Construction Statistics Annual

2000 Edition

October 2000

Department of the Environment, Transport and the Regions: London

Department of the Environment, Transport and the Regions
Eland House
Bressenden Place
London SW1E 5DU
Telephone 020 7944 3000
Fax: 020 7944 4242
Internet service http://www.detr.gov.uk/

Further copies of this report are available from:

Department of the Environment, Transport and the Regions
Publications Sale Centre
Unit 21
Goldthorpe Industrial Estate
Goldthorpe
Rotherham S63 9BL
Tel: 01709 891318
Fax: 01709 881673

ISBN 1 85112 432 2

You may also access this document at the Department's web site.

Printed in Great Britain. Text printed on material containing 100% post-consumer waste.
Cover printed on material containing 75% post-consumer waste and 25% ECF pulp.
October 2000

CONTENTS

INTRODUCTION   5

LIST OF TABLES   6

LIST OF FIGURES   12

CHAPTER 1
New Orders   14

CHAPTER 2
Output   29

CHAPTER 3
Structure of the Industry   45

CHAPTER 4
Price Indices   71

CHAPTER 5
Cost Indices   84

CHAPTER 6
Investment   90

CHAPTER 7
Commercial Floorspace   93

CHAPTER 8
Government Departments' Construction Plans   96

CHAPTER 9
Local Authority Expenditure   135

CHAPTER 10
National Lottery Funded Projects   140

CHAPTER 11
Planning Applications and Decisions   149

CHAPTER 12

Trends in Employment and the Professions 156

CHAPTER 13

Northern Ireland, Wales and Scotland 182

CHAPTER 14

Building Materials 197

CHAPTER 15

International Comparison 212

CHAPTER 16

DETR Construction Innovation and Research Business Plans 236

CHAPTER 17

Construction Industry Key Performance Indicators and Benchmarking 240

SUPPLEMENTARY INFORMATION

CHAPTER 8

Public/Private Partnerships and the Private Finance Initiative 96

CHAPTER 8

Northern Ireland PFI Plans 133

CHAPTER 11

Thames Gateway 154

CHAPTER 15

Export Promotion 235

APPENDIX 1

List of DETR, DOE and Standard Regions 246

APPENDIX 2

List of SSI Inspection Regions 247

APPENDIX 3

Notes and Definitions 248

INDEX **261**

Introduction

This is the first edition of the new *Construction Statistics Annual* which conveniently brings together under one cover the construction statistics previously included in the annual *Housing and Construction Statistics* and the construction related data previously included in the annual *Digest of Data for the Construction Industry*.

This publication marks the beginning of the separate annual dissemination of the statistical data for housing and construction, which was announced in Housing and Construction Statistics 1988-1998. In future the housing data will be included in Housing Statistics, the 1989-1999 edition to be published later this year.

This *2000 Edition of Construction Statistics Annual* gives a broad perspective of statistical trends in the construction industry in Great Britain through the last decade together with some international comparisons and features on leading initiatives which may influence the future.

We believe the compendium of statistics and other data in this edition make it essential reading for all those engaged in planning the future workload and other activities which go to make up the Construction Industry in Great Britain.

Our thanks go to the contributors, too many to list here, and in particular to our colleagues in the Scottish Executive, the National Assembly for Wales and the Northern Ireland Executive without whose contributions publication of *Construction Statistics Annual* would not have been possible.

Our thanks also go to the members of the Consultative Committee on Construction Industry Statistics (the forum for joint consultation on construction by the industry and the DETR). Special thanks are due to Ms Andrea Solomou of DETR for her hard work in putting together this edition.

Comments and suggestions from readers are welcome and should be addressed to:

Bob Packham
Construction Market Intelligence
Department of the Environment, Transport and the Regions
Zone 3/B4, Eland House
Bressenden Place, London SW1E 5DU
Tel: 020 7944 5764

Bob Davies
Head of Construction Market Intelligence
Department of the Environment, Transport and the Regions
September 2000

LIST OF TABLES

Table	Title	Contact	Phone	Page
One	**New Orders**			
1.1	New Orders obtained by Contractors (Current Prices)	Neville Price	020 7944 5587	16
1.2	New Orders obtained by Contractors (1995 Prices)			17
1.3	New Orders obtained by Contractors (Index)			18
1.4	New Orders obtained by Contractors by Type of Work			19
1.5	New Orders obtained by Contractors by Region			21
1.6	New Orders obtained by Contractors by Value Range			26
1.7	New Orders obtained by Contractors by Duration			28
Two	**Output**			
2.1	Output (Current Prices)	Neville Price	020 7944 5587	31
2.2	Output (1995 Prices)			32
2.3	Output (Index)			33
2.4	Contractors' Output (Current Prices)			34
2.5	Contractors' Output (1995 Prices)			35
2.6	Direct Labour Output (Current Prices)			36
2.7	Direct Labour Output (1995 Prices)			37
2.8	Contractors' Output by Type of Work			38
2.9	Contractors' Output by Region			40
Three	**Structure of the Industry**			
3.1	Private Contractors: Number of Firms	Neville Price	020 7944 5587	45
3.2	Insolvencies and Bankruptcies of Construction Firms			47
3.3	Private Contractors: Work Done			48
3.4	Private Contractors: Total Employment			50
3.5	Private Contractors: Employment of Operatives			52
3.6	Private Contractors: Number of Firms by Size and Trade of Firm			54
3.7	Private Contractors: Number of Firms by Size of Firm and Region of Registration			56
3.8	Private Contractors: Value of Work Done by Size of Firm and Type of Work			57
3.9	Private Contractors: Value of Work Done by Trade of Firm and Type of Work			58

Table	Title	Contact	Phone	Page
3.10	Private Contractors: Value of Work Done by Size and Trade of Firm	Neville Price	020 7944 5587	59
3.11	Private Contractors: Employment of Operatives by Size and Trade of Firm			62
3.12	Private Contractors: Employment of APTCs by Size and Trade of Firm			65
3.13	Private Contractors: Total Employment by Size of Firm			68
3.14	Private Contractors: Total Employment by Trade of Firm			69
3.15	Private Contractors: Total Employment by Region of Registration			70

Four Price Indices

4.1	Tender Price Indices	Marcella Douglas	020 7944 5594	71
4.2	Regional Factors for the Social Housing Tender Price Index			73
4.3	Regional Factors for the Public Sector Building (Non-Housing) Tender Price Index			74
4.4	Function Factors for the Public Sector Building (Non-Housing) Tender Price Index			75
4.5	Regional Factors for the Road Construction Tender Price Index			76
4.6	Road Type Factors for the Road Construction Tender Price Index			77
4.7	Value Factors for the Road Construction Tender Price Index			78
4.8	Comparison of Output Price Indices			80
4.9	Output Price Indices			81
4.10	Public Works Output Price Indices			82
4.11	Direct Labour Output Deflators and Contractors Output Deflators			83

Five Cost Indices

5.1	Resource Cost Index of Building Non-Housing	Marcella Douglas	020 7944 5594	84
5.2	Resource Cost Index of House Building			85
5.3	Resource Cost Index of Road Construction			86
5.4	Resource Cost Index of Infrastructure			87
5.5	Resource Cost Index of Maintenance of Building Non-Housing			88
5.6	Resource Cost Index of Maintenance of House Building			89

Six Investment

| 6.1 | Gross Fixed Capital Formation at Current Purchaser's Prices: Analysis by Type of Asset and Sector | Neville Price | 020 7944 5587 | 90 |

Table	Title	Contact	Phone	Page
6.2	Gross Domestic Fixed Capital Formation at Current Purchaser's Prices: Analysis by Broad Sector and Type of Asset	Neville Price	020 7944 5587	91
6.3	Gross Domestic Fixed Capital Formation at 1995 Purchaser's Prices: Analysis by Broad Sector and Type of Asset: Total Economy			92

Seven   Commercial Floorspace

| 7.1 | Take-up and Availability of Floorspace in the City of London and its Surrounding Area | Malcolm Trice | 020 7606 7461 | 94 |
| 7.2 | New Office Designs Awarded BREEAM Certificates | Hilary John | 01923 664462 | 95 |

Eight   Government Departments' Construction Plans

8.1	Department for Education and Employment PFI Projects	Sue Meehan	020 7925 6348	100
8.2	Department of the Environment, Transport and the Regions PFI Projects	Rachel Edwards	020 7944 5015	102
8.3	Department of Health PFI Projects	Christian Richardson	01132 547385	103
8.4	Capital Programme for Non-PFI Health Building Construction Projects over £2.5m	Simon Wright	0113 254 7185	107
8.5	Number of Homes for the Elderly and Younger Physically Disabled People	Laurent Orthmans	020 7972 5600	108
8.6	Highways Agency PFI Projects	Ian Farrand	020 7921 4169	109
8.7	Major (£1m and Over) Non-PFI Road Contracts Awarded by the Highways Agency	Adrian McCabe	020 7921 4333	110
8.8	Home Office PFI Projects	Donna Sanders	020 7273 2353	113
8.9	Local Authorities PFI Projects			114
8.10	Inland Revenue PFI Projects	Alexandria Barling	020 7438 6859	116
8.11	Lord Chancellor's Department PFI Projects	Mark Armstrong	020 7210 8578	117
8.12	Ministry of Defence PFI Projects	Angela Barrett	020 7218 5951	118
8.13	Ministry of Defence Non-PPI Expenditure	Eric Crane	020 7218 0781	126
8.14	Scottish Executive PFI Projects	Fiona McLellan	0131 244 7499	127
8.15	National Assembly for Wales PFI Projects	Lisa Thomas	029 2082 5213	131
8.16	Northern Ireland PFI Projects	James McAleer	01247 279279	133

Nine   Local Authority Expenditure

9.1	English Local Authority Capital Expenditure on Construction, Conversion and Renovation	Mervion Kirwood	020 7944 4074	136
9.2	Local Authority Housing Repair & Maintenance Revenue Expenditure	Dan Varey	020 7944 3598	138
9.3	Local Authorities and New Towns: Value of Construction Work Done by Type of Work	Neville Price	020 7944 5587	139

Table	Title	Contact	Phone	Page
Ten	**National Lottery Funded Projects**			
10.1	National Lottery Projects – Fund Application	Sarah Wood	01202 432121	140
10.2	National Lottery Projects – Funding Received			
Eleven	**Planning Applications and Decisions**			
11.1	Planning Applications and Decisions by District Planning Authorities by Speed of Decision	Ian Rowe	020 7944 5502	145
11.2	Planning Decisions by District Planning Authorities by Speed of Decision, Region and Type of Authority			150
11.3	Planning Decisions by District Planning Authorities by Speed of Decision and Type and Size of Development			151
11.4	Mineral Planning Decisions by 'County Matters' Authorities			152
11.5	Decisions on Mineral Applications by All 'County Matters' Authorities by Type of Development			153
Twelve	**Trends in Employment and the Professions**			
12.1	Manpower – Seasonally Adjusted	Neville Price	020 7944 5587	157
12.2	Manpower			158
12.3	Earnings and Hours in the Construction Industry			159
12.4	Earnings in the Construction Industry by Craft			160
12.5	Earnings and Hours in the Construction Industry and in All Industries and Services			161
12.6	Employment in the Construction Industry by Gender	Labour Market Statistics Helpline	020 7533 6094	162
12.7	Employees and Self-Employed in the Construction Industry by Ethnic Origin	Andrew Risdon	020 7533 6145	163
12.8	Royal Institute of Chartered Surveyors Total Workload Survey	Milan Khatri	020 7334 3774	164
12.9	Architects' Workload – Estimated Value of New Commissions Analysed by Building Type	Aziz Mirza	01243 551302	165
12.10	Architects' Workload – Estimated Value of Production Drawings Analysed by Building Type			166
12.11	Consulting Engineers' Fees for Work in the UK	Craig Beaumont	020 7222 6557	167
12.12	Consulting Engineers' Source of New Commissions			167
12.13	CITB Trainee Numbers Survey: First Year Intake for 1998/99	Martin Turner	01485 577640	168
12.14	CITB Trainee Numbers Survey: First Year Intake for 1997/98			168
12.15	Claimant Unemployed for Skilled Construction Trades	David Taylor	020 8926 0473	169

Table	Title	Contact	Phone	Page
12.16	Job Centre Vacancies in Skilled Construction Trades	David Taylor	020 8926 0473	171
12.17	Claimant Unemployed for Carpenters and Joiners			173
12.18	Overseas Service Trade by the Type of Service being Supplied or Purchased from Overseas	Chris Peggie	01633 813458	174
12.19	Overseas Trade in Services by the Industrial Classification of the Business			174
12.20	Professional and Scientific Services	Keri Hodson	01633 812264	175
12.21	Stoppages in the Construction Industry	Jackie Davies	01928 792825	176
12.22	Injuries in the Construction Industry Reported to All Enforcement Authorities	Operations Unit	0151 951 4841	179
12.23	Injuries to Employees in the Construction Sector Reported to HSE's Field Operations Division Inspectorates and Local Authorities			180
12.24	Injuries to Employees Reported to all Enforcement Authorities: Construction v All Industries			180
12.25	HSE Enforcement Actions			181

Thirteen   Northern Ireland, Wales and Scotland

13.1	Northern Ireland – Index of Construction Output	Rodney Redmond	028 9054 0799	183
13.2	Northern Ireland – Housing Executive Starts by District Council			184
13.3	Northern Ireland – Private Sector Housing Starts by District Council			185
13.4	Northern Ireland – Housing Association Starts by District Council			186
13.5	Wales – New Orders by Contractors	Neville Price	020 7944 5587	187
13.6	Wales – Output obtained by Contractors1			188
13.7	Wales – Index of Construction	Kathryn Hughes	01222 825017	190
13.8	Wales – Planning Applications			191
13.9	Wales – Mineral Planning Applications			192
13.10	Expenditure for Wales – Detailed Analysis of Capital Account			193
13.11	Wales – 1996 Based Household Projections			194
13.12	Scotland – New Orders obtained by Contractors	Neville Price	020 7944 5587	195
13.13	Scotland – Output obtained by Contractors			196

Fourteen   Building Materials

14.1	Price Indices of Construction Materials	David Williams	020 7944 5593	199
14.2	Bricks			201
14.3	Concrete Building Blocks			202
14.4	Sales of Aggregate by Use from Standard Regions, Wales and Scotland			203
14.5	Other Building Materials			204

Table	Title	Contact	Phone	Page
14.6	Value of Overseas Trade in Selected Materials and Components for Constructional Use: Imports and Exports	David Williams	020 7944 5593	205
14.7	Steelwork Prices and Lead Times	Gillian Mitchell	020 7839 8566	211

Fifteen   International Comparison

Table	Title	Contact	Phone	Page
15.1	Key Data	David Hart	020 7209 3000	212
15.2	Building Labour Rates			213
15.3	Building Material Supply Prices			214
15.4	Building Costs: Residential, Retail & Hotels			215
15.5	Building Costs: Offices & Industrial			217
15.6	Inflation Rates and Forecasts			219
15.7	Production of Constructional Steelwork	Gillian Mitchell	020 7839 8566	221
15.8	Direct Investment of UK Companies in Overseas Subsidiary and Associate Companies and Earnings from UK Direct Investment Overseas	Simon Harrington	01633 813314	222
15.9	Level of Direct Investment Assets held Overseas by the UK in Subsidiary and Associate Companies			226
15.10	Direct Investment by Foreign Companies in UK Subsidiary and Associate Companies and Earnings from Overseas Direct Investment in the UK			228
15.11	Level of Direct Investment Assets held in the UK by Foreign Investors			231
15.12	Overseas Construction Activity by British Companies	David Williams	020 7944 5593	232

Sixteen   DETR Construction Innovation and Research Business Plans

Table	Title	Contact	Phone	Page
16.1	The Plan	John Troughton	020 7944 5689	237

Seventeen   Construction Industry Key Performance Indicators and Benchmarking

Table	Title	Contact	Phone	Page
17.1	Definitions used for Project and Company Performance KPIs	Bob Packham	020 7944 5764	242
17.2	Industry Average Performances			243
17.3	Summary of Industry Performance 1998 and 1999			243

LIST OF FIGURES

Figure	Title	Page

One   New Orders

1.1	New Housing Orders	14
1.2	New Non-Housing Orders: Including Infrastructure	14
1.3	New Non-Housing Orders: Excluding Infrastructure	15
1.4	New Orders	15
1.5	New Orders obtained by Contractors by Region for 1999	25

Two   Output

2.1	Output: New Work and Repair & Maintenance	29
2.2	Output: Housing and Non-Housing	29
2.3	Output: All Agencies, Contractors and DLO	30
2.4	Output obtained by Contractors by Region for 1999	44

Three   Structure of the Industry

3.1	Insolvencies and Bankruptcies of Construction Firms	47

Four   Price Indices

4.1	Comparison of Tender Price Indices	72
4.2	Comparison of Output Price Indices	79

Seven   Commercial Floorspace

7.1	Take-up of Floorspace in the City and Fringe	93
7.2	Availability of Floorspace in the City and Fringe	93

Eight   Government Departments' Construction Plans

8.1	Number of Homes for Elderly and for Younger Physically Disabled People	107

Nine   Local Authority Expenditure

9.1	English Local Authority Capital Expenditure on Construction, Conversion and Renovation	135
9.2	Breakdown by Sector of Local Authority Capital Expenditure on Construction, Conversion and Renovation	135
9.3	Local Authority Housing Repair & Maintenance Revenue Expenditure	138
9.4	Local Authorities and New Towns: Value of Construction Work Done by Type of Work	139

Figure	Title	Page
Twelve	**Trends in Employment and the Professions**	
12.1	Manpower – Seasonally Adjusted	156
12.2	Employment in the Construction Industry by Gender	162
12.3	Architects' Workload – Value of New Commissions	165
Thirteen	**Northern Ireland, Wales and Scotland**	
13.1	Northern Ireland Index of Construction	182
13.2	Wales Index of Construction	189
Fourteen	**Building Materials**	
14.1	Deliveries of Roofing Materials	197
14.2	Sales of Aggregates and Ready Mix Concrete	197
14.3	Deliveries of Bricks, Cement and Blocks	198
14.4	External Trade: Building Materials	198
14.5	Steelwork Prices	210
14.6	Lead Times	210
Fifteen	**International Comparison**	
15.1	Overseas Construction by British Companies: Value of New Contracts	233
15.2	Overseas Construction by British Companies: Value of Work Done	233
15.3	Overseas Construction by British Companies: Value of Work Outstanding	234
Sixteen	**DETR Construction Innovation and Research Business Plans**	
16.1	Baseline Trend Scenario	239
Seventeen	**Construction Industry Key Performance Indicators and Benchmarking**	
17.1	Family Tree of KPI Wallcharts	241
17.2	Key Projects Stages	241
17.3	Year on Year Comparisons of Industry Performance 1998 and 1999	244
17.4	Supply Chain Cube	245

CHAPTER 1

New Orders

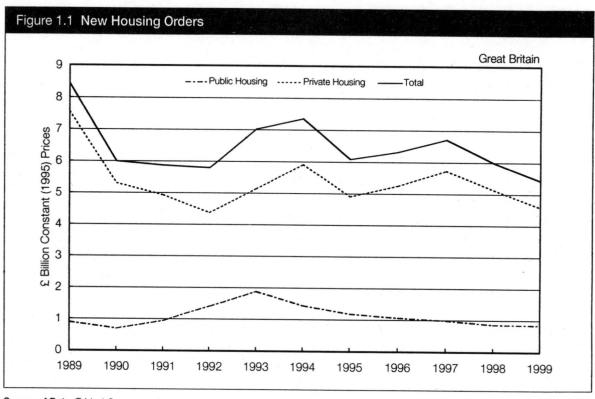

Figure 1.1 New Housing Orders

Source of Data: Table 1.2

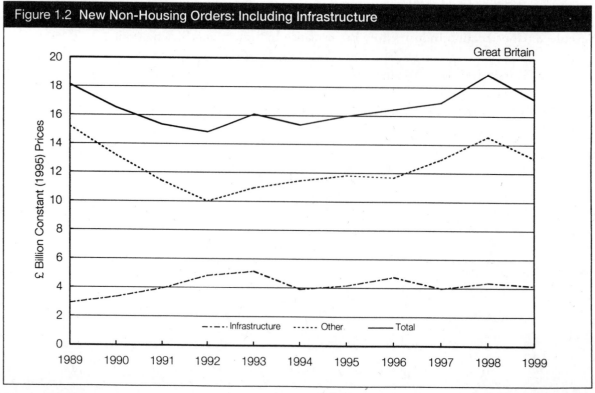

Figure 1.2 New Non-Housing Orders: Including Infrastructure

Source of Data: Table 1.2

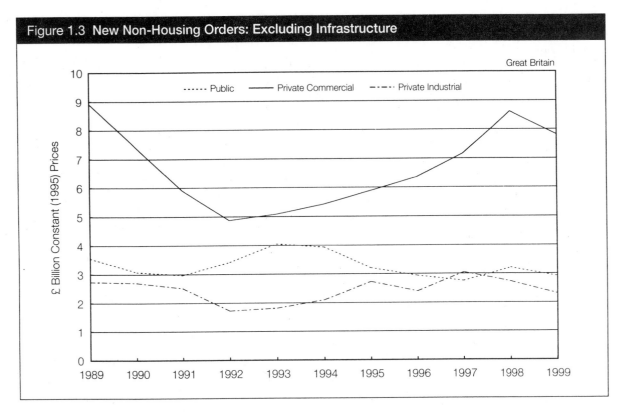

Figure 1.3 New Non-Housing Orders: Excluding Infrastructure

Source of Data: Table 1.2

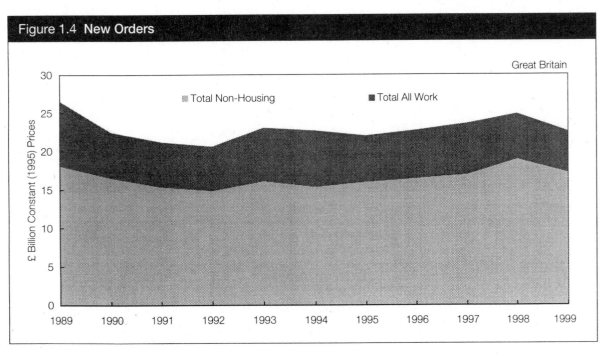

Figure 1.4 New Orders

Source of Data: Table 1.2

Table 1.1 New Orders obtained by Contractors[1]

Current Prices (£ Million)   Great Britain

| Year & Quarter | | New Housing | | | Other New Work | | | | | |
| | | | | | Infra-structure | Other New Work Exc. Infrastructure | | | | All New Work |
		Public	Private	All New Housing		Public	Private Industrial	Private Commercial	All Other New Work	
1989		872	6,497	7,370	2,960	3,836	3,049	9,927	19,772	27,142
1990	Q1	244	1,336	1,580	839	833	775	2,113	4,559	6,139
	Q2	152	1,403	1,555	832	726	917	2,211	4,686	6,241
	Q3	153	1,168	1,320	733	853	649	1,809	4,043	5,363
	Q4	134	949	1,083	787	705	509	1,663	3,665	4,748
1991	Q1	181	1,102	1,283	899	620	434	1,461	3,414	4,697
	Q2	213	1,310	1,523	744	677	965	1,544	3,930	5,453
	Q3	216	1,213	1,429	950	599	507	1,319	3,375	4,804
	Q4	266	926	1,192	829	781	442	1,257	3,309	4,501
1992	Q1	365	1,008	1,373	1,168	714	439	1,225	3,545	4,919
	Q2	277	1,144	1,421	776	598	372	1,029	2,775	4,196
	Q3	289	1,102	1,391	844	766	376	1,083	3,068	4,459
	Q4	314	762	1,076	958	685	317	882	2,842	3,918
1993	Q1	446	1,128	1,575	1,106	957	421	925	3,409	4,984
	Q2	468	1,315	1,783	762	590	412	1,128	2,892	4,675
	Q3	402	1,265	1,667	784	942	440	1,171	3,336	5,003
	Q4	351	1,166	1,517	1,347	908	404	1,127	3,786	5,303
1994	Q1	414	1,445	1,859	1,191	1,025	445	1,081	3,741	5,600
	Q2	383	1,595	1,978	955	786	569	1,307	3,616	5,594
	Q3	269	1,530	1,799	638	934	573	1,492	3,637	5,436
	Q4	319	1,151	1,470	667	910	413	1,195	3,184	4,654
1995	Q1	366	1,363	1,729	1,333	812	698	1,269	4,112	5,841
	Q2	348	1,338	1,686	818	824	521	1,347	3,510	5,196
	Q3	244	1,223	1,466	961	745	599	1,425	3,730	5,196
	Q4	224	981	1,205	1,058	826	908	1,836	4,627	5,832
1996	Q1	363	1,181	1,544	1,289	838	555	1,526	4,208	5,753
	Q2	260	1,360	1,620	964	690	586	1,772	4,013	5,633
	Q3	250	1,343	1,594	1,094	778	585	1,488	3,944	5,538
	Q4	198	1,532	1,730	1,316	650	612	1,603	4,181	5,911
1997	Q1	356	1,491	1,848	1,083	731	736	1,799	4,349	6,197
	Q2	246	1,575	1,820	1,100	664	811	1,892	4,467	6,287
	Q3	168	1,611	1,780	917	752	820	1,928	4,417	6,196
	Q4	225	1,575	1,800	892	747	781	1,906	4,326	6,126
1998	Q1	370	1,692	2,062	1,033	786	735	2,357	4,910	6,972
	Q2	216	1,562	1,777	1,032	886	795	2,063	4,776	6,553
	Q3	168	1,568	1,736	1,262	832	734	2,582	5,410	7,146
	Q4	180	1,175	1,355	1,152	1,001	707	2,591	5,451	6,806
1999	Q1	328	1,500	1,828	992	928	621	2,367	4,908	6,736
	Q2	196	1,451	1,647	1,061	839	676	2,537	5,112	6,759
	Q3	197	1,525	1,722	1,045	764	629	2,193	4,630	6,352
	Q4	247	1,425	1,672	1,097	743	633	2,087	4,560	6,233

Notes

1. Classified to construction in the 1992 Revised Standard Industrial Clasification.

Source of Data: Construction Market Intelligence, Department of Environment, Transport and the Regions
Contact: Neville Price 020 7944 5587

Table 1.2 New Orders obtained by Contractors[1]

Constant (1995) Prices Seasonally Adjusted (£ Million) Great Britain

| Year & Quarter | | New Housing | | | Other New Work | | | | | |
| | | | | | | Other New Work Exc. Infrastructure | | | | |
		Public	Private	All New Housing	Infra-structure	Public	Private Industrial	Private Commercial	All Other New Work	All New Work
1989		893	7,548	8,441	2,921	3,548	2,738	8,911	18,118	26,559
1990	Q1	212	1,516	1,728	806	776	695	1,905	4,182	5,910
	Q2	150	1,396	1,547	871	698	846	1,983	4,398	5,944
	Q3	180	1,225	1,405	828	863	614	1,766	4,071	5,476
	Q4	152	1,164	1,316	828	738	544	1,752	3,862	5,178
1991	Q1	156	1,229	1,386	877	653	442	1,492	3,464	4,849
	Q2	218	1,280	1,498	861	736	1,004	1,513	4,114	5,612
	Q3	259	1,260	1,519	1,212	673	557	1,402	3,846	5,365
	Q4	310	1,155	1,466	991	908	507	1,503	3,909	5,375
1992	Q1	347	1,125	1,472	1,336	853	501	1,384	4,074	5,546
	Q2	297	1,128	1,426	1,097	740	419	1,110	3,366	4,792
	Q3	362	1,155	1,517	1,201	974	437	1,262	3,874	5,391
	Q4	407	970	1,377	1,200	847	364	1,112	3,524	4,900
1993	Q1	454	1,200	1,654	1,271	1,151	450	1,125	3,996	5,650
	Q2	509	1,237	1,746	1,078	699	427	1,219	3,423	5,169
	Q3	494	1,268	1,761	1,102	1,121	463	1,280	3,966	5,727
	Q4	426	1,434	1,861	1,692	1,077	473	1,456	4,698	6,559
1994	Q1	361	1,506	1,867	1,261	1,099	482	1,243	4,085	5,952
	Q2	402	1,523	1,924	1,174	940	579	1,316	4,009	5,933
	Q3	318	1,532	1,850	738	949	590	1,582	3,859	5,709
	Q4	363	1,349	1,712	728	944	448	1,276	3,396	5,107
1995	Q1	276	1,366	1,642	1,183	768	655	1,319	3,925	5,567
	Q2	342	1,248	1,590	873	909	482	1,262	3,525	5,115
	Q3	296	1,202	1,498	1,024	697	603	1,403	3,728	5,226
	Q4	268	1,089	1,357	1,090	833	984	1,893	4,801	6,158
1996	Q1	283	1,145	1,428	1,142	783	542	1,570	4,036	5,464
	Q2	245	1,243	1,488	1,072	777	566	1,659	4,075	5,562
	Q3	310	1,280	1,590	1,208	729	607	1,473	4,016	5,607
	Q4	227	1,587	1,814	1,342	647	676	1,651	4,316	6,131
1997	Q1	271	1,379	1,650	938	667	716	1,796	4,117	5,767
	Q2	227	1,358	1,585	1,168	707	750	1,698	4,323	5,908
	Q3	223	1,445	1,669	983	672	777	1,831	4,263	5,931
	Q4	264	1,544	1,808	888	703	801	1,835	4,228	6,036
1998	Q1	268	1,486	1,754	862	682	683	2,204	4,431	6,185
	Q2	192	1,274	1,466	1,042	905	715	1,737	4,398	5,864
	Q3	196	1,314	1,510	1,338	711	663	2,340	5,053	6,562
	Q4	197	1,061	1,258	1,129	909	661	2,320	5,019	6,278
1999	Q1	190	1,148	1,338	997	796	565	2,010	4,369	5,707
	Q2	176	1,117	1,292	1,023	761	605	2,075	4,464	5,757
	Q3	202	1,137	1,339	1,026	703	554	1,875	4,158	5,497
	Q4	274	1,183	1,456	1,095	648	584	1,859	4,186	5,642

Notes

1. Classified to construction in the 1992 Revised Standard Industrial Clasification.

Source of Data: Construction Market Intelligence, Department of Environment, Transport and the Regions
Contact: Neville Price 020 7944 5587

Table 1.3 New Orders obtained by Contractors[1]

Index at 1995=100 Prices   Great Britain

Year & Quarter		New Housing			Other New Work					
						Other New Work Exc. Infrastructure				
		Public	Private	All New Housing	Infra-structure	Public	Private Industrial	Private Commercial	All Other New Work	All New Work
1989		75.5	153.9	138.7	70.1	110.7	100.5	151.6	113.4	120.4
1990	Q1	71.6	123.6	113.5	77.3	96.8	102.0	129.6	104.7	107.1
	Q2	50.9	113.9	101.6	83.6	87.0	124.1	135.0	110.1	107.8
	Q3	61.0	99.9	92.4	79.4	107.6	90.2	120.2	101.9	99.3
	Q4	51.4	94.9	86.5	79.4	92.0	79.9	119.3	96.7	93.9
1991	Q1	52.9	100.3	91.0	84.1	81.4	64.9	101.5	86.7	87.9
	Q2	73.9	104.4	98.5	82.6	91.9	147.3	103.0	103.0	101.7
	Q3	87.8	102.7	99.8	116.3	84.0	81.8	95.4	96.3	97.2
	Q4	104.9	94.2	96.3	95.1	113.2	74.5	102.3	97.9	97.4
1992	Q1	117.3	91.8	96.7	128.1	106.4	73.5	94.2	102.0	100.5
	Q2	100.6	92.0	93.7	105.2	92.3	61.6	75.5	84.3	86.9
	Q3	122.4	94.2	99.7	115.2	121.5	64.1	85.9	97.0	97.7
	Q4	137.6	79.1	90.5	115.1	105.6	53.5	75.7	88.2	88.8
1993	Q1	153.6	97.9	108.7	121.9	143.6	66.0	76.6	100.0	102.4
	Q2	172.3	100.8	114.7	103.4	87.1	62.8	83.0	85.7	93.7
	Q3	167.0	103.4	115.7	105.7	139.9	67.9	87.1	99.3	103.8
	Q4	144.2	117.0	122.3	162.3	134.3	69.5	99.1	117.6	118.9
1994	Q1	122.2	122.8	122.7	120.9	137.1	70.8	84.6	102.3	107.9
	Q2	135.9	124.2	126.5	112.6	117.2	85.0	89.6	100.3	107.6
	Q3	107.6	124.9	121.6	70.8	118.4	86.6	107.7	96.6	103.5
	Q4	122.8	110.0	112.5	69.8	117.7	65.8	86.9	85.0	92.6
1995	Q1	93.5	111.4	107.9	113.5	95.8	96.2	89.8	98.3	100.9
	Q2	115.7	101.8	104.5	83.7	113.4	70.8	85.9	88.2	92.7
	Q3	100.1	98.0	98.4	98.3	87.0	88.6	95.5	93.3	94.7
	Q4	90.8	88.8	89.2	104.6	103.9	144.5	128.8	120.2	111.6
1996	Q1	95.8	93.4	93.8	109.6	97.6	79.5	106.9	101.0	99.1
	Q2	82.8	101.4	97.8	102.8	96.9	83.1	112.9	102.0	100.8
	Q3	105.1	104.4	104.5	115.9	90.9	89.1	100.2	100.5	101.6
	Q4	77.0	129.4	119.2	128.7	80.7	99.2	112.4	108.1	111.1
1997	Q1	91.8	112.4	108.4	90.0	83.2	105.2	122.2	103.1	104.5
	Q2	76.8	110.8	104.2	112.1	88.2	110.1	115.6	108.2	107.1
	Q3	75.6	117.9	109.7	94.3	83.9	114.0	124.7	106.7	107.5
	Q4	89.3	125.9	118.8	85.2	87.7	117.6	124.9	105.8	109.4
1998	Q1	91.2	121.0	115.2	82.7	85.0	100.3	150.2	111.0	112.2
	Q2	65.0	103.8	96.3	100.0	112.9	105.0	118.4	110.2	106.4
	Q3	66.4	107.1	99.2	128.4	88.7	97.4	159.5	126.6	119.0
	Q4	67.0	86.4	82.7	108.4	113.4	97.1	158.1	125.7	113.9
1999	Q1	64.4	93.6	88.0	95.7	99.4	83.0	136.8	109.4	103.5
	Q2	59.5	91.1	84.9	98.1	94.9	88.8	141.2	111.8	104.4
	Q3	68.4	92.7	88.0	98.4	87.7	81.3	127.6	104.1	99.7
	Q4	92.6	96.5	95.7	105.0	80.9	85.7	126.6	104.8	102.3

Notes
1. Classified to construction in the 1992 Revised Standard Industrial Clasification.

Source of Data: Construction Market Intelligence, Department of Environment, Transport and the Regions
Contact: Neville Price 020 7944 5587

Table 1.4 New Orders obtained by Contractors[1] by Type of Work

(a) New Orders (£ Million) obtained from Public Sector Great Britain

Year	New Housing	Infra-structure	Other New Work Excluding Infrastructure												Other New Work	All Public Sector
			Fac-tories	Ware-houses	Oil, Steel & Coal	Schools & Colleges	Univ-ersities	Health	Offices	Enter-tainment	Garages	Shops	Agri-culture	Miscell-aneous		
1989	872	2,369	188	48	115	471	139	824	519	382	75	26	6	1,043	3,836	7,077
1990	683	2,029	149	14	81	527	146	663	570	283	56	24	5	601	3,117	5,829
1991	875	2,090	89	14	13	584	150	578	492	250	44	25	17	421	2,677	5,642
1992	1,246	1,690	128	54	47	584	203	644	499	189	28	23	14	351	2,763	5,699
1993	1,668	2,472	111	23	30	655	353	697	684	281	42	26	44	450	3,397	7,538
1994	1,386	2,211	111	38	12	658	376	752	469	308	49	14	22	844	3,654	7,251
1995	1,182	2,327	94	29	13	710	373	717	393	285	51	21	12	508	3,206	6,716
1996	1,073	1,663	91	14	4	708	355	674	381	253	28	11	8	419	2,945	5,681
1997	995	1,352	72	27	4	749	273	491	391	342	34	35	33	441	2,894	5,241
1998	933	1,505	84	20	2	770	405	769	292	432	19	35	17	660	3,504	5,942
1999	969	1,495	72	24	5	791	345	635	390	435	36	29	9	503	3,273	5,737

(b) New Orders (£ Million) obtained from Private Sector Great Britain

Year	New Housing	Infra-structure	Industrial				Commercial								All Comm-ercial	All Private Sector
			Fac-tories	Ware-houses	Oil, Steel & Coal	All In-dustrial	Schools & Universities	Health	Offices	Enter-tainment	Garages	Shops	Agri-culture	Miscell-aneous		
1989	6,497	592	2,269	706	73	3,049	179	295	5,271	1,418	405	2,086	127	149	9,927	20,065
1990	4,855	1,161	2,094	648	108	2,850	169	271	4,215	1,195	364	1,345	127	109	7,796	16,662
1991	4,552	1,333	1,830	438	79	2,347	175	369	2,215	1,111	261	1,222	105	123	5,581	13,813
1992	4,016	2,056	1,006	426	72	1,503	121	193	1,691	747	234	1,034	94	103	4,218	11,794
1993	4,874	1,525	1,221	429	27	1,677	134	179	1,471	751	308	1,278	108	122	4,351	12,427
1994	5,721	1,240	1,451	498	51	1,999	115	255	1,777	928	300	1,453	120	127	5,075	14,034
1995	4,905	1,843	2,055	594	76	2,725	105	288	2,123	940	301	1,871	124	126	5,877	15,350
1996	5,416	2,993	1,603	663	71	2,337	156	277	2,169	1,407	265	1,795	123	198	6,390	17,135
1997	6,253	2,639	2,184	901	64	3,149	189	356	2,506	1,847	344	1,937	148	198	7,525	19,565
1998	5,997	2,974	1,878	1,014	79	2,971	351	651	3,472	2,244	315	2,154	146	259	9,593	21,535
1999	5,901	2,700	1,698	821	38	2,558	393	411	3,566	2,224	266	1,901	100	321	9,184	20,342

Table 1.4 New Orders obtained by Contractors¹ by Type of Work (continued)

(c) New Orders (£ Million) obtained from Public and Private Sectors

Great Britain

Year	New Housing	Infrastructure							Total Infra-structure
		Water	Sewerage	Elec-tricity	Commu-nications, Gas & Air	Railways	Harbours	Roads	
1989	7,370	251	332	271	339	362	180	1,226	2,960
1990	5,538	321	491	187	391	160	216	1,425	3,190
1991	5,427	509	429	301	441	125	203	1,415	3,423
1992	5,263	669	469	281	554	200	252	1,322	3,746
1993	6,542	421	447	211	642	623	220	1,435	3,998
1994	7,106	412	389	170	494	412	218	1,356	3,451
1995	6,087	500	394	218	904	351	273	1,531	4,170
1996	6,487	640	481	294	745	524	270	1,710	4,664
1997	7,248	733	656	382	693	416	182	928	3,991
1998	6,930	957	737	359	745	573	287	821	4,479
1999	6,869	760	789	254	713	471	250	957	4,195

Year	Other Non-Housing Excluding Infrastructure											Total Other Non-Housing	Total All New Work
	Fac-tories	Ware-houses	Oil, Steel & Coal	Schools & Univer-sities	Health	Offices	Entertain-ment	Garages	Shops	Agri-culture	Miscel-laneous		
1989	2,457	754	188	789	1,118	5,789	1,800	479	2,112	132	1,192	16,812	27,142
1990	2,243	662	188	842	934	4,786	1,477	420	1,368	131	709	13,762	22,491
1991	1,919	452	92	909	947	2,707	1,361	305	1,247	122	544	10,606	19,455
1992	1,134	480	119	907	837	2,190	937	263	1,057	108	453	8,484	17,493
1993	1,332	452	57	1,143	876	2,155	1,032	350	1,304	152	572	9,425	19,965
1994	1,561	535	63	1,149	1,007	2,246	1,235	349	1,467	143	971	10,727	21,285
1995	2,148	623	89	1,188	1,005	2,516	1,225	352	1,892	136	634	11,808	22,065
1996	1,694	676	75	1,218	958	2,548	1,666	293	1,807	131	616	11,683	22,834
1997	2,256	928	68	1,211	847	2,466	1,831	379	1,971	181	639	13,567	24,806
1998	1,961	1,033	81	1,526	1,420	3,765	2,677	334	2,189	164	919	16,068	27,477
1999	1,771	845	43	1,529	1,046	3,956	2,659	302	1,930	110	824	15,015	26,079

Notes

1. Classified to construction in the 1992 Revised Standard Industrial Clasification.

Source of Data: Construction Market Intelligence, Department of Environment, Transport and the Regions
Contact: Neville Price 020 7944 5587

Table 1.5 New Orders obtained by Contractors[1] by Region

£ Million

Year	New Housing			Other New Work					All New Work
	Public	Private	All New Housing	Infra-structure	Public	Private Industrial	Private Commercial	All Other New Work	
North									
1989	29	357	386	190	182	147	339	858	1,245
1990	31	338	370	167	98	176	241	681	1,051
1991	62	261	323	215	101	118	188	622	945
1992	41	222	263	134	120	71	160	486	750
1993	52	293	345	223	180	127	170	700	1,044
1994	40	354	394	119	179	132	287	717	1,110
1995	54	197	251	210	142	534	274	1,159	1,410
1996	26	272	298	292	154	184	275	905	1,203
1997	39	347	386	310	123	285	312	1,030	1,416
1998	27	291	318	331	135	163	597	1,226	1,544
1999	17	257	274	232	189	116	330	866	1,140
Yorkshire & Humberside									
1989	36	690	726	154	306	333	558	1,350	2,077
1990	38	459	497	278	471	293	482	1,523	2,020
1991	56	371	427	321	153	204	440	1,118	1,545
1992	63	361	424	285	241	172	305	1,003	1,426
1993	131	433	564	291	317	124	332	1,064	1,629
1994	117	476	593	193	277	160	414	1,044	1,637
1995	93	448	541	278	272	228	441	1,219	1,760
1996	71	400	471	488	257	203	406	1,354	1,825
1997	80	469	549	326	194	283	540	1,343	1,890
1998	63	507	570	527	263	276	644	1,710	2,280
1999	83	444	527	385	190	264	644	1,482	2,009
East Midlands									
1989	50	606	655	164	222	220	429	1,035	1,690
1990	30	421	451	263	200	467	362	1,291	1,742
1991	70	427	497	199	191	204	268	863	1,360
1992	52	424	477	377	219	119	243	958	1,434
1993	69	557	626	213	221	176	210	820	1,446
1994	86	617	703	131	229	160	250	770	1,474
1995	56	448	504	329	194	199	342	1,064	1,568
1996	43	634	677	187	148	268	381	984	1,660
1997	50	647	697	228	204	359	392	1,183	1,880
1998	33	576	609	270	164	316	479	1,228	1,837
1999	41	568	609	229	207	264	432	1,132	1,741
East Anglia									
1989	40	382	422	91	151	163	468	874	1,296
1990	20	207	227	99	108	104	250	560	787
1991	29	259	288	156	100	51	198	505	793
1992	48	198	246	108	111	62	98	379	624
1993	46	256	302	177	218	69	116	580	881
1994	26	296	322	119	174	95	143	531	852
1995	46	296	342	162	147	78	193	579	921
1996	33	280	313	213	148	67	155	583	896
1997	23	319	342	154	150	78	205	587	929
1998	30	278	308	116	159	72	371	718	1,025
1999	29	255	284	157	183	99	246	686	970

Table 1.5 New Orders obtained by Contractors[1] by Region (continued)

£ Million

	New Housing			Other New Work					
Year	Public	Private	All New Housing	Infra-structure	Public	Private Industrial	Private Commercial	All Other New Work	All New Work
South East (part): Beds, Essex, Herts									
1989	19	359	378	136	81	221	621	1,058	1,437
1990	28	249	277	147	138	135	470	890	1,167
1991	25	265	291	160	55	517	258	990	1,281
1992	57	210	267	125	76	63	189	454	721
1993	82	280	362	111	119	117	245	592	955
1994	74	349	423	134	127	167	231	659	1,082
1995	82	357	439	143	105	127	224	600	1,039
1996	52	355	407	201	145	140	264	750	1,158
1997	54	346	400	150	121	152	307	730	1,130
1998	37	403	440	189	102	95	352	738	1,179
1999	43	346	389	178	91	138	268	676	1,065
South East (part): Greater London									
1989	126	325	452	590	671	209	2,666	4,136	4,588
1990	68	185	253	359	437	152	2,454	3,402	3,655
1991	138	176	314	365	501	145	1,175	2,185	2,499
1992	240	166	406	406	366	87	883	1,742	2,148
1993	501	201	702	920	552	77	897	2,446	3,145
1994	256	268	524	683	581	83	1,160	2,507	3,031
1995	161	273	434	742	473	127	1,316	2,658	3,092
1996	173	459	632	941	462	88	1,605	3,096	3,728
1997	151	612	763	463	394	124	1,744	2,725	3,488
1998	188	505	693	460	717	254	2,280	3,711	4,403
1999	207	442	649	593	528	118	2,456	3,695	4,344
South East (part): Kent, Surrey, Sussex									
1989	89	448	536	267	301	213	776	1,557	2,094
1990	56	263	320	306	129	158	580	1,173	1,492
1991	42	253	295	329	137	91	466	1,024	1,319
1992	84	243	327	265	169	52	299	786	1,113
1993	112	261	373	361	151	128	244	884	1,257
1994	115	342	457	338	190	75	216	819	1,275
1995	79	304	383	341	144	80	674	1,238	1,621
1996	79	380	459	199	195	108	349	851	1,310
1997	82	442	524	370	187	129	528	1,214	1,738
1998	58	430	488	502	198	210	608	1,518	2,006
1999	73	558	631	365	253	96	498	1,212	1,843
South East (part): Berks, Bucks, Hants, Oxon									
1989	77	481	557	279	321	262	876	1,738	2,296
1990	75	376	451	296	268	132	547	1,243	1,694
1991	39	330	369	294	196	116	401	1,007	1,376
1992	82	338	420	225	208	111	258	802	1,222
1993	73	435	508	151	214	73	334	772	1,280
1994	80	458	538	183	367	182	325	1,057	1,595
1995	85	332	417	184	322	155	334	995	1,412
1996	115	402	517	362	207	101	450	1,120	1,637
1997	79	429	508	265	198	168	682	1,313	1,821
1998	85	349	434	193	310	225	759	1,487	1,921
1999	54	406	460	330	259	159	842	1,589	2,049

Table 1.5 New Orders obtained by Contractors[1] by Region (continued)

£ Million

Year	New Housing			Other New Work					All New Work
	Public	Private	All New Housing	Infra-structure	Public	Private Industrial	Private Commercial	All Other New Work	
South East: All									
1989	311	1,613	1,924	1,272	1,374	905	4,939	8,490	10,414
1990	228	1,073	1,301	1,107	972	578	4,051	6,708	8,009
1991	245	1,025	1,269	1,148	889	869	2,300	5,206	6,475
1992	463	957	1,420	1,021	820	313	1,630	3,784	5,203
1993	766	1,177	1,943	1,542	1,036	396	1,720	4,694	6,637
1994	525	1,417	1,942	1,338	1,264	507	1,932	5,041	6,983
1995	406	1,267	1,673	1,411	1,043	488	2,549	5,491	7,164
1996	419	1,596	2,015	1,704	1,009	437	2,668	5,818	7,833
1997	365	1,830	2,195	1,248	901	573	3,261	5,983	8,177
1998	368	1,686	2,054	1,345	1,328	785	3,998	7,456	9,509
1999	376	1,753	2,129	1,465	1,131	511	4,064	7,172	9,301
South West									
1989	83	803	886	158	342	240	833	1,572	2,458
1990	55	439	494	213	235	190	495	1,134	1,628
1991	65	508	573	250	231	130	564	1,174	1,747
1992	114	354	468	664	234	116	377	1,390	1,859
1993	107	473	580	288	352	128	444	1,212	1,790
1994	116	536	652	244	327	175	382	1,128	1,778
1995	108	435	543	282	239	188	349	1,057	1,600
1996	53	454	507	368	252	133	370	1,123	1,630
1997	71	572	643	219	304	273	430	1,226	1,869
1998	62	560	622	271	347	187	563	1,368	1,989
1999	53	486	539	340	259	262	660	1,521	2,060
West Midlands									
1989	88	614	702	182	385	359	670	1,596	2,297
1990	79	461	539	300	305	291	594	1,490	2,029
1991	101	462	564	288	221	174	462	1,145	1,709
1992	95	460	555	251	210	195	374	1,030	1,587
1993	116	429	545	329	255	196	339	1,119	1,663
1994	85	514	599	220	256	211	429	1,116	1,716
1995	86	508	594	383	219	309	364	1,276	1,870
1996	67	496	563	246	234	275	443	1,198	1,761
1997	69	605	674	390	282	469	603	1,744	2,418
1998	85	631	716	324	247	365	752	1,688	2,405
1999	77	708	785	272	317	369	805	1,763	2,548
North West									
1989	59	598	658	261	330	316	744	1,651	2,308
1990	55	501	555	354	245	347	539	1,485	2,040
1991	89	497	586	236	271	252	479	1,239	1,825
1992	159	410	569	347	292	223	446	1,308	1,877
1993	122	514	636	334	341	194	444	1,313	1,950
1994	148	665	813	431	303	229	523	1,486	2,300
1995	99	545	644	471	317	218	687	1,694	2,338
1996	115	500	615	568	258	325	595	1,746	2,360
1997	97	531	628	466	228	297	750	1,741	2,369
1998	95	595	690	605	309	368	903	2,185	2,874
1999	100	541	641	372	309	285	745	1,710	2,351

Table 1.5 New Orders obtained by Contractors[1] by Region (continued)

£ Million

Year	New Housing			Other New Work					All New Work
	Public	Private	All New Housing	Infra-structure	Public	Private Industrial	Private Commercial	All Other New Work	
England									
1989	697	5,662	6,359	2,472	3,292	2,683	8,979	17,426	23,785
1990	535	3,899	4,434	2,781	2,634	2,444	7,013	14,872	19,306
1991	717	3,810	4,527	2,812	2,158	2,002	4,900	11,872	16,399
1992	1,035	3,386	4,421	3,187	2,248	1,269	3,633	10,337	14,759
1993	1,409	4,133	5,542	3,397	2,919	1,409	3,775	11,500	17,042
1994	1,143	4,875	6,018	2,795	3,009	1,668	4,360	11,832	17,851
1995	948	4,144	5,092	3,526	2,572	2,242	5,199	13,539	18,631
1996	827	4,633	5,460	4,065	2,461	1,892	5,293	13,711	19,171
1997	794	5,315	6,109	3,353	2,387	2,616	6,491	14,847	20,957
1998	762	5,124	5,886	3,788	2,952	2,532	8,306	17,578	23,464
1999	776	5,012	5,788	3,451	2,785	2,169	7,927	16,332	22,120
Wales									
1989	42	310	352	189	154	161	219	722	1,073
1990	61	335	396	154	190	149	233	725	1,121
1991	45	193	239	372	163	171	186	893	1,132
1992	68	202	270	179	162	76	108	526	796
1993	55	209	264	234	158	76	147	615	879
1994	86	208	294	344	296	131	198	969	1,263
1995	49	212	261	297	171	93	172	734	995
1996	59	215	274	194	141	178	269	782	1,057
1997	37	252	289	217	163	190	353	923	1,212
1998	50	219	269	240	151	141	298	830	1,100
1999	42	215	257	257	102	139	240	737	994
Scotland									
1989	133	526	659	300	390	206	729	1,624	2,283
1990	88	621	709	256	293	257	550	1,355	2,064
1991	113	548	661	238	357	174	494	1,264	1,924
1992	143	428	571	381	353	156	475	1,365	1,936
1993	204	532	736	367	320	193	429	1,309	2,045
1994	156	638	794	312	349	200	517	1,378	2,171
1995	185	548	733	347	464	389	505	1,706	2,439
1996	186	568	754	404	354	268	828	1,854	2,607
1997	164	685	849	422	344	343	681	1,790	2,638
1998	121	654	775	450	401	298	989	2,138	2,913
1999	151	674	825	487	387	250	1,017	2,141	2,966

Notes

1. Classified to construction in the 1992 Revised Standard Industrial Clasification.

Source of Data: Construction Market Intelligence, Department of Environment, Transport and the Regions
Contact: Neville Price 020 7944 5587

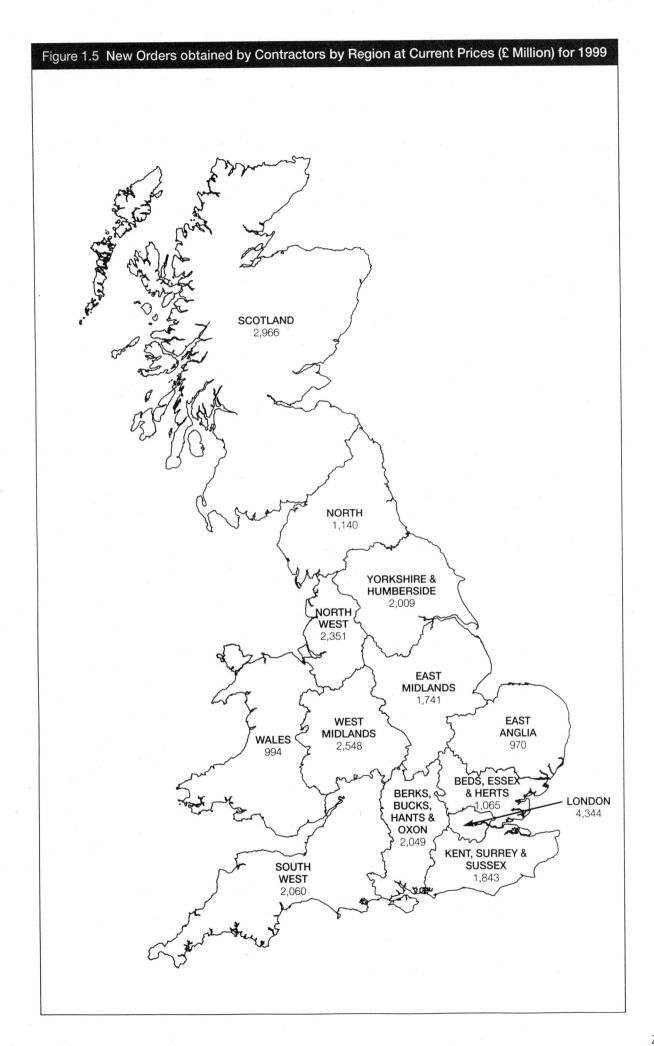

Figure 1.5 New Orders obtained by Contractors by Region at Current Prices (£ Million) for 1999

Table 1.6 New Orders obtained by Contractors by Value Range

Percentage

Value Range (£000s)	New Housing				Infrastructure		Other New Work (Excluding Infrastructure)					
	Public		Private				Public		Private			
									Industrial		Commercial	
	Number	Value	Number	Value	Number	Value	Number	Value	Number	Value	Number	Value
(a) 1996												
25 and under 50	12.3	0.6	11.8	2.0	21.7	1.1	30.2	3.5	30.7	2.9	32.8	2.9
50 and under 100	9.0	1.0	28.5	8.4	23.6	2.4	24.7	5.6	21.8	4.2	24.5	4.3
100 and under 150	6.8	1.3	16.4	8.5	13.8	2.5	11.2	4.5	11.6	3.7	10.7	3.2
150 and under 200	4.6	1.2	11.3	8.3	8.2	2.1	6.9	3.8	6.6	3.1	6.7	2.9
200 and under 250	6.0	2.1	7.2	7.0	4.7	1.5	4.3	3.1	4.1	2.4	3.5	2.0
250 and under 300	3.8	1.6	5.2	6.1	3.3	1.3	3.3	2.9	3.0	2.2	2.8	1.9
300 and under 400	8.4	4.4	6.7	10.0	5.0	2.6	4.6	5.1	4.1	3.8	3.6	3.1
400 and under 500	7.1	4.9	3.7	7.2	2.9	1.9	2.6	3.7	2.8	3.4	2.2	2.5
500 and under 750	15.2	14.3	4.4	11.5	4.6	4.3	4.0	8.0	4.3	7.2	4.2	6.5
750 and under 1,000	6.9	9.1	1.7	6.4	2.4	3.1	2.4	6.6	3.0	7.0	2.2	4.8
1,000 and under 2,000	14.6	30.3	2.6	14.8	4.3	9.2	3.3	14.0	4.3	16.0	3.3	11.2
2,000 and Over	5.5	29.2	0.5	9.4	4.9	35.9	2.7	36.0	3.8	42.1	3.3	39.8
All	100.0	100.0	100.0	100.0	100.0	100.0	100.0	100.0	100.0	100.0	100.0	100.0
Number (Thousand) £25,000 and Over	1.6		22.7		6.1		9.0		4.6		13.8	
Value (£ Million) £25,000 and Over		1,072		5,395		4,627		2,886		2,248		6,170
(b) 1997												
25 and under 50	5.7	0.3	10.6	1.6	26.2	1.7	28.5	3.0	29.2	2.3	30.6	2.8
50 and under 100	8.7	0.9	28.5	8.2	20.5	2.7	24.7	5.2	24.8	3.9	25.0	4.5
100 and under 150	6.1	1.0	16.3	8.0	10.4	2.4	11.5	4.2	9.2	2.6	10.5	3.3
150 and under 200	7.5	1.9	10.8	7.5	6.8	2.2	6.5	3.3	5.2	2.1	6.8	3.0
200 and under 250	8.3	2.7	8.0	7.2	5.4	2.3	4.7	3.1	3.6	1.8	3.8	2.2
250 and under 300	6.4	2.5	5.5	6.0	3.5	1.8	3.5	2.8	3.8	2.4	3.2	2.2
300 and under 400	7.1	3.2	7.0	9.6	5.2	3.4	4.5	4.6	4.8	3.8	4.0	3.5
400 and under 500	8.1	5.2	3.7	6.6	3.6	3.1	3.1	4.1	2.5	2.6	2.7	3.1
500 and under 750	11.5	9.9	4.3	10.3	5.3	6.1	4.1	7.4	4.7	6.6	3.8	5.9
750 and under 1,000	8.8	10.9	1.7	6.0	3.2	5.2	2.5	6.4	2.4	4.7	2.3	5.2
1,000 and under 2,000	14.9	29.7	2.9	15.8	5.2	13.3	3.4	13.3	5.2	16.0	3.7	12.8
2,000 and Over	6.9	31.9	0.7	13.1	4.8	55.7	2.9	42.5	4.6	51.2	3.6	51.5
All	100.0	100.0	100.0	100.0	100.0	100.0	100.0	100.0	100.0	100.0	100.0	100.0
Number (Thousand) £25,000 and Over	1.4		24.5		6.4		8.1		5.3		16.2	
Value (£ Million) £25,000 and Over		990		6,237		3,896		2,815		3,056		7,136

Table 1.6 New Orders obtained by Contractors by Value Range (continued)

Percentage

Value Range (£000s)	New Housing Public Number	Value	Private Number	Value	Infrastructure Number	Value	Other New Work (Excluding Infrastructure) Public Number	Value	Private Industrial Number	Value	Commercial Number	Value
(c) 1998												
25 and under 50	8.6	0.5	8.5	1.2	26.1	1.5	29.5	2.8	28.3	2.2	30.3	2.1
50 and under 100	10.3	1.2	26.0	7.0	19.2	2.2	25.3	4.8	20.5	3.2	21.8	3.0
100 and under 150	10.0	1.9	18.3	7.9	10.0	1.9	10.7	3.5	10.4	2.7	11.1	2.6
150 and under 200	6.5	1.8	11.5	6.9	7.5	2.1	6.6	3.1	6.1	2.4	7.1	2.4
200 and under 250	6.5	2.3	8.3	6.8	4.6	1.6	4.9	2.9	4.2	2.1	4.1	1.8
250 and under 300	4.8	2.1	5.4	5.6	3.7	1.6	3.0	2.2	3.3	2.0	2.9	1.5
300 and under 400	8.2	4.5	6.6	9.0	5.2	3.0	3.9	3.6	4.1	3.2	4.3	3.0
400 and under 500	5.9	4.2	4.2	7.4	3.3	2.4	2.9	3.5	3.0	3.0	2.5	2.2
500 and under 750	13.7	13.2	5.0	12.1	4.8	4.7	4.3	7.0	5.6	7.4	4.1	5.0
750 and under 1,000	6.8	9.2	2.1	7.0	3.6	5.1	2.1	5.0	2.1	4.2	2.0	3.4
1,000 and under 2,000	13.1	28.0	3.5	17.9	5.8	12.3	3.7	12.8	7.6	23.6	4.2	11.1
2,000 and Over	5.6	31.1	0.8	11.3	6.2	61.7	3.1	48.8	4.8	44.2	5.7	61.9
All	**100.0**	**100.0**	**100.0**	**100.0**	**100.0**	**100.0**	**100.0**	**100.0**	**100.0**	**100.0**	**100.0**	**100.0**
Number (Thousand) £25,000 and Over	1.5		23.7		6.3		8.9		5.0		16.6	
Value (£ Million) £25,000 and Over		932		5,972		4,379		3,415		2,862		9,267
(d) 1999												
25 and under 50	5.7	0.3	7.6	0.9	26.9	1.7	26.3	2.4	28.3	2.1	30.2	1.2
50 and under 100	6.5	0.6	24.4	5.5	20.4	2.5	24.9	4.6	20.7	3.0	21.7	1.7
100 and under 150	5.9	1.1	16.5	6.1	10.5	2.2	10.4	3.2	10.3	2.6	9.9	1.3
150 and under 200	8.7	2.1	10.4	5.5	6.0	1.8	6.6	2.9	5.1	1.8	6.7	1.3
200 and under 250	6.1	2.0	8.0	5.5	4.9	1.9	4.9	2.8	4.2	2.0	4.0	1.0
250 and under 300	6.6	2.6	5.9	4.9	4.2	2.0	3.1	2.1	2.7	1.5	3.1	1.0
300 and under 400	8.5	4.2	8.0	8.5	5.2	3.2	4.5	4.0	4.6	3.4	3.9	1.5
400 and under 500	8.6	5.5	4.3	5.9	3.0	2.4	3.5	4.1	3.9	3.6	3.0	1.5
500 and under 750	13.6	11.7	5.7	10.8	5.7	6.0	5.2	8.1	4.9	6.2	5.0	3.4
750 and under 1,000	7.1	8.4	2.6	6.9	3.0	4.4	2.8	6.2	3.0	5.5	2.6	2.5
1,000 and under 2,000	14.2	27.8	5.3	22.4	4.9	11.9	4.6	15.9	7.2	20.7	4.5	6.8
2,000 and Over	8.4	33.9	1.3	17.2	5.3	59.8	3.1	43.6	5.1	47.6	5.4	36.9
All	**100.0**	**100.0**	**100.0**	**100.0**	**100.0**	**100.0**	**100.0**	**100.0**	**100.0**	**100.0**	**100.0**	**100.0**
Number (Thousand) £25,000 and Over	1.3		17.5		6.3		7.7		3.9		14.8	
Value (£ Million) £25,000 and Over		907		5,848		4,086		3,116		2,422		14,753

Source of Data: Construction Market Intelligence, Department of Environment, Transport and the Regions
Contact: Neville Price 020 7944 5587

Table 1.7 New Orders obtained by Contractors by Duration[1]

Percentage

Duration (Months)	New Housing				Infrastructure		Other New Work (Excluding Infrastructure)					
	Public		Private				Public		Private			
									Industrial		Commercial	
	Number	Value	Number	Value	Number	Value	Number	Value	Number	Value	Number	Value
(a) 1996												
0-6	41.5	15.8	80.3	65.3	83.8	24.6	86.1	37.1	89.7	50.7	90.8	40.3
7-12	48.7	53.0	18.1	27.0	11.7	23.8	11.8	35.4	9.4	40.7	7.8	29.5
13-18	8.0	24.1	0.9	3.4	2.1	14.5	1.6	17.7	0.7	7.1	0.9	14.2
19-24	1.0	4.2	0.6	3.4	1.4	11.5	0.3	5.2	0.2	1.5	0.3	8.8
25-30	0.4	1.4	0.0	0.3	0.3	7.4	0.1	3.2	0.0	0.0	0.0	1.0
31-36	0.2	1.1	0.0	0.1	0.3	7.8	0.1	1.0	0.0	0.0	0.0	1.0
Over 36	0.1	0.4	0.0	0.5	0.4	10.4	0.0	0.4	0.0	0.1	0.0	5.2
All New Work	100.0	100.0	100.0	100.0	100.0	100.0	100.0	100.0	100.0	100.0	100.0	100.0
Number (Thousand) £25,000 and Over	1.6		22.7		6.1		9.0		4.6		13.8	
Value (£ Million) £25,000 and Over		1,072		5,395		4,627		2,886		2,248		6,170
(b) 1997												
0-6	37.0	14.7	75.1	56.8	83.6	33.7	85.5	36.8	88.0	43.4	90.7	44.9
7-12	53.1	56.0	22.2	33.6	12.2	31.6	12.1	35.5	10.9	42.9	8.1	29.3
13-18	7.8	20.3	1.7	4.4	1.8	8.3	1.6	17.0	0.8	8.8	0.8	13.9
19-24	1.5	6.4	0.8	3.2	1.3	10.9	0.7	8.9	0.1	3.2	0.2	5.2
25-30	0.4	0.9	0.1	1.0	0.3	5.5	0.1	0.8	0.1	1.1	0.1	3.8
31-36	0.1	0.7	0.1	0.6	0.4	6.4	0.0	0.7	0.1	0.4	0.1	1.0
Over 36	0.2	1.0	0.1	0.5	0.4	3.7	0.0	0.4	0.0	0.1	0.0	1.9
All New Work	100.0	100.0	100.0	100.0	100.0	100.0	100.0	100.0	100.0	100.0	100.0	100.0
Number (Thousand) £25,000 and Over	1.4		24.5		6.4		8.1		5.3		16.2	
Value (£ Million) £25,000 and Over		999		6,237		3,896		2,815		3,056		7,136

Notes

1. Figures for 1998 and 1999 will be published in the issue covering 1990-2000.

Source of Data: Construction Market Intelligence, Department of Environment, Transport and the Regions
Contact: Neville Price 020 7944 5587

CHAPTER 2

Output

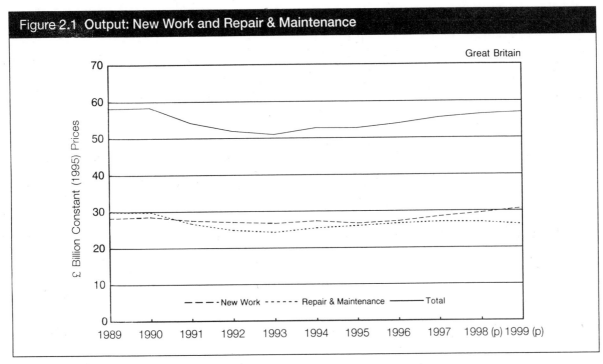

Figure 2.1 Output: New Work and Repair & Maintenance

Source of Data: Table 2.2

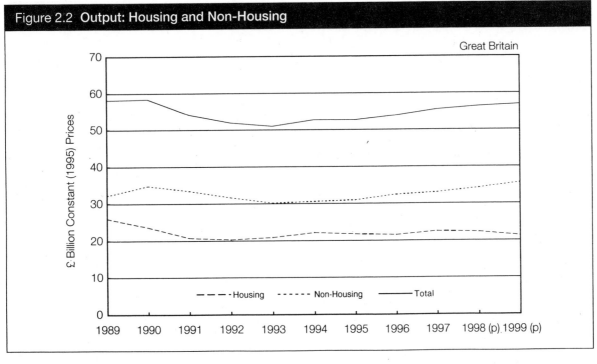

Figure 2.2 Output: Housing and Non-Housing

Source of Data: Table 2.2

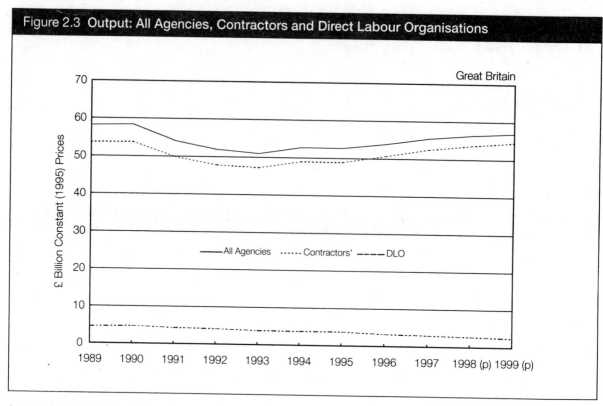

Figure 2.3 Output: All Agencies, Contractors and Direct Labour Organisations

Great Britain

Source of Data: Table 2.2 and 2.7

Table 2.1 Output[1]

Current Prices (£ Million)   Great Britain

Year & Quarter		New Housing Public	New Housing Private	Infrastructure	Other New Work Exc. Infrastructure Public	Private Indus- trial	Private Comm- ercial	All New Work	Repair & Maintenance Housing Public	Housing Private	Other Work Public	Other Work Private	All Repair & Mainte- nance	All Work
1989		966	7,117	4,017	3,903	3,425	9,892	29,320	4,943	8,149	4,982	4,755	22,830	52,150
1990	Q1	258	1,434	1,080	1,096	830	2,716	7,414	1,415	2,087	1,394	1,283	6,180	13,594
	Q2	233	1,417	1,174	1,136	885	2,949	7,795	1,387	2,088	1,299	1,281	6,055	13,850
	Q3	236	1,446	1,397	1,160	801	2,945	7,986	1,327	2,148	1,417	1,299	6,190	14,175
	Q4	207	1,449	1,313	1,022	878	2,699	7,568	1,255	2,132	1,378	1,355	6,120	13,688
1991	Q1	202	1,139	1,381	1,133	664	2,430	6,949	1,313	1,972	1,400	1,302	5,988	12,937
	Q2	187	1,350	1,439	1,055	633	2,355	7,019	1,179	1,928	1,291	1,340	5,739	12,757
	Q3	193	1,288	1,623	1,007	701	2,250	7,063	1,272	2,122	1,330	1,220	5,944	13,006
	Q4	211	1,226	1,619	947	624	2,067	6,695	1,173	2,039	1,270	1,236	5,719	12,414
1992	Q1	252	1,163	1,348	1,074	556	1,821	6,215	1,327	1,798	1,380	1,226	5,731	11,946
	Q2	299	1,215	1,446	1,069	586	1,773	6,388	1,186	1,745	1,202	1,243	5,376	11,764
	Q3	333	1,286	1,469	1,044	583	1,624	6,339	1,263	2,011	1,276	1,214	5,763	12,102
	Q4	359	1,177	1,452	993	510	1,382	5,873	1,216	2,041	1,229	1,302	5,788	11,661
1993	Q1	328	1,124	1,479	1,057	482	1,203	5,674	1,392	1,760	1,305	1,206	5,663	11,337
	Q2	366	1,313	1,457	1,008	571	1,258	5,974	1,339	1,744	1,130	1,241	5,455	11,429
	Q3	361	1,353	1,392	1,034	567	1,335	6,042	1,308	1,887	1,282	1,295	5,772	11,814
	Q4	359	1,423	1,215	946	587	1,335	5,865	1,401	1,978	1,199	1,300	5,877	11,742
1994	Q1	409	1,357	1,162	1,029	546	1,240	5,743	1,598	1,869	1,332	1,271	6,069	11,812
	Q2	432	1,500	1,301	1,058	636	1,357	6,283	1,411	1,909	1,251	1,302	5,872	12,156
	Q3	449	1,451	1,360	1,214	611	1,487	6,571	1,468	1,975	1,317	1,352	6,111	12,683
	Q4	381	1,438	1,325	1,083	696	1,564	6,488	1,486	2,052	1,312	1,451	6,301	12,789
1995	Q1	393	1,440	1,254	1,167	676	1,358	6,289	1,723	1,909	1,418	1,357	6,407	12,697
	Q2	425	1,446	1,481	1,170	718	1,527	6,767	1,525	2,014	1,231	1,462	6,232	12,999
	Q3	442	1,329	1,415	1,208	817	1,642	6,853	1,629	2,062	1,354	1,535	6,580	13,433
	Q4	399	1,259	1,510	1,115	797	1,682	6,762	1,588	2,145	1,395	1,624	6,752	13,514
1996	Q1	348	1,187	1,495	1,200	705	1,575	6,509	1,786	1,901	1,540	1,614	6,840	13,349
	Q2	387	1,384	1,616	1,102	803	1,732	7,024	1,530	2,053	1,188	1,765	6,535	13,560
	Q3	350	1,458	1,596	1,143	828	1,872	7,247	1,762	2,062	1,243	1,770	6,837	14,083
	Q4	337	1,563	1,630	997	783	1,836	7,146	1,560	2,382	1,281	1,881	7,105	14,251
1997	Q1	290	1,510	1,495	979	770	1,934	6,977	1,720	2,160	1,343	1,761	6,985	13,961
	Q2	318	1,678	1,616	897	907	2,112	7,528	1,623	2,137	1,202	1,911	6,873	14,400
	Q3	351	1,658	1,596	967	908	2,212	7,691	1,606	2,348	1,235	1,891	7,080	14,771
	Q4	273	1,905	1,604	914	907	2,131	7,733	1,679	2,480	1,299	2,027	7,486	15,219
1998	Q1	285	1,739	1,647	969	898	2,109	7,648	1,725	2,337	1,351	2,151	7,564	15,211
	Q2	276	1,953	1,489	1,026	959	2,329	8,034	1,423	2,461	1,216	1,994	7,095	15,129
	Q3	268	1,829	1,460	1,077	969	2,821	8,423	1,633	2,462	1,288	1,967	7,350	15,773
	Q4 (p)	240	1,839	1,586	1,079	984	2,658	8,386	1,725	2,437	1,364	2,035	7,561	15,947
1999	Q1 (p)	238	1,741	1,441	1,197	928	2,630	8,175	1,699	2,386	1,415	1,963	7,463	15,638
	Q2 (p)	234	1,952	1,496	1,170	1,015	2,973	8,841	1,508	2,430	1,236	2,017	7,191	16,032
	Q3 (p)	261	1,887	1,536	1,280	979	3,149	9,092	1,620	2,498	1,348	2,101	7,567	16,658
	Q4 (p)	266	1,896	1,690	1,162	1,004	3,155	9,173	1,632	2,511	1,365	2,253	7,760	16,933

Notes

p = provisional

1. Output by contractors, including estimates of unrecorded output by small firms and self-employed workers, and output by public sector direct labour departments, classified to construction in the 1992 Revised Standard Industrial Classification.

Source of Data: Construction Market Intelligence, Department of Environment, Transport and the Regions
Contact: Neville Price 020 7944 5587

Table 2.2 Output[1]

Constant (1995) Prices Seasonally Adjusted (£ Million)   Great Britain

Year & Quarter		New Housing		Infra-structure	Other New Work Non-Housing			All New Work	Repair & Maintenance				All Repair & Mainte-nance	All Work
						Other New Work Exc. Infrastructure			Housing		Other Work			
		Public	Private		Public	Private Indus-trial	Private Comm-ercial		Public	Private	Public	Private		
1989		1,016	7,803	3,853	3,557	3,217	8,777	28,224	6,490	10,658	6,541	6,223	29,913	58,137
1990	Q1	261	1,540	1,149	946	790	2,456	7,143	1,747	2,647	1,680	1,617	7,690	14,833
	Q2	227	1,423	1,131	971	807	2,574	7,134	1,778	2,636	1,668	1,559	7,641	14,775
	Q3	235	1,407	1,265	1,001	716	2,524	7,147	1,559	2,473	1,693	1,579	7,304	14,451
	Q4	217	1,482	1,224	935	820	2,435	7,112	1,473	2,509	1,643	1,581	7,205	14,316
1991	Q1	204	1,193	1,524	1,014	685	2,331	6,951	1,490	2,307	1,548	1,513	6,858	13,808
	Q2	189	1,349	1,460	972	645	2,216	6,831	1,413	2,270	1,544	1,511	6,738	13,569
	Q3	202	1,267	1,572	964	705	2,121	6,832	1,411	2,316	1,494	1,413	6,634	13,466
	Q4	233	1,286	1,641	979	660	2,070	6,869	1,316	2,285	1,447	1,372	6,420	13,289
1992	Q1	270	1,255	1,653	1,087	651	1,948	6,866	1,447	2,032	1,469	1,376	6,325	13,190
	Q2	319	1,258	1,670	1,104	666	1,849	6,866	1,375	1,991	1,391	1,354	6,112	12,978
	Q3	365	1,311	1,611	1,112	637	1,698	6,735	1,347	2,114	1,379	1,359	6,199	12,933
	Q4	417	1,289	1,636	1,131	574	1,524	6,571	1,310	2,204	1,347	1,393	6,254	12,825
1993	Q1	375	1,282	1,895	1,147	587	1,419	6,705	1,459	1,928	1,325	1,305	6,017	12,722
	Q2	413	1,380	1,779	1,115	623	1,408	6,718	1,495	1,914	1,260	1,308	5,977	12,695
	Q3	408	1,399	1,603	1,156	588	1,458	6,610	1,366	1,964	1,339	1,419	6,088	12,698
	Q4	405	1,530	1,455	1,123	641	1,509	6,663	1,467	2,069	1,307	1,358	6,201	12,865
1994	Q1	465	1,516	1,484	1,123	663	1,487	6,738	1,645	1,990	1,306	1,378	6,319	13,057
	Q2	455	1,519	1,497	1,171	673	1,492	6,808	1,554	2,047	1,394	1,365	6,359	13,168
	Q3	465	1,475	1,507	1,267	632	1,534	6,879	1,518	2,033	1,371	1,416	6,337	13,216
	Q4	405	1,468	1,453	1,197	720	1,631	6,873	1,498	2,071	1,359	1,449	6,377	13,251
1995	Q1	418	1,533	1,388	1,176	742	1,478	6,735	1,706	1,951	1,329	1,420	6,405	13,140
	Q2	411	1,396	1,461	1,199	688	1,526	6,682	1,614	2,071	1,332	1,470	6,487	13,169
	Q3	425	1,309	1,361	1,150	788	1,559	6,591	1,614	2,032	1,361	1,531	6,538	13,129
	Q4	406	1,237	1,449	1,136	790	1,645	6,663	1,532	2,075	1,377	1,557	6,541	13,204
1996	Q1	367	1,240	1,527	1,139	767	1,622	6,662	1,715	1,882	1,396	1,639	6,632	13,294
	Q2	372	1,313	1,522	1,092	775	1,663	6,737	1,573	2,061	1,255	1,727	6,616	13,352
	Q3	333	1,420	1,521	1,056	814	1,741	6,885	1,707	1,996	1,236	1,728	6,667	13,552
	Q4	338	1,494	1,548	993	792	1,784	6,948	1,471	2,250	1,232	1,764	6,717	13,665
1997	Q1	303	1,522	1,524	915	846	1,965	7,075	1,605	2,087	1,185	1,747	6,625	13,700
	Q2	298	1,527	1,504	879	869	1,978	7,055	1,627	2,096	1,236	1,820	6,779	13,834
	Q3	326	1,544	1,505	869	857	1,996	7,097	1,520	2,230	1,201	1,813	6,764	13,860
	Q4	267	1,728	1,495	872	865	1,998	7,224	1,540	2,265	1,204	1,841	6,850	14,074
1998	Q1	289	1,664	1,647	867	928	2,047	7,441	1,562	2,196	1,159	2,072	6,990	14,431
	Q2	249	1,679	1,351	959	869	2,075	7,182	1,382	2,347	1,212	1,841	6,782	13,964
	Q3	236	1,599	1,335	923	863	2,408	7,364	1,461	2,192	1,183	1,767	6,604	13,968
	Q4 (p)	220	1,562	1,445	984	886	2,360	7,457	1,485	2,115	1,194	1,755	6,549	14,007
1999	Q1 (p)	226	1,549	1,432	1,033	905	2,409	7,554	1,461	2,134	1,153	1,804	6,553	14,106
	Q2 (p)	199	1,552	1,365	1,058	879	2,498	7,550	1,402	2,236	1,183	1,802	6,624	14,174
	Q3 (p)	218	1,579	1,421	1,065	850	2,540	7,672	1,394	2,159	1,196	1,834	6,584	14,256
	Q4 (p)	233	1,475	1,550	1,032	898	2,656	7,844	1,358	2,118	1,157	1,891	6,523	14,367

Notes

p = provisional

1. Output by contractors, including estimates of unrecorded output by small firms and self-employed workers, and output by public sector direct labour departments, classified to construction in the 1992 Revised Standard Industrial Classification.

Source of Data: Construction Market Intelligence, Department of Environment, Transport and the Regions

Contact: Neville Price 020 7944 5587

Table 2.3 Output[1]

Index at 1995=100 Prices   Great Britain

| Year & Quarter | | New Housing | | Infra-structure | Other New Work Non-Housing Other New Work Exc. Infrastructure | | | All New Work | Repair & Maintenance Housing | | Other Work | | All Repair & Mainte-nance | All Work |
		Public	Private		Public	Private Indus-trial	Private Comm-ercial		Public	Private	Public	Private		
1989		61.2	142.5	68.1	76.3	107.0	141.4	105.8	100.4	131.1	121.2	104.1	115.2	110.4
1990	Q1	63.0	112.5	81.2	81.2	105.1	158.2	107.1	108.1	130.2	124.5	108.2	118.4	112.7
	Q2	54.8	104.0	80.0	83.4	107.3	165.8	107.0	110.0	129.7	123.6	104.3	117.7	112.3
	Q3	56.8	102.8	89.4	85.9	95.2	162.6	107.2	96.4	121.7	125.5	105.6	112.5	109.8
	Q4	52.2	108.3	86.5	80.2	109.0	156.9	106.7	91.1	123.4	121.7	105.8	111.0	108.8
1991	Q1	49.1	87.2	107.7	87.1	91.2	150.2	104.2	92.2	113.5	114.7	101.2	105.6	104.9
	Q2	45.6	98.6	103.2	83.4	85.7	142.8	102.4	87.4	111.7	114.4	101.1	103.8	103.1
	Q3	48.8	92.6	111.1	82.7	93.8	136.7	102.5	87.3	114.0	110.7	94.5	102.2	102.3
	Q4	56.2	94.0	116.0	84.0	87.8	133.3	103.0	81.4	112.4	107.2	91.8	98.9	101.0
1992	Q1	65.1	91.7	116.8	93.3	86.6	125.5	103.0	89.5	100.0	108.9	92.1	97.4	100.2
	Q2	77.0	91.9	118.0	94.7	88.5	119.1	103.0	85.1	98.0	103.1	90.6	94.1	98.6
	Q3	88.0	95.8	113.9	95.4	84.7	109.4	101.0	83.3	104.0	102.2	90.9	95.5	98.3
	Q4	100.6	94.2	115.6	97.1	76.3	98.2	98.5	81.0	108.4	99.8	93.2	96.3	97.5
1993	Q1	90.3	93.7	133.9	98.4	78.0	91.4	100.6	90.3	94.8	98.2	87.4	92.7	96.7
	Q2	99.5	100.8	125.7	95.7	82.8	90.7	100.8	92.5	94.2	93.4	87.5	92.1	96.5
	Q3	98.2	102.2	113.3	99.2	78.1	93.9	99.1	84.5	96.6	99.2	94.9	93.8	96.5
	Q4	97.7	111.8	102.8	96.4	85.2	97.2	99.9	90.7	101.8	96.9	90.9	95.5	97.7
1994	Q1	112.2	110.8	104.9	96.4	88.1	95.8	101.1	101.8	97.9	96.8	92.2	97.3	99.2
	Q2	109.8	111.0	105.8	100.5	89.5	96.1	102.1	96.1	100.7	103.3	91.3	97.9	100.1
	Q3	112.1	107.8	106.5	108.7	84.0	98.8	103.2	93.9	100.0	101.6	94.8	97.6	100.4
	Q4	97.6	107.2	102.7	102.7	95.8	105.0	103.1	92.6	101.9	100.7	96.9	98.2	100.7
1995	Q1	100.7	112.0	98.1	100.9	98.7	95.2	101.0	105.5	96.0	98.5	95.0	98.6	99.8
	Q2	99.1	102.0	103.3	102.9	91.6	98.3	100.2	99.8	101.9	98.7	98.4	99.9	100.1
	Q3	102.4	95.6	96.2	98.7	104.8	100.5	98.8	99.9	100.0	100.8	102.4	100.7	99.8
	Q4	97.8	90.4	102.4	97.5	105.0	106.0	99.9	94.8	102.1	102.0	104.2	100.7	100.3
1996	Q1	88.5	90.6	107.9	97.7	102.0	104.5	99.9	106.1	92.6	103.5	109.7	102.1	101.0
	Q2	89.6	95.9	107.6	93.7	103.1	107.1	101.0	97.3	101.4	93.0	115.6	101.9	101.5
	Q3	80.3	103.7	107.5	90.6	108.3	112.2	103.3	105.6	98.2	91.6	115.6	102.7	103.0
	Q4	81.5	109.1	109.4	85.2	105.4	114.9	104.2	91.0	110.7	91.3	118.1	103.4	103.8
1997	Q1	73.0	111.2	107.7	78.5	112.5	126.6	106.1	99.3	102.7	87.8	116.9	102.0	104.1
	Q2	71.8	111.6	106.3	75.4	115.6	127.5	105.8	100.7	103.1	91.6	121.8	104.4	105.1
	Q3	78.6	112.8	106.4	74.6	113.9	128.6	106.4	94.0	109.7	89.0	121.3	104.2	105.3
	Q4	64.2	126.2	105.7	74.8	115.1	128.7	108.3	95.3	111.4	89.2	123.2	105.5	106.9
1998	Q1	69.6	121.6	116.4	74.4	123.4	131.9	111.6	96.6	108.0	85.9	138.7	107.7	109.7
	Q2	59.9	122.7	95.5	82.3	115.6	133.7	107.7	85.5	115.5	89.8	123.2	104.5	106.1
	Q3	57.0	116.8	94.4	79.2	114.8	155.1	110.4	90.4	107.9	87.7	118.2	101.7	106.1
	Q4 (p)	53.1	114.1	102.1	84.5	117.9	152.0	111.8	91.9	104.1	88.5	117.4	100.9	106.4
1999	Q1 (p)	54.6	113.2	101.2	88.6	120.3	155.2	113.3	90.4	105.0	85.4	120.7	100.9	107.2
	Q2 (p)	47.9	113.4	96.5	90.8	116.9	160.9	113.2	86.7	110.0	87.7	120.6	102.0	107.7
	Q3 (p)	52.5	115.3	100.4	91.4	113.1	163.6	115.1	86.3	106.3	88.6	122.7	101.4	108.3
	Q4 (p)	56.1	107.7	109.5	88.6	119.5	171.1	117.6	84.0	104.2	85.8	126.5	100.5	109.2

Notes

p = provisional

1. Output by contractors, including estimates of unrecorded output by small firms and self-employed workers, and output by public sector direct labour departments, classified to construction in the 1992 Revised Standard Industrial Classification.

Source of Data: Construction Market Intelligence, Department of Environment, Transport and the Regions
Contact: Neville Price 020 7944 5587

Table 2.4 Contractors' Output[1]

Current Prices (£ Million)   Great Britain

Year & Quarter		New Housing Public	New Housing Private	Infra-structure	Other New Work Exc. Infrastructure Public	Private Indus-trial	Private Comm-ercial	All New Work	Repair & Maintenance Housing Public	Repair & Maintenance Housing Private	Other Work Public	Other Work Private	All Repair & Mainte-nance	All Work
1989		944	7,117	3,899	3,695	3,425	9,892	28,972	3,661	8,149	3,210	4,755	19,775	48,747
1990	Q1	250	1,434	1,026	1,029	830	2,716	7,285	1,037	2,087	898	1,283	5,306	12,591
	Q2	227	1,417	1,133	1,101	885	2,949	7,713	1,057	2,088	864	1,281	5,289	13,002
	Q3	231	1,446	1,342	1,111	801	2,945	7,876	969	2,148	910	1,299	5,326	13,201
	Q4	203	1,449	1,256	976	878	2,699	7,461	901	2,132	892	1,355	5,280	12,740
1991	Q1	196	1,139	1,338	1,080	664	2,430	6,847	911	1,972	853	1,302	5,038	11,886
	Q2	182	1,350	1,406	1,017	633	2,355	6,944	854	1,928	862	1,340	4,984	11,928
	Q3	189	1,288	1,582	968	701	2,250	6,978	918	2,122	866	1,220	5,126	12,105
	Q4	207	1,226	1,578	904	624	2,067	6,607	797	2,039	790	1,236	4,863	11,470
1992	Q1	248	1,163	1,303	1,008	556	1,821	6,099	910	1,798	838	1,226	4,772	10,871
	Q2	296	1,215	1,403	1,033	586	1,773	6,306	863	1,745	788	1,243	4,640	10,946
	Q3	329	1,286	1,424	1,000	583	1,624	6,247	882	2,011	828	1,214	4,935	11,182
	Q4	356	1,177	1,407	952	510	1,382	5,784	814	2,041	796	1,302	4,953	10,737
1993	Q1	325	1,124	1,443	998	482	1,203	5,576	948	1,760	858	1,206	4,773	10,348
	Q2	365	1,313	1,427	972	571	1,258	5,906	977	1,744	787	1,241	4,750	10,656
	Q3	358	1,353	1,367	992	567	1,335	5,972	904	1,887	883	1,295	4,969	10,942
	Q4	357	1,423	1,191	904	587	1,335	5,795	970	1,978	809	1,300	5,056	10,852
1994	Q1	406	1,357	1,137	962	546	1,240	5,648	1,126	1,869	906	1,271	5,171	10,819
	Q2	430	1,500	1,285	1,013	636	1,357	6,221	1,041	1,909	910	1,302	5,162	11,382
	Q3	446	1,451	1,344	1,164	611	1,487	6,503	1,036	1,975	934	1,352	5,297	11,800
	Q4	379	1,438	1,309	1,029	696	1,564	6,416	1,015	2,052	936	1,451	5,454	11,870
1995	Q1	391	1,440	1,235	1,103	676	1,358	6,204	1,211	1,909	1,014	1,357	5,491	11,695
	Q2	424	1,446	1,462	1,126	718	1,527	6,703	1,113	2,014	872	1,462	5,461	12,164
	Q3	441	1,329	1,400	1,160	817	1,642	6,789	1,182	2,062	937	1,535	5,715	12,504
	Q4	397	1,259	1,497	1,064	797	1,682	6,696	1,136	2,145	978	1,624	5,884	12,580
1996	Q1	346	1,187	1,478	1,150	705	1,575	6,441	1,299	1,901	1,085	1,614	5,899	12,340
	Q2	384	1,384	1,611	1,072	803	1,732	6,987	1,119	2,053	940	1,765	5,876	12,863
	Q3	347	1,458	1,593	1,113	828	1,872	7,211	1,293	2,062	972	1,770	6,097	13,308
	Q4	335	1,563	1,628	964	783	1,836	7,109	1,082	2,382	1,004	1,881	6,349	13,458
1997	Q1	287	1,510	1,493	949	770	1,934	6,942	1,201	2,160	1,019	1,761	6,141	13,084
	Q2	315	1,678	1,612	867	907	2,112	7,491	1,193	2,137	959	1,911	6,200	13,691
	Q3	348	1,658	1,594	933	908	2,212	7,652	1,136	2,348	969	1,891	6,344	13,996
	Q4	269	1,905	1,602	878	907	2,131	7,691	1,207	2,480	1,043	2,027	6,758	14,449
1998	Q1	281	1,739	1,645	933	898	2,109	7,605	1,214	2,337	1,039	2,151	6,741	14,346
	Q2	274	1,953	1,486	995	959	2,329	7,999	1,016	2,461	985	1,994	6,457	14,456
	Q3	267	1,829	1,457	1,040	969	2,821	8,384	1,182	2,462	1,024	1,967	6,634	15,018
	Q4 (p)	240	1,839	1,581	1,047	984	2,658	8,349	1,276	2,437	1,110	2,035	6,859	15,208
1999	Q1 (p)	234	1,741	1,437	1,163	928	2,630	8,133	1,240	2,386	1,112	1,963	6,701	13,492
	Q2 (p)	234	1,952	1,493	1,152	1,015	2,973	8,819	1,101	2,430	1,009	2,017	6,558	13,540
	Q3 (p)	259	1,887	1,534	1,261	979	3,149	9,069	1,199	2,498	1,113	2,101	6,911	13,672
	Q4 (p)	262	1,896	1,687	1,139	1,004	3,155	9,141	1,205	2,511	1,114	2,253	7,083	13,794

Notes

p = provisional

1. Output by contractors, including estimates of unrecorded output by small firms and self-employed workers, classified to construction in the 1992 Revised Standard Industrial Classification.

Source of Data: Construction Market Intelligence, Department of Environment, Transport and the Regions
Contact: Neville Price 020 7944 5587

Table 2.5 Contractors' Output[1]

Constant (1995) Prices Seasonally Adjusted (£ Million)   Great Britain

Year & Quarter		New Housing		Infra-structure	Other New Work Exc. Infrastructure			All New Work	Repair & Maintenance Housing		Other Work		All Repair & Mainte-nance	All Work
		Public	Private		Public	Private Indus-trial	Private Comm-ercial		Public	Private	Public	Private		
1989		988	7,803	3,700	3,287	3,217	8,777	27,772	4,791	10,658	4,199	6,223	25,870	53,642
1990	Q1	253	1,540	1,091	874	790	2,456	7,004	1,312	2,647	1,116	1,617	6,692	13,696
	Q2	220	1,423	1,071	920	807	2,574	7,015	1,331	2,636	1,070	1,559	6,596	13,610
	Q3	228	1,407	1,195	940	716	2,524	7,010	1,113	2,473	1,070	1,579	6,235	13,245
	Q4	212	1,482	1,158	880	820	2,435	6,986	1,057	2,509	1,070	1,581	6,216	13,203
1991	Q1	198	1,193	1,481	961	685	2,331	6,849	1,066	2,307	975	1,513	5,860	12,709
	Q2	183	1,349	1,415	921	645	2,216	6,729	1,006	2,270	996	1,511	5,783	12,512
	Q3	198	1,267	1,524	918	705	2,121	6,734	1,002	2,316	966	1,413	5,697	12,431
	Q4	229	1,286	1,596	930	660	2,070	6,771	893	2,285	904	1,372	5,454	12,225
1992	Q1	265	1,255	1,610	1,023	651	1,948	6,753	1,028	2,032	923	1,376	5,360	12,112
	Q2	316	1,258	1,614	1,055	666	1,849	6,758	985	1,991	883	1,354	5,213	11,971
	Q3	362	1,311	1,561	1,062	637	1,698	6,631	928	2,114	889	1,359	5,288	11,919
	Q4	415	1,289	1,587	1,086	574	1,524	6,475	879	2,204	879	1,393	5,354	11,829
1993	Q1	372	1,282	1,861	1,092	587	1,419	6,613	1,038	1,928	898	1,305	5,169	11,781
	Q2	411	1,380	1,741	1,069	623	1,408	6,633	1,072	1,914	852	1,308	5,147	11,779
	Q3	405	1,399	1,575	1,110	588	1,458	6,534	941	1,964	923	1,419	5,247	11,782
	Q4	403	1,530	1,428	1,077	641	1,509	6,588	1,015	2,069	894	1,358	5,337	11,924
1994	Q1	462	1,516	1,463	1,066	663	1,487	6,658	1,199	1,990	905	1,378	5,472	12,130
	Q2	453	1,519	1,476	1,114	673	1,492	6,728	1,117	2,047	990	1,365	5,519	12,247
	Q3	463	1,475	1,489	1,211	632	1,534	6,803	1,066	2,033	976	1,416	5,490	12,293
	Q4	403	1,468	1,436	1,141	720	1,631	6,797	1,025	2,071	977	1,449	5,522	12,319
1995	Q1	416	1,533	1,373	1,123	742	1,478	6,666	1,237	1,951	964	1,420	5,571	12,237
	Q2	410	1,396	1,439	1,146	688	1,526	6,605	1,145	2,071	922	1,470	5,608	12,213
	Q3	424	1,309	1,345	1,098	788	1,559	6,523	1,163	2,032	943	1,531	5,669	12,192
	Q4	403	1,237	1,437	1,085	790	1,645	6,598	1,098	2,075	971	1,557	5,702	12,300
1996	Q1	366	1,240	1,513	1,098	767	1,622	6,606	1,286	1,882	996	1,639	5,803	12,409
	Q2	369	1,313	1,516	1,057	775	1,663	6,692	1,123	2,061	976	1,727	5,887	12,579
	Q3	330	1,420	1,517	1,025	814	1,741	6,847	1,252	1,996	968	1,728	5,945	12,792
	Q4	336	1,494	1,546	961	792	1,784	6,912	1,022	2,250	965	1,764	6,001	12,913
1997	Q1	301	1,522	1,522	891	846	1,965	7,048	1,161	2,087	912	1,747	5,906	12,955
	Q2	295	1,527	1,500	845	869	1,978	7,015	1,171	2,096	973	1,820	6,060	13,074
	Q3	323	1,544	1,503	835	857	1,996	7,058	1,079	2,230	948	1,813	6,069	13,127
	Q4	263	1,728	1,493	838	865	1,998	7,185	1,102	2,265	965	1,841	6,173	13,358
1998	Q1	286	1,664	1,645	839	928	2,047	7,408	1,141	2,196	905	2,072	6,314	13,722
	Q2	247	1,679	1,348	926	869	2,075	7,143	965	2,347	971	1,841	6,124	13,268
	Q3	236	1,599	1,333	886	863	2,408	7,325	1,052	2,192	940	1,767	5,952	13,277
	Q4 (p)	219	1,562	1,441	957	886	2,360	7,425	1,108	2,115	976	1,755	5,954	13,379
1999	Q1 (p)	223	1,549	1,428	1,008	905	2,409	7,522	1,109	2,134	922	1,804	5,970	13,492
	Q2 (p)	198	1,552	1,362	1,038	879	2,498	7,527	1,013	2,236	961	1,802	6,013	13,540
	Q3 (p)	216	1,579	1,419	1,047	850	2,540	7,650	1,037	2,159	992	1,834	6,022	13,672
	Q4 (p)	228	1,475	1,546	1,013	898	2,656	7,817	1,017	2,118	953	1,891	5,977	13,794

Notes

p = provisional

1. Output by contractors, including estimates of unrecorded output by small firms and self-employed workers, classified to construction in the 1992 Revised Standard Industrial Classification.

Source of Data: Construction Market Intelligence, Department of Environment, Transport and the Regions
Contact: Neville Price 020 7944 5587

Table 2.6 Direct Labour Output[1]

Current Prices (£ Million)   Great Britain

Year & Quarter		New Work			All New Work	Repair & Maintenance			All Work
		Housing	Public Works	Infra-structure		Housing	Non-Housing	All Repair & Maintenance	
1989		22	207	118	348	1,283	1,772	3,055	3,403
1990	Q1	8	67	55	129	378	496	874	1,003
	Q2	5	35	42	82	331	435	766	848
	Q3	6	49	56	110	357	507	864	974
	Q4	5	46	57	108	354	486	840	948
1991	Q1	6	54	43	102	402	547	949	1,051
	Q2	5	37	33	75	325	429	755	829
	Q3	4	39	41	84	354	463	817	902
	Q4	4	43	40	88	376	480	856	944
1992	Q1	5	66	45	116	417	543	959	1,075
	Q2	3	36	43	81	323	413	736	818
	Q3	3	44	45	92	380	448	829	921
	Q4	3	41	45	89	402	433	835	924
1993	Q1	3	59	37	98	444	447	891	989
	Q2	1	36	30	68	362	343	705	773
	Q3	3	42	26	70	403	399	803	873
	Q4	2	43	25	70	431	390	821	891
1994	Q1	3	67	25	96	472	426	898	994
	Q2	2	45	17	63	370	341	711	773
	Q3	2	50	16	68	432	382	814	883
	Q4	2	54	16	72	471	376	846	919
1995	Q1	2	64	19	85	513	404	916	1,002
	Q2	1	44	19	64	412	359	771	835
	Q3	1	49	15	65	447	417	865	929
	Q4	2	51	12	66	452	417	869	935
1996	Q1	2	50	17	68	487	455	941	1,010
	Q2	3	29	5	38	411	248	659	697
	Q3	3	30	3	36	469	271	739	775
	Q4	2	33	2	37	479	277	756	793
1997	Q1	2	30	2	35	519	324	843	878
	Q2	3	30	3	36	429	244	673	709
	Q3	3	34	2	39	470	266	736	774
	Q4	3	36	2	42	472	256	728	770
1998	Q1	3	37	3	43	510	312	822	865
	Q2	2	31	3	35	407	231	638	673
	Q3	0	37	2	40	451	264	716	756
	Q4 (p)	1	32	5	37	448	254	702	739
1999	Q1 (p)	4	34	4	42	459	303	762	805
	Q2 (p)	0	18	3	22	406	226	633	654
	Q3 (p)	2	19	2	23	420	235	655	678
	Q4 (p)	5	23	4	31	427	250	677	709

Notes

p = provisional

1. Output of public sector direct labour departments classified to construction in the 1992 Revised Standard Industrial Classification.

Source of Data: Construction Market Intelligence, Department of Environment, Transport and the Regions
Contact: Neville Price 020 7944 5587

Table 2.7 Direct Labour Output[1]

Constant (1995) Prices Seasonally Adjusted (£ Million)   Great Britain

Year & Quarter		New Work			All New Work	Repair & Maintenance			All Work
		Housing	Public Works	Infra-structure		Housing	Non-Housing	All Repair & Maintenance	
1989		29	270	153	452	1,700	2,343	4,042	4,495
1990	Q1	8	72	58	139	435	563	998	1,137
	Q2	8	51	60	119	447	599	1,046	1,165
	Q3	7	61	69	137	446	623	1,069	1,206
	Q4	5	54	66	125	416	572	988	1,113
1991	Q1	6	54	43	102	424	573	997	1,099
	Q2	6	51	45	102	407	548	955	1,057
	Q3	5	46	48	99	409	528	937	1,036
	Q4	4	49	45	98	423	543	965	1,064
1992	Q1	5	65	43	113	419	546	965	1,078
	Q2	3	48	56	108	390	508	898	1,006
	Q3	4	50	50	104	420	491	910	1,015
	Q4	3	45	48	97	432	469	900	997
1993	Q1	3	55	34	92	421	428	849	941
	Q2	2	46	38	85	423	408	830	916
	Q3	3	46	28	76	425	416	840	916
	Q4	3	46	26	75	452	413	865	940
1994	Q1	3	56	21	80	446	401	848	928
	Q2	2	57	21	80	437	403	840	921
	Q3	2	56	18	76	452	395	847	923
	Q4	2	56	17	76	473	383	855	932
1995	Q1	2	52	15	69	469	365	834	903
	Q2	1	53	23	77	469	410	879	956
	Q3	1	52	16	69	451	417	868	937
	Q4	2	51	12	65	434	405	840	905
1996	Q1	1	40	14	56	429	400	829	885
	Q2	3	35	7	45	450	279	729	773
	Q3	3	32	4	38	455	268	722	760
	Q4	2	32	2	36	449	267	716	752
1997	Q1	2	23	2	27	444	274	718	745
	Q2	3	34	4	40	456	263	719	760
	Q3	3	34	2	39	442	253	694	733
	Q4	4	34	2	39	438	240	677	716
1998	Q1	3	28	2	33	421	255	676	709
	Q2	2	34	3	38	417	241	658	696
	Q3	0	37	2	39	409	243	652	691
	Q4 (p)	1	27	4	32	377	218	595	627
1999	Q1 (p)	3	25	3	32	352	231	583	614
	Q2 (p)	0	20	3	23	388	222	610	633
	Q3 (p)	2	18	2	22	358	204	562	584
	Q4 (p)	5	19	3	27	341	205	546	573

Notes

p = provisional

1. Output of public sector direct labour departments classified to construction in the 1992 Revised Standard Industrial Classification.

Source of Data: Construction Market Intelligence, Department of Environment, Transport and the Regions
Contact: Neville Price 020 7944 5587

Table 2.8 Contractors' Output[1] by Type of Work

(a) New Work (£ Million) for Public Sector — Great Britain

Year	New Housing	Infra-structure	Other New Work Excluding Infrastructure												Other New Work	All Public Sector
			Fac-tories	Ware-houses	Oil, Steel & Coal	Schools & Colleges	Univ-ersities	Health	Offices	Enter-tainment	Garages	Shops	Agri-culture	Miscell-aneous		
1989	944	2,716	264	34	146	486	94	731	595	387	81	38	15	824	3,695	7,355
1990	911	3,058	185	39	136	577	164	853	720	471	80	30	7	956	4,217	8,185
1991	775	3,408	150	22	89	576	151	836	616	343	80	27	13	1,066	3,969	8,152
1992	1,230	3,125	142	36	53	729	226	956	579	286	51	34	22	882	3,994	8,348
1993	1,405	3,063	141	39	48	735	345	869	632	250	58	30	26	693	3,866	8,334
1994	1,662	3,067	135	47	23	780	407	817	709	336	47	25	31	811	4,168	8,896
1995	1,653	3,401	132	46	16	819	433	960	662	370	49	24	35	907	4,453	9,508
1996	1,412	3,510	118	25	5	876	503	967	658	353	42	17	28	706	4,299	9,221
1997	1,220	2,592	72	26	6	821	368	822	457	347	45	17	20	626	3,626	7,439
1998 (p)	1,063	2,226	92	34	4	962	399	764	545	457	41	23	26	668	4,015	7,304
1999 (p)	988	2,258	116	43	6	1,185	464	860	624	537	42	24	27	785	4,714	7,959

(b) New Work (£ Million) for Private Sector — Great Britain

Year	New Housing	Infra-structure	Industrial				Commercial									All Comm-ercial	All Private Sector
			Fac-tories	Ware-houses	Oil, Steel & Coal	All In-dustrial	Schools & Universities	Health	Offices	Enter-tainment	Garages	Shops	Agri-culture	Miscell-aneous		
1989	7,117	1,183	2,477	868	81	3,425	209	272	5,061	1,385	391	2,232	144	199	9,892	21,617
1990	5,746	1,699	2,430	838	127	3,394	204	374	6,043	1,639	453	2,252	173	172	11,310	22,148
1991	5,003	2,496	1,993	545	84	2,622	206	327	4,835	1,335	416	1,713	134	137	9,103	19,225
1992	4,841	2,413	1,648	473	114	2,234	195	344	2,820	1,160	292	1,503	130	157	6,600	16,088
1993	5,213	2,364	1,588	525	95	2,208	170	300	1,839	884	312	1,395	115	116	5,131	14,915
1994	5,746	2,008	1,813	579	96	2,489	120	281	1,791	1,098	352	1,731	137	138	5,648	15,890
1995	5,475	2,193	2,235	691	81	3,008	141	320	2,240	1,110	369	1,743	135	151	6,209	16,884
1996	5,592	2,801	2,247	778	93	3,119	144	335	2,626	1,368	337	1,932	128	144	7,015	78,526
1997	6,751	3,709	2,426	997	68	3,491	201	307	3,055	1,977	347	2,118	159	224	8,388	22,338
1998 (p)	7,361	3,943	2,546	1,163	101	3,810	266	389	3,486	2,393	407	2,559	156	261	9,917	25,031
1999 (p)	7,476	3,893	2,574	1,249	104	3,926	338	466	4,087	2,816	469	3,250	169	312	11,907	27,203

Table 2.8 Contractors' Output[1] by Type of Work (continued)

(c) New work (£ Million) for Public and Private Sectors   Great Britain

Year	New Housing	Infrastructure								Total Infra-structure
		Water	Sewerage	Elec-tricity	Commu-nications, Gas & Air	Railways	Harbours	Roads	
1989	8,061	218	395	342	350	795	351	1,448	3,899
1990	6,656	269	461	429	439	1,149	314	1,695	4,756
1991	5,778	490	588	531	456	1,353	384	2,102	5,904
1992	6,071	683	623	445	548	843	373	2,023	5,538
1993	6,618	664	573	457	582	774	426	1,951	5,427
1994	7,407	630	541	330	732	426	262	2,154	5,075
1995	7,128	629	522	303	849	744	349	2,199	5,594
1996	7,004	820	571	363	1,066	1,057	344	2,089	6,311
1997	7,971	830	648	407	1,285	989	315	1,827	6,301
1998 (p)	8,423	915	759	450	1,283	868	257	1,638	6,170
1999 (p)	8,464	969	806	473	1,146	831	268	1,659	6,151

Year	Other Non-Housing Excluding Infrastructure											Other Non-Housing Total	Total All New Work
	Fac-tories	Ware-houses	Oil, Steel & Coal	Schools & Univer-sities	Health	Offices	Entertain-ment	Garages	Shops	Agri-culture	Miscel-laneous		
1989	2,741	902	227	789	1,003	5,656	1,772	472	2,270	159	1,023	17,013	28,972
1990	2,615	876	263	944	1,226	6,763	2,110	533	2,283	180	1,128	18,921	30,334
1991	2,143	567	173	933	1,163	5,451	1,678	497	1,740	147	1,203	15,695	27,377
1992	1,790	509	166	1,150	1,299	3,399	1,446	342	1,537	152	1,038	12,828	24,437
1993	1,729	564	142	1,250	1,169	2,471	1,134	370	1,425	141	809	11,205	23,250
1994	1,948	626	119	1,308	1,097	2,500	1,433	400	1,757	168	948	12,304	24,787
1995	2,368	737	97	1,394	1,280	2,901	1,481	418	1,766	169	1,059	13,669	26,392
1996	2,366	803	98	1,523	1,302	3,284	1,720	379	1,950	157	851	14,432	27,747
1997	2,498	1,024	73	1,390	1,129	3,512	2,323	391	2,135	179	850	15,505	29,776
1998 (p)	2,638	1,197	106	1,627	1,152	4,031	2,850	448	2,582	182	929	17,743	32,336
1999 (p)	2,690	1,292	109	1,986	1,326	4,711	3,354	511	3,275	196	1,097	20,547	35,162

Notes

p = provisional

1. Output of contractors, including estimates of unrecorded output of small firms and self-employed workers, classified to construction in the 1992 Revised Standard Industrial Classification.

Source of Data: Construction Market Intelligence, Department of Environment, Transport and the Regions
Contact: Neville Price 020 7944 5587

Table 2.9 Contractors' Output[1] by Region[2]

£ Million

| | New Housing | | | Other New Work Non-Housing | | | | All New Work | Repair & Maintenance | | | All Repair & Mainte-nance | All Work |
| | | | | | Other New Work Exc. Infrastructure | | | | | | Other Work | | | |
Year	Public	Private	Infra-structure	Public	Private Indus-trial	Private Comm-ercial			Housing	Public	Private		
North													
1989	36	392	261	166	181	221	1,257	456	122	224	802	2,059
1990	33	395	272	178	167	292	1,336	471	142	260	873	2,209
1991	44	290	302	148	172	270	1,226	444	126	261	832	2,058
1992	63	270	260	141	145	238	1,117	431	125	273	829	1,946
1993	53	276	244	168	123	186	1,050	436	126	251	813	1,863
1994	56	363	230	197	160	246	1,252	479	137	271	887	2,139
1995	59	273	219	180	172	340	1,243	504	150	289	943	2,186
1996	41	283	278	172	394	362	1,531	511	147	317	975	2,506
1997	38	387	351	148	302	357	1,584	550	153	350	1,052	2,636
1998 (p)	38	425	379	185	310	413	1,750	558	189	390	1,137	2,887
1999 (p)	39	386	406	222	293	491	1,837	546	150	350	1,046	2,883
Yorkshire & Humberside													
1989	49	601	197	333	311	543	2,032	954	296	449	1,699	3,732
1990	36	496	265	414	370	684	2,265	987	355	486	1,828	4,093
1991	50	355	360	395	270	553	1,983	928	301	462	1,691	3,673
1992	67	351	390	372	259	506	1,946	886	294	440	1,621	3,566
1993	104	425	393	338	182	392	1,834	910	286	438	1,634	3,468
1994	138	463	342	375	186	468	1,972	984	309	439	1,732	3,703
1995	121	465	286	363	272	431	1,937	1,069	336	519	1,924	3,862
1996	106	474	418	388	268	581	2,235	1,100	379	610	2,089	4,324
1997	91	478	491	252	393	624	2,330	1,139	348	738	2,225	4,555
1998 (p)	74	549	580	330	338	681	2,552	1,180	311	789	2,281	4,833
1999 (p)	66	538	665	412	352	796	2,829	1,201	270	935	2,405	5,235
East Midlands													
1989	59	670	176	233	283	401	1,823	809	230	312	1,351	3,174
1990	49	499	276	263	298	503	1,888	843	232	320	1,395	3,283
1991	44	462	282	235	340	364	1,728	787	213	331	1,332	3,060
1992	71	489	282	262	213	292	1,610	772	219	333	1,324	2,934
1993	58	605	322	263	186	267	1,702	768	203	346	1,317	3,019
1994	84	636	305	342	196	303	1,865	821	236	378	1,436	3,301
1995	90	551	416	252	265	329	1,903	878	241	424	1,543	3,446
1996	67	562	609	236	267	423	2,165	894	256	512	1,662	3,827
1997	63	732	609	199	351	489	2,443	952	298	521	1,771	4,213
1998 (p)	60	744	530	239	438	541	2,551	985	307	538	1,830	4,382
1999 (p)	54	695	486	270	466	630	2,601	946	352	527	1,825	4,425
East Anglia													
1989	52	497	186	167	180	376	1,457	512	104	173	789	2,246
1990	30	260	200	193	159	430	1,271	513	118	195	826	2,098
1991	25	275	273	124	82	335	1,115	485	110	184	779	1,895
1992	36	249	259	117	74	257	991	522	108	180	810	1,801
1993	49	284	215	149	69	140	905	594	109	177	881	1,786
1994	40	288	225	234	94	159	1,042	627	116	192	935	1,977
1995	49	309	237	210	89	198	1,091	653	116	204	973	2,064
1996	44	297	207	196	88	198	1,030	644	117	229	990	2,021
1997	36	341	236	202	93	215	1,123	615	124	257	996	2,119
1998 (p)	25	335	212	198	132	278	1,180	687	123	273	1,083	2,263
1999 (p)	23	331	226	220	135	346	1,281	680	117	256	1,053	2,334

Table 2.9 Contractors Output[1] by Region[2] (continued)

£ Million

| | New Housing | | | Other New Work Non-Housing | | | | | Repair & Maintenance | | | | |
| | | | | Other New Work Exc. Infrastructure | | | | | | Other Work | | All Repair & | |
Year	Public	Private	Infra-structure	Public	Private Indus-trial	Private Comm-ercial	All New Work	Housing	Public	Private	Mainte-nance	All Work
South East (part): Beds, Essex, Herts												
1989	42	461	163	129	272	576	1,643	785	201	305	1,290	2,933
1990	35	296	182	118	252	685	1,568	833	217	328	1,378	2,945
1991	28	301	300	119	149	508	1,404	778	224	327	1,328	2,733
1992	50	271	238	85	177	304	1,126	742	211	322	1,275	2,401
1993	69	335	216	120	361	248	1,349	746	194	317	1,257	2,606
1994	86	348	140	127	307	291	1,298	801	244	342	1,388	2,686
1995	90	369	188	162	225	291	1,326	860	236	372	1,469	2,795
1996	70	393	282	138	204	341	1,428	884	287	477	1,648	3,076
1997	75	400	236	147	225	334	1,417	955	263	468	1,686	3,103
1998 (p)	56	421	217	168	158	390	1,410	988	267	495	1,750	3,160
1999 (p)	52	409	208	188	132	479	1,468	1,035	342	590	1,967	3,436
South East (part): Greater London												
1989	131	493	412	675	236	3,177	5,124	1,456	476	634	2,566	7,690
1990	116	307	518	675	205	3,613	5,435	1,567	539	714	2,820	8,255
1991	97	249	738	719	172	3,027	5,002	1,436	527	724	2,688	7,690
1992	202	194	740	742	156	1,728	3,762	1,356	476	702	2,534	6,296
1993	269	200	740	674	100	1,121	3,104	1,307	503	732	2,542	5,646
1994	391	233	800	621	96	1,274	3,414	1,390	535	779	2,703	6,118
1995	343	295	1,083	689	130	1,459	3,999	1,472	582	864	2,918	6,917
1996	218	342	1,318	733	166	1,617	4,394	1,519	521	993	3,034	7,428
1997	178	597	1,212	583	126	2,023	4,720	1,608	533	1,096	3,237	7,957
1998 (p)	165	693	1,092	701	193	2,568	5,412	1,741	548	1,253	3,542	8,954
1999 (p)	150	972	998	824	210	2,995	6,148	1,956	592	1,126	3,674	9,823
South East (part): Kent, Surrey, Sussex												
1989	100	500	888	206	224	751	2,669	1,071	283	391	1,745	4,414
1990	90	346	1,254	266	180	770	2,906	1,118	304	444	1,865	4,771
1991	56	305	1,451	224	136	709	2,882	1,019	348	419	1,787	4,668
1992	77	276	927	227	106	479	2,091	966	298	405	1,669	3,760
1993	96	271	806	192	107	402	1,874	973	350	459	1,782	3,657
1994	110	310	513	223	168	296	1,620	1,001	402	460	1,863	3,483
1995	108	347	480	201	123	313	1,573	1,059	417	463	1,938	3,512
1996	98	381	405	193	128	424	1,630	1,116	442	544	2,102	3,732
1997	100	612	384	251	155	500	2,002	1,190	450	574	2,214	4,216
1998 (p)	83	565	424	228	161	785	2,246	1,231	481	633	2,346	4,592
1999 (p)	74	501	473	263	174	1,077	2,561	1,258	596	749	2,603	5,163
South East (part): Berks, Bucks, Hants, Oxon												
1989	105	604	335	317	339	940	2,640	903	267	441	1,610	4,250
1990	94	448	373	338	266	1,016	2,535	959	290	470	1,719	4,254
1991	58	380	386	310	147	573	1,854	899	300	434	1,632	3,486
1992	79	403	319	277	131	382	1,590	846	291	434	1,570	3,160
1993	72	445	333	282	104	319	1,555	838	314	355	1,507	3,062
1994	77	451	287	342	158	394	1,708	891	344	369	1,603	3,312
1995	117	406	228	418	216	392	1,776	951	316	423	1,690	3,467
1996	119	422	300	415	152	442	1,850	995	345	540	1,880	3,729
1997	110	475	389	250	164	616	2,004	1,070	303	586	1,958	3,963
1998 (p)	89	461	383	283	223	770	2,208	1,102	314	616	2,032	4,241
1999 (p)	76	479	367	352	250	924	2,449	1,172	332	642	2,146	4,595

Table 2.9 Contractors Output[1] by Region[2] (continued)

£ Million

Year & Region	New Housing		Infra-structure	Other New Work Non-Housing Other New Work Exc. Infrastructure			All New Work	Repair & Maintenance			All Repair & Mainte-nance	All Work
	Public	Private		Public	Private Indus-trial	Private Comm-ercial		Housing	Public	Private		
South East: All												
1989	378	2,058	1,798	1,327	1,071	5,443	12,076	4,215	1,226	1,771	7,211	19,287
1990	335	1,396	2,327	1,397	903	6,084	12,443	4,477	1,349	1,956	7,782	20,225
1991	238	1,236	2,875	1,372	604	4,816	11,142	4,132	1,399	1,903	7,435	18,576
1992	407	1,143	2,224	1,331	570	2,893	8,569	3,910	1,275	1,863	7,048	15,617
1993	506	1,251	2,095	1,268	671	2,091	7,882	3,863	1,361	1,864	7,088	14,970
1994	664	1,342	1,739	1,312	729	2,254	8,041	4,083	1,525	1,949	7,557	15,598
1995	658	1,417	1,979	1,471	695	2,455	8,675	4,342	1,551	2,123	8,015	16,690
1996	505	1,538	2,306	1,480	650	2,824	9,302	4,513	1,595	2,555	8,663	17,965
1997	463	2,084	2,220	1,231	671	3,474	10,143	4,823	1,548	2,724	9,095	19,238
1998 (p)	393	2,140	2,117	1,379	735	4,513	11,277	5,062	1,610	2,997	9,669	20,946
1999 (p)	351	2,360	2,045	1,627	767	5,475	12,626	5,421	1,863	3,107	10,391	23,017
South West												
1989	82	970	236	299	282	742	2,610	1,142	290	425	1,857	4,467
1990	84	601	228	390	247	883	2,433	1,200	301	451	1,953	4,386
1991	61	507	273	306	161	674	1,984	1,141	270	451	1,862	3,846
1992	103	530	480	325	161	543	2,142	1,073	284	415	1,772	3,914
1993	109	491	476	352	167	471	2,064	1,043	275	399	1,717	3,781
1994	124	533	497	391	187	474	2,205	1,128	296	426	1,850	4,055
1995	143	488	480	479	251	491	2,333	1,189	313	482	1,984	4,317
1996	93	538	416	313	205	467	2,032	1,268	361	532	2,161	4,183
1997	65	574	451	295	253	513	2,151	1,334	347	566	2,247	4,398
1998 (p)	68	709	384	328	298	582	2,370	1,368	383	599	2,349	4,719
1999 (p)	63	738	382	388	308	662	2,541	1,365	384	631	2,380	4,921
West Midlands												
1989	87	605	219	319	341	663	2,233	1,067	274	403	1,744	3,977
1990	100	514	237	400	385	713	2,349	1,127	302	428	1,857	4,207
1991	82	527	376	368	212	664	2,229	1,034	262	412	1,708	3,937
1992	108	536	391	299	190	547	2,071	983	269	407	1,660	3,730
1993	121	552	338	294	210	431	1,948	1,016	301	453	1,769	3,716
1994	137	580	416	299	256	441	2,129	1,097	309	505	1,911	4,039
1995	87	568	398	304	337	458	2,153	1,158	314	531	2,003	4,157
1996	106	495	443	278	363	522	2,206	1,219	338	635	2,192	4,399
1997	69	617	412	303	401	662	2,465	1,291	331	687	2,308	4,773
1998 (p)	77	742	460	340	489	739	2,846	1,333	346	723	2,402	5,248
1999 (p)	75	707	467	385	517	919	3,071	1,200	392	740	2,322	5,404
North West												
1989	45	546	250	246	369	711	2,167	1,141	333	465	1,938	4,106
1990	69	649	326	311	388	802	2,545	1,210	398	534	2,141	4,687
1991	59	568	423	324	335	608	2,316	1,137	323	496	1,957	4,273
1992	142	527	342	366	260	529	2,167	1,050	295	473	1,817	3,984
1993	166	580	437	374	270	443	2,271	1,068	285	493	1,846	4,117
1994	135	655	493	395	259	581	2,519	1,184	328	546	2,057	4,576
1995	154	592	634	417	312	702	2,811	1,300	337	624	2,261	5,072
1996	148	548	780	395	355	699	2,923	1,344	368	777	2,489	5,412
1997	152	549	651	315	386	858	2,912	1,392	393	861	2,646	5,557
1998 (p)	116	669	623	326	423	886	3,044	1,432	431	879	2,742	5,785
1999 (p)	114	647	592	387	431	1,104	3,276	1,472	383	854	2,708	5,984

Table 2.9 Contractors Output[1] by Region[2] (continued)

£ Million

Year	New Housing Public	New Housing Private	Infra-structure	Other New Work Exc. Infrastructure Public	Private Indus-trial	Private Comm-ercial	All New Work	Repair & Maintenance Housing	Other Work Public	Other Work Private	All Repair & Mainte-nance	All Work
England												
1989	787	6,338	3,323	3,091	3,018	9,099	25,656	10,295	2,874	4,222	17,392	43,048
1990	735	4,810	4,131	3,547	2,917	10,391	26,531	10,829	3,198	4,629	18,655	45,187
1991	604	4,221	5,165	3,274	2,176	8,284	23,723	10,090	3,004	4,502	17,596	41,319
1992	998	4,095	4,628	3,213	1,872	5,805	20,611	9,626	2,869	4,385	16,880	37,492
1993	1,164	4,465	4,519	3,206	1,879	4,421	19,655	9,698	2,946	4,420	17,065	36,720
1994	1,377	4,860	4,247	3,545	2,067	4,927	21,024	10,403	3,255	4,707	18,365	39,389
1995	1,362	4,664	4,650	3,676	2,392	5,403	22,146	11,092	3,359	5,196	19,647	41,794
1996	1,111	4,734	5,455	3,458	2,591	6,076	23,425	11,494	3,562	6,166	21,222	44,647
1997	977	5,762	5,421	2,948	2,850	7,193	25,150	12,095	3,542	6,703	22,340	47,490
1998 (p)	851	6,313	5,285	3,326	3,163	8,632	27,569	12,605	3,701	7,187	23,494	51,063
1999 (p)	786	6,403	5,270	3,911	3,270	10,423	30,063	12,831	3,910	7,399	24,140	54,203
Wales												
1989	42	313	187	189	208	236	1,175	517	102	172	791	1,966
1990	58	303	222	186	210	262	1,240	548	110	200	857	2,097
1991	52	189	304	191	195	207	1,137	512	114	195	822	1,959
1992	75	250	389	220	149	187	1,270	501	118	194	812	2,082
1993	66	198	327	188	110	133	1,023	496	112	195	803	1,826
1994	79	226	387	216	181	212	1,301	541	125	205	871	2,172
1995	64	221	490	290	158	233	1,455	576	121	225	922	2,377
1996	65	226	402	270	164	219	1,345	595	138	252	986	2,331
1997	65	258	355	233	245	334	1,490	629	137	284	1,094	2,539
1998 (p)	49	292	342	233	225	419	1,559	637	149	297	1,082	2,641
1999 (p)	42	286	292	262	224	493	1,598	608	147	245	1,000	2,599
Scotland												
1989	115	466	388	416	199	557	2,141	997	234	361	1,592	3,733
1990	118	632	403	484	268	657	2,562	1,042	257	390	1,688	4,250
1991	118	593	436	505	252	613	2,517	941	252	401	1,594	4,111
1992	157	496	521	561	213	607	2,555	937	263	406	1,606	4,161
1993	174	550	581	471	218	577	2,572	975	278	427	1,680	4,251
1994	205	660	441	406	240	509	2,462	1,078	307	463	1,848	4,310
1995	227	591	455	488	458	572	2,790	1,103	321	556	1,981	4,771
1996	236	631	454	571	364	720	2,977	1,101	301	612	2,014	4,991
1997	178	731	525	446	396	861	3,137	1,139	311	604	2,054	5,191
1998 (p)	164	756	543	457	422	866	3,208	1,144	309	662	2,115	5,323
1999 (p)	160	788	588	541	433	991	3,501	1,132	291	690	2,113	5,614

Notes

p = provisional

1. Output of contractors, including estimates of unrecorded output of small firms and self-employed workers, classified to construction in the 1992 Revised Standard Industrial Classification.
2. For new work – the region of the site; for repair & maintenance – the region in which the reporting unit is based.

Source of Data: Construction Market Intelligence, Department of Environment, Transport and the Regions
Contact: Neville Price 020 7944 5587

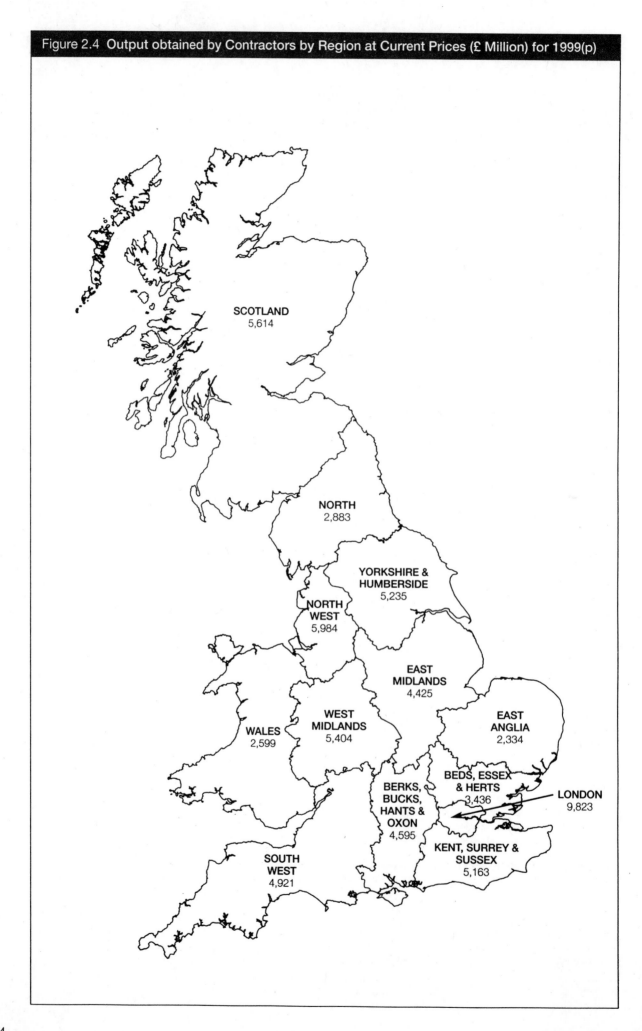

Figure 2.4 Output obtained by Contractors by Region at Current Prices (£ Million) for 1999(p)

SCOTLAND
5,614

NORTH
2,883

YORKSHIRE &
HUMBERSIDE
5,235

NORTH
WEST
5,984

EAST
MIDLANDS
4,425

WEST
MIDLANDS
5,404

WALES
2,599

EAST
ANGLIA
2,334

BEDS, ESSEX
& HERTS
3,436

BERKS,
BUCKS,
HANTS &
OXON
4,595

LONDON
9,823

KENT, SURREY &
SUSSEX
5,163

SOUTH
WEST
4,921

CHAPTER 3

Structure of the Industry

Table 3.1 **Private Contractors: Number of Firms** [1,2]										

Great Britain										3rd Quarter Each Year
	1989	**1990**	**1991**	**1992[3]**	**1993**	**1994**	**1995**	**1996[4]**	**1997**	**1998**	**1999**
(a) By Size of Firm											
1	94,218	101,223	103,169	94,452	93,585	97,141	99,099	81,363	86,269	87,837	88,018
2-3	67,189	71,498	70,452	68,486	64,438	65,188	64,837	56,106	47,644	47,918	49,350
4-7	24,984	23,403	21,664	30,395	26,072	22,145	20,288	15,317	15,737	16,391	16,969
8-13	5,869	5,362	4,981	5,240	4,630	4,221	4,021	4,366	3,787	3,988	4,148
14-24	4,212	3,935	3,429	3,574	3,129	2,881	2,828	2,952	3,101	3,274	3,271
25-34	1,478	1,420	1,186	1,146	1,066	956	938	1,103	1,176	1,201	1,332
35-59	1,458	1,305	1,100	1,148	1,098	1,008	968	984	1,156	1,263	1,188
60-79	450	442	382	361	294	325	307	325	396	419	397
80-114	421	432	372	317	283	262	258	263	296	319	304
115-299	530	507	431	387	330	356	337	348	381	405	379
300-599	153	150	137	103	96	92	105	101	107	125	105
600-1,199	66	69	58	59	53	50	51	54	60	56	58
1,200 and Over	48	47	39	36	33	32	33	33	38	40	42
All Firms	**201,076**	**209,793**	**207,400**	**205,704**	**195,107**	**194,657**	**194,070**	**163,315**	**160,148**	**163,236**	**165,561**
(b) By Trade of Firm											
General Builders	77,222	78,981	76,991	74,393	70,765	69,160	68,502	..	..	..	..
Building and Civil Engineering Contractors	4,522	4,773	4,772	6,180	6,264	6,845	7,043	..	..	..	..
Civil Engineers[5]	3,468	3,742	3,760	4,312	4,070	4,182	4,298	..	..	..	..
Plumbers	16,774	17,046	16,752	14,647	13,880	13,181	13,111	11,698	12,045	12,519	13,600
Carpenters and Joiners	14,033	15,244	15,499	14,199	13,302	12,614	12,385	10,202	9,974	10,016	9,725
Painters	16,118	16,511	16,142	10,788	9,774	8,974	8,939	8,284	8,634	8,969	8,921
Roofers	7,294	7,767	7,722	7,524	6,891	6,470	6,461	5,457	5,374	5,599	5,636
Plasterers	4,567	4,834	4,761	3,893	3,549	3,160	3,129	2,475	2,443	2,538	2,741
Glaziers	6,231	6,531	6,435	7,001	6,599	6,918	7,015	4,174	3,484	3,128	3,841
Demolition Contractors	656	667	635	717	708	685	712	740	750	793	1,021
Scaffolding Specialists	1,401	1,524	1,597	1,779	1,645	1,733	1,791	1,270	1,112	1,009	1,262
Reinforced Concrete Specialists	733	910	999	859	729	637	615	415	357	321	351
Heating and Ventilating Engineers	9,485	9,624	9,375	9,774	9,355	9,136	8,892	6,697	5,981	5,500	6,161
Electrical Contractors	19,215	20,752	21,020	21,780	20,589	21,004	21,033	19,463	19,077	19,385	19,036
Asphalt and Tar Sprayers	1,036	1,080	1,081	1,163	1,071	1,077	1,086	866	772	711	584
Plant Hirers[6]	4,473	4,626	4,579	5,621	5,567	5,940	5,886	4,607	4,182	3,882	3,549
Flooring Contractors	1,841	1,999	2,033	2,387	2,248	2,320	2,288	2,249	2,412	2,684	2,805
Constructional Engineers	2,286	2,592	2,640	2,713	2,375	2,168	1,976	1,216	1,002	864	1,042
Insulating Specialists	1,399	1,376	1,281	1,265	1,147	1,131	1,133	977	926	934	832
Suspended Ceiling Specialists	1,381	1,554	1,659	1,757	1,597	1,509	1,522	..	..	..	..
Floor and Wall Tiling Specialists	1,516	1,657	1,675	1,607	1,492	1,430	1,394	1,011	872	770	803
Miscellaneous	5,425	6,003	5,992	11,345	11,490	14,383	14,859	13,016	13,934	15,535	17,560
All Trades	**201,076**	**209,793**	**207,400**	**205,704**	**195,107**	**194,657**	**194,070**	**163,315**	**160,148**	**163,236**	**165,561**

Table 3.1 Private Contractors: Number of Firms [1,2] (continued)

Great Britain 3rd Quarter Each Year

	1989	1990	1991	1992[3]	1993	1994	1995	1996[4]	1997	1998	1999
(c) By Region of Registration											
North	7,324	7,544	7,649	7,629	7,304	7,333	7,261	5,940	5,741	5,724	6,183
Yorkshire & Humberside	15,429	15,772	15,763	15,903	15,336	15,290	15,366	12,676	12,581	12,543	12,442
East Midlands	14,192	14,182	14,943	14,455	13,772	13,877	13,926	11,567	11,489	11,956	11,396
East Anglia	9,673	10,376	9,923	9,617	8,968	8,953	8,908	7,559	7,589	7,668	7,220
South East:											
Beds, Essex, Herts	15,279	15,788	15,927	15,591	14,648	14,835	14,560	12,746	12,606	13,281	13,388
Greater London	22,106	23,193	22,984	21,975	20,539	20,053	20,174	16,977	16,530	17,142	17,222
Kent, Surrey, Sussex	20,194	20,864	20,847	19,810	18,423	17,874	17,541	14,750	14,799	15,348	15,897
Berks, Bucks, Hants, Oxon	16,813	17,935	17,317	17,390	16,198	15,903	15,623	13,188	13,025	13,458	14,810
South East: All	74,392	77,780	77,075	74,766	69,808	68,665	67,898	57,661	56,960	59,229	61,317
South West	22,060	23,928	23,051	22,567	20,852	20,032	19,773	16,689	16,462	16,705	16,985
West Midlands	17,472	18,471	17,919	18,301	17,754	17,740	18,058	15,213	14,656	14,844	13,974
North West	18,713	19,393	18,837	17,999	17,471	17,800	17,953	14,784	14,022	14,162	14,576
England	179,255	187,446	185,160	181,237	171,265	169,690	169,143	142,089	139,500	142,804	144,093
Wales	9,287	9,612	9,754	10,187	9,930	10,269	10,274	8,585	8,392	8,273	8,382
Scotland	12,534	12,735	12,486	14,280	13,912	14,698	14,653	12,641	12,256	12,159	13,179
Great Britain	**201,076**	**209,793**	**207,400**	**205,704**	**195,107**	**194,657**	**194,070**	**163,315**	**160,148**	**163,236**	**165,561**

Notes

. . = not available due to technical problems with the trade type breakdown
1. Information relates to the number of private contractors' firms on the Department's register.
2. The number of firms include some which were temporarily inactive.
3. There is a discontinuity between 1991 and 1992 when the Department's register was enhanced by integration with the ONS register.
4. There is a discontinuity in the series between 1995 and 1996 as improved survey techniques, resulting in better coverage and classification of businesses during the previous 24 months, have been incorporated in a single update. This affects the trade classification of all businesses and the number of businesses with 7 or fewer employees.
5. Excluding open-cast coal specialists.

Source of Data: Construction Market Intelligence, Department of the Environment, Transport and the Regions
Contact: Neville Price 020 7944 5587

Figure 3.1 Insolvencies and Bankruptcies of Construction Firms[1]

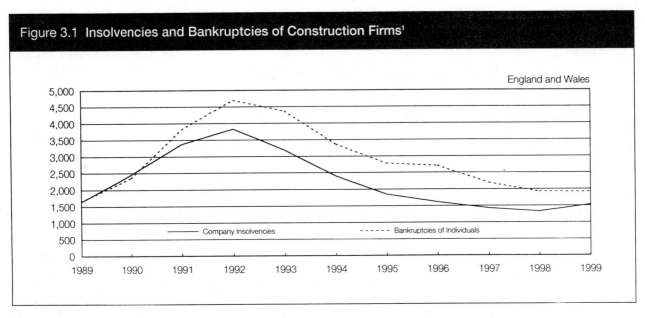

Source of Data: Table 3.2

Table 3.2 Insolvencies and Bankruptcies of Construction Firms[1]

England and Wales

	Bankruptcies & Deeds of Arrangement Construction Self-Employed		Compulsory and Creditors Voluntary Liquidations Construction Companies	
	Number	Percentage of Total[2]	Number	Percentage of Total
1989	1,652	20.3	1,638	15.7
1990	2,348	19.5	2,445	16.2
1991 [3]	3,812	16.8	3,373	15.5
1992 [3]	4,692	14.6	3,830	15.7
1993 [3]	4,361	14.1	3,189	15.4
1994 [3]	3,362	13.1	2,401	14.4
1995 [3]	2,783	12.7	1,844	12.7
1996 [3]	2,713	12.4	1,610	12.0
1997 [3]	2,182	11.0	1,419	11.3
1998 [3]	1,919	9.8	1,325	10.0
1999 [3]	1,911	8.8	1,529	10.7

Notes
1. Based on the definition of the construction industry as given in the Revised 1992 Standard Industrial Classification.
2. Includes employees, unemployed, directors of companies as well as individuals whose occupation is unknown.
3. From 1991 excludes deeds of arrangement.

Source of Data: Construction Market Intelligence, Department of Environment, Transport and the Regions
Contact: Neville Price 020 7944 5587

Table 3.3 Private Contractors: Work Done[1]

Great Britain 3rd Quarter Each Year: £ Million

	1989	1990	1991	1992[2]	1993	1994	1995	1996[3]	1997	1998	1999
(a) By Size of Firm											
1	700.2	784.5	809.3	675.1	729.4	903.8	1,005.5	870.0	1,095.4	1,106.2	1,471.1
2-3	918.2	955.0	883.5	942.7	868.0	1,050.4	1,145.2	1,054.9	1,030.1	1,069.4	1,364.8
4-7	814.0	783.0	754.5	914.1	878.7	879.7	964.3	786.9	860.3	889.8	1,173.7
8-13	477.1	467.5	435.8	521.0	508.9	524.9	525.0	622.7	530.9	574.4	554.4
14-24	727.2	696.6	615.8	700.8	670.8	681.3	753.9	808.9	770.1	865.0	773.2
25-34	421.8	421.7	365.6	375.6	342.5	370.3	412.7	493.5	514.6	533.0	557.9
35-59	658.9	632.8	556.2	597.3	678.1	673.3	676.5	710.7	812.2	1,004.1	1,059.3
60-79	350.1	377.3	354.4	343.2	301.9	368.1	368.4	410.0	468.3	503.8	574.1
80-114	516.7	508.7	472.3	379.9	396.7	391.9	478.7	465.5	519.3	630.9	627.4
115-299	1,146.8	1,178.5	974.8	898.6	824.7	1,004.7	981.0	1,106.1	1,224.5	1,388.5	1,405.2
300-599	743.3	811.6	835.6	703.1	565.6	639.4	729.6	671.1	758.4	934.9	868.6
600-1,199	827.4	857.8	810.8	695.7	727.7	760.5	826.8	920.4	1,019.8	1,120.4	1,099.3
1,200 and Over	1,532.0	1,812.8	1,369.2	1,172.5	1,098.1	1,113.2	1,313.3	1,255.1	1,219.3	1,488.4	1,730.0
All Firms	**9,833.8**	**10,287.9**	**9,237.6**	**8,919.4**	**8,591.1**	**9,361.5**	**10,180.9**	**10,175.7**	**10,823.3**	**12,163.0**	**13,259.0**
(b) By Trade of Firm											
General Builders	3,205.8	3,192.7	2,842.4	2,657.4	2,534.4	2,827.8	3,077.4	..	..	..	..
Building and Civil Engineering Contractors	1,855.0	2,057.6	1,789.0	1,597.7	1,416.2	1,599.7	1,566.0	..	..	..	..
Civil Engineers[4]	725.9	783.6	780.2	775.4	745.5	797.9	973.5	..	..	..	..
Plumbers	268.4	268.8	251.8	233.0	228.8	252.4	269.7	286.2	329.2	374.1	380.5
Carpenters and Joiners	207.6	222.3	207.7	214.1	206.3	228.7	241.1	237.2	260.9	292.7	336.4
Painters	332.3	337.7	284.4	249.8	236.1	244.6	261.8	272.7	302.1	329.8	374.8
Roofers	286.8	295.7	251.6	234.5	226.7	241.7	262.4	248.4	280.7	302.0	290.9
Plasterers	85.6	87.9	80.2	74.4	71.5	71.8	73.9	68.4	76.9	93.0	105.7
Glaziers	253.4	257.1	238.6	273.9	280.4	295.9	286.4	219.2	203.6	228.6	153.7
Demolition Contractors	33.6	32.2	27.4	34.7	34.3	37.2	51.1	48.8	55.9	59.3	126.2
Scaffolding Specialists	147.6	154.7	126.4	121.4	133.4	150.6	167.9	192.3	140.0	184.9	212.5
Reinforced Concrete Specialists	73.6	63.6	51.6	46.5	41.7	35.1	35.1	34.8	31.6	34.8	45.3
Heating and Ventilating Engineers	583.3	643.4	566.6	533.4	564.7	549.0	653.9	607.2	628.2	575.9	639.6
Electrical Contractors	751.2	839.5	800.0	813.3	819.3	862.9	954.4	1,066.8	1,122.5	1,221.8	1,439.2
Asphalt and Tar Sprayers	168.5	162.0	120.2	111.0	129.6	184.9	165.0	161.1	178.9	177.0	185.5
Plant Hirers[5]	205.4	198.4	181.3	188.9	188.6	199.7	220.3	190.9	178.0	226.3	239.5
Flooring Contractors	76.1	77.3	70.8	84.1	85.3	87.3	93.7	94.8	112.3	118.0	115.8
Constructional Engineers	197.6	191.7	188.7	192.8	162.1	173.0	154.9	156.5	164.1	164.0	121.9
Insulating Specialists	89.9	95.3	79.7	104.3	88.0	79.0	89.6	115.9	127.3	98.7	101.0
Suspended Ceiling Specialists	48.7	52.9	53.8	53.5	46.8	47.8	48.4	..	..	..	..
Floor and Wall Tiling Specialists	36.9	39.3	36.0	35.9	31.4	32.0	37.8	30.9	33.6	32.8	36.3
Miscellaneous	200.5	234.3	209.2	289.3	319.9	362.7	496.4	533.1	632.9	744.5	1,042.4
All Trades	**9,833.8**	**10,287.9**	**9,237.6**	**8,919.4**	**8,591.1**	**9,361.5**	**10,180.9**	**10,175.7**	**10,823.3**	**12,163.0**	**13,259.0**

Table 3.3 Private Contractors: Work Done [1]

Great Britain 3rd Quarter Each Year: £ Million

	1989	1990	1991	1992 [2]	1993	1994	1995	1996 [3]	1997	1998	1999
(c) By Region of Registration											
North	434.4	427.0	405.3	401.2	362.5	395.0	445.4	412.2	449.5	522.6	563.1
Yorkshire & Humberside	811.4	806.8	754.2	750.0	738.6	816.1	892.7	972.0	981.0	1,111.0	1,168.6
East Midlands	590.5	627.8	574.4	545.0	533.1	574.0	617.6	629.9	679.6	838.9	808.8
East Anglia	312.4	321.4	283.1	314.2	291.6	323.4	349.0	310.6	348.9	398.0	393.6
South East:											
Beds, Essex, Herts	557.6	591.8	508.5	483.3	455.9	561.9	567.2	617.1	703.3	791.9	1,146.8
Greater London	1,750.2	1,955.2	1,836.0	1,555.7	1,424.5	1,439.6	1,610.7	1,557.8	1,663.0	1,849.3	1,777.7
Kent, Surrey, Sussex	911.9	894.8	732.6	816.6	725.1	747.5	794.2	779.9	815.7	981.6	1,113.2
Berks, Bucks, Hants, Oxon	808.6	864.2	739.5	672.4	655.1	655.8	715.5	702.9	769.7	923.9	1,041.1
South East: All	4,028.3	4,306.0	3,816.6	3,528.0	3,260.6	3,404.8	3,687.7	3,657.6	3,951.6	4,546.8	5,078.7
South West	730.9	714.3	666.5	598.4	565.4	609.3	712.4	703.4	770.8	800.3	931.4
West Midlands	994.9	1,033.5	844.2	769.8	811.7	991.9	1,000.6	1,113.4	1,098.5	1,223.0	1,319.1
North West	830.9	919.6	826.1	849.4	895.2	1,009.9	1,121.0	1,027.1	1,136.6	1,225.0	1,289.6
England	8,733.6	9,156.3	8,170.4	7,756.1	7,458.8	8,124.4	8,826.3	8,826.3	9,416.5	10,655.5	11,552.9
Wales	303.8	310.6	293.3	293.9	295.6	329.1	364.9	351.4	390.9	405.2	437.8
Scotland	796.4	821.0	773.9	869.5	836.6	908.0	989.6	998.1	1,015.9	1,092.2	1,268.3
Great Britain	9,833.8	10,287.8	9,237.6	8,919.4	8,591.1	9,361.5	10,180.9	10,175.7	10,823.3	12,163.0	13,259.0

Notes

. . = not available due to technical problems with the trade type breakdown

1. Information relates to work done by firms on the Department's register.

2. There is a discontinuity between 1991 and 1992 when the Department's register was enhanced by integration with the ONS register.

3. There is a discontinuity in the series between 1995 and 1996 as improved survey techniques, resulting in better coverage and classification of businesses during the previous 24 months, have been incorporated in a single update. This affects the trade classification of all businesses and the number of businesses with 7 or fewer employees.

4. Excluding open-cast coal specialists.

5. Excluding firms whose major activity is plant hire without operatives.

Source of Data: Construction Market Intelligence, Department of the Environment, Transport and the Regions
Contact: Neville Price 020 7944 5587

Table 3.4 Private Contractors: Total Employment [1]

Great Britain

3rd Quarter Each Year: Thousands

	1989	1990	1991	1992 [2]	1993	1994	1995	1996 [3]	1997	1998	1999
(a) By Size of Firm											
1	88.6	96.5	94.5	87.2	82.0	88.6	89.5	72.0	77.4	77.4	183.2
2-3	155.3	167.3	147.5	145.4	130.0	146.0	153.2	131.0	145.2	148.3	120.5
4-7	111.3	103.4	91.9	115.9	97.9	90.0	86.5	68.5	77.8	81.2	119.3
8-13	59.3	54.0	49.8	52.5	46.8	42.8	40.6	44.2	39.7	41.6	48.0
14-24	75.5	70.3	61.0	62.7	55.3	51.2	50.4	52.5	56.0	59.5	63.2
25-34	42.7	41.0	34.2	32.9	30.8	27.6	27.1	32.0	34.5	35.1	36.0
35-59	64.8	58.2	49.0	50.6	49.0	44.9	42.6	43.9	51.9	56.6	58.1
60-79	31.0	30.5	25.9	24.4	20.0	22.2	20.8	22.1	27.1	28.8	29.3
80-114	40.0	41.1	35.4	29.5	26.7	24.6	24.3	24.9	28.3	30.9	31.7
115-299	93.5	89.1	73.5	67.5	58.9	62.4	59.3	62.3	68.1	71.1	78.5
300-599	63.8	63.1	57.0	41.6	39.0	37.6	43.0	41.0	45.5	52.1	47.3
600-1,199	55.3	56.7	46.0	47.8	43.0	43.1	41.9	44.2	50.8	49.1	51.6
1,200 and Over	129.6	125.3	100.6	84.1	73.8	72.2	71.8	68.5	76.3	82.0	92.3
All Firms	**1,010.6**	**996.7**	**866.6**	**842.2**	**753.2**	**753.3**	**750.9**	**707.1**	**778.5**	**813.6**	**958.8**
(b) By Trade of Firm											
General Builders	337.6	322.9	279.8	263.9	232.0	232.1	231.2	..	..	..	..
Building and Civil Engineering Contractors	143.1	140.6	116.7	107.3	90.6	91.6	83.9	..	..	..	..
Civil Engineers [4]	55.2	57.0	51.3	50.7	44.1	44.8	46.9	..	..	..	..
Plumbers	36.6	36.4	32.3	29.7	26.8	26.8	26.8	26.0	30.3	32.8	39.5
Carpenters and Joiners	30.6	32.1	28.7	29.3	25.2	25.3	24.9	22.5	25.0	26.2	31.8
Painters	53.6	52.0	43.5	36.2	32.7	30.9	30.9	30.1	34.5	35.3	41.0
Roofers	29.7	29.7	25.1	23.6	20.2	19.8	19.6	17.9	20.2	20.7	24.4
Plasterers	12.9	12.9	11.4	9.8	8.7	8.2	8.4	7.1	8.0	8.5	10.5
Glaziers	28.1	28.2	24.8	27.2	26.3	26.7	25.2	17.2	17.4	16.6	13.9
Demolition Contractors	4.1	3.9	3.2	3.9	3.5	3.2	3.6	3.8	4.4	4.6	9.6
Scaffolding Specialists	18.9	18.8	16.2	16.0	15.5	15.4	15.8	12.5	13.7	16.4	19.9
Reinforced Concrete Specialists	6.5	6.5	5.2	3.9	2.9	2.3	2.1	1.7	1.7	1.8	4.5
Heating and Ventilating Engineers	57.1	56.3	50.3	49.6	47.2	44.1	45.3	40.5	40.9	39.5	44.7
Electrical Contractors	91.4	94.4	86.2	85.9	82.6	82.4	81.8	86.4	94.3	98.0	118.7
Asphalt and Tar Sprayers	12.6	11.5	8.0	7.7	7.7	8.7	8.5	7.8	8.7	7.9	9.0
Plant Hirers [5]	23.5	22.7	20.3	21.8	19.7	20.2	20.3	17.1	16.5	17.0	20.1
Flooring Contractors	7.4	7.5	6.7	7.8	7.3	7.4	7.1	7.1	8.0	8.4	10.4
Constructional Engineers	17.3	17.7	16.7	16.7	12.7	12.7	11.4	9.3	9.0	8.2	8.1
Insulating Specialists	11.4	11.2	10.0	10.2	8.2	7.4	7.8	7.3	7.3	7.7	7.6
Suspended Ceiling Specialists	4.6	4.5	4.6	4.7	4.0	3.7	3.7	..	..	..	..
Floor and Wall Tiling Specialists	4.8	4.9	4.5	4.5	3.5	3.4	3.3	2.6	2.7	2.5	3.0
Miscellaneous	23.6	25.0	21.2	32.1	31.6	36.3	42.8	43.0	51.1	56.3	87.8
All Trades	**1,010.6**	**996.7**	**866.6**	**842.2**	**753.2**	**753.3**	**750.9**	**707.1**	**778.5**	**813.6**	**958.8**

Table 3.4 Private Contractors: Total Employment [1] (continued)

Great Britain								3rd Quarter Each Year: Thousands			
	1989	1990	1991	1992 [2]	1993	1994	1995	1996 [3]	1997	1998	1999
(c) By Region of Registration											
North	50.3	48.6	44.2	41.1	37.1	37.8	39.2	35.2	37.6	37.5	45.1
Yorkshire & Humberside	87.3	85.3	77.9	77.4	69.8	70.4	71.4	68.1	75.3	81.9	90.0
East Midlands	64.2	61.4	55.5	53.0	46.1	46.5	47.5	43.4	51.6	55.0	61.5
East Anglia	36.5	37.5	31.2	32.9	28.7	28.0	27.8	25.3	29.5	31.4	32.6
South East:											
Beds, Essex, Herts	57.6	58.7	49.8	47.6	40.4	44.0	44.2	42.0	48.4	52.5	65.0
Greater London	165.7	160.0	135.2	125.1	109.3	101.0	101.9	92.0	98.8	103.8	106.4
Kent, Surrey, Sussex	80.0	78.4	68.8	66.8	57.3	56.2	55.2	52.0	58.4	63.6	80.8
Berks, Bucks, Hants, Oxon	73.9	75.6	60.3	60.2	52.0	50.2	48.8	47.8	52.1	54.2	77.6
South East: All	377.3	372.7	314.1	299.7	258.9	251.4	250.1	233.8	257.6	274.1	329.8
South West	80.0	81.2	70.2	65.2	56.9	56.2	57.9	53.1	61.2	61.8	76.6
West Midlands	89.0	87.6	76.5	70.7	66.0	70.0	70.0	71.1	73.0	74.4	87.4
North West	94.3	92.0	80.7	80.9	75.6	76.4	72.3	70.5	77.7	81.4	97.4
England	878.9	866.3	750.5	721.0	639.4	636.8	636.2	600.6	663.7	697.6	820.4
Wales	35.7	35.9	32.0	31.5	28.4	29.5	30.1	27.0	30.6	32.0	37.5
Scotland	95.9	94.4	84.1	89.7	85.3	87.1	84.6	79.4	84.2	83.9	100.3
Great Britain	**1,010.6**	**996.6**	**866.6**	**842.2**	**753.2**	**753.3**	**750.9**	**707.1**	**778.5**	**813.6**	**958.8**

Notes

. . = not available due to technical problems with the trade type breakdown

1. Information relates to employment by firms on the Department's register.

2. There is a discontinuity between 1991 and 1992 when the Department's register was enhanced by integration with the ONS register.

3. There is a discontinuity in the series between 1995 and 1996 as improved survey techniques, resulting in better coverage and classification of businesses during the previous 24 months, have been incorporated in a single update. This affects the trade classification of all businesses and the number of businesses with 7 or fewer employees.

4. Excluding open-cast coal specialists.

5. Excluding firms whose major activity is plant hire without operatives.

Source of Data: Construction Market Intelligence, Department of the Environment, Transport and the Regions
Contact: Neville Price 020 7944 5587

Table 3.5 Private Contractors: Employment of Operatives [1]

3rd Quarter Each Year: Thousands

	1989	1990	1991	1992 [2]	1993	1994	1995	1996 [3]	1997	1998	1999
(a) By Size of Firm											
2-3	54.3	58.8	46.2	45.7	40.9	50.1	54.4	45.3	66.4	68.7	59.6
4-7	61.0	56.6	48.7	59.9	49.2	46.0	42.7	34.1	44.1	46.6	83.2
8-13	39.8	35.5	32.5	33.7	29.7	27.1	25.0	27.2	25.6	27.2	32.1
14-24	54.1	49.8	42.6	43.0	37.6	34.6	33.3	34.2	38.5	41.6	44.6
25-34	31.5	29.6	24.4	23.3	21.7	19.6	18.8	22.0	24.5	25.2	26.0
35-59	47.4	41.7	34.5	35.5	33.9	31.0	29.2	30.4	37.1	40.8	42.7
60-79	22.5	21.3	18.1	16.5	13.5	14.9	13.5	13.7	18.3	20.0	19.5
80-114	27.7	28.5	23.4	19.6	17.1	15.3	14.4	15.2	18.1	20.4	21.7
115-299	63.3	58.8	47.2	42.9	37.6	39.4	36.8	38.4	43.7	45.7	49.2
300-599	40.3	37.9	34.4	25.2	23.7	22.8	26.0	25.1	28.7	33.2	29.5
600-1,199	35.0	36.2	27.9	26.8	23.6	24.7	23.0	23.2	28.7	28.3	30.3
1,200 and Over	73.6	71.1	57.6	47.2	41.0	38.4	38.0	39.4	44.5	44.8	54.7
All Firms	**550.5**	**525.6**	**437.5**	**419.2**	**369.4**	**363.7**	**354.9**	**348.3**	**418.2**	**442.4**	**492.8**
(b) By Trade of Firm											
General Builders	170.2	157.8	129.9	121.1	101.7	100.4	97.2	..	..	..	..
Building and Civil Engineering Contractors	81.3	76.5	62.2	55.3	45.3	45.8	39.9	..	..	..	..
Civil Engineers [4]	37.0	37.4	32.2	30.7	26.4	26.2	27.2	..	..	..	..
Plumbers	14.5	13.9	11.4	10.7	9.6	9.6	9.6	10.1	13.7	15.3	14.4
Carpenters and Joiners	12.4	12.4	9.9	10.9	9.0	9.1	8.9	8.7	11.3	12.3	12.8
Painters	30.0	27.8	21.4	19.5	17.8	16.7	16.4	16.5	20.3	20.9	23.4
Roofers	15.7	15.2	11.8	10.9	9.2	9.1	8.7	8.2	10.4	10.8	12.3
Plasterers	6.2	5.9	4.9	4.2	3.8	3.6	3.8	3.3	4.1	4.4	4.5
Glaziers	14.5	13.9	11.6	12.4	11.8	11.9	10.6	7.4	7.8	7.5	6.2
Demolition Contractors	2.8	2.5	2.0	2.5	2.2	1.9	2.2	2.4	2.8	3.0	5.7
Scaffolding Specialists	13.1	12.8	11.3	11.0	10.9	10.1	10.1	8.8	10.1	11.9	12.7
Reinforced Concrete Specialists	4.3	4.2	3.1	2.1	1.6	1.2	1.0	0.9	0.9	1.1	3.0
Heating and Ventilating Engineers	31.8	30.8	27.0	26.2	25.6	23.5	24.2	22.5	23.7	23.2	22.3
Electrical Contractors	54.2	54.7	48.5	46.8	45.4	44.0	43.0	46.5	54.4	57.3	65.7
Asphalt and Tar Sprayers	8.6	7.7	5.2	4.8	5.0	5.6	5.3	4.9	5.9	5.4	6.2
Plant Hirers [5]	15.0	14.3	12.4	12.4	11.1	11.2	11.0	9.6	9.9	10.5	11.1
Flooring Contractors	3.9	3.8	3.2	3.7	3.4	3.4	3.1	3.2	3.9	4.0	4.1
Constructional Engineers	10.6	10.3	8.9	9.3	7.1	7.3	6.3	5.2	5.4	4.7	4.9
Insulating Specialists	7.6	7.2	6.8	6.8	5.5	4.8	5.1	4.9	4.9	5.2	5.0
Suspended Ceiling Specialists	2.1	1.9	1.7	1.9	1.5	1.3	1.3	..	..	..	..
Floor and Wall Tiling Specialists	2.3	2.3	2.0	1.9	1.4	1.3	1.2	1.1	1.3	1.3	1.5
Miscellaneous	12.4	12.5	10.3	14.1	14.0	15.6	18.8	20.4	26.5	27.8	41.4
All Trades	**550.5**	**525.6**	**437.5**	**419.2**	**369.4**	**363.7**	**354.9**	**348.3**	**418.2**	**442.4**	**492.9**
(c) By Region of Registration											
North	33.8	31.9	28.2	25.1	22.7	23.2	23.6	21.9	24.7	24.7	28.2
Yorkshire & Humberside	54.1	51.2	45.3	44.6	39.6	38.7	38.4	38.3	45.4	49.9	52.5
East Midlands	33.5	32.2	27.3	25.3	20.9	21.0	21.1	19.9	26.7	28.8	31.3
East Anglia	19.2	18.7	14.7	14.8	12.5	11.6	11.6	10.8	14.1	15.3	15.4
South East:											
Beds, Essex, Herts	27.2	27.1	21.3	19.8	16.4	17.9	17.0	17.1	22.3	25.3	28.8
Greater London	84.6	78.9	64.7	59.5	52.3	46.2	45.0	43.4	49.2	50.7	49.5
Kent, Surrey, Sussex	39.3	37.3	31.2	29.6	24.5	23.9	23.1	23.0	28.9	31.9	38.7
Berks, Bucks, Hants, Oxon	34.5	34.7	25.9	26.9	21.5	20.5	19.0	19.6	24.3	26.3	33.5
South East: All	185.6	178.0	143.1	135.8	114.6	108.5	104.2	103.1	125.7	134.2	150.5

Table 3.5 Private Contractors: Employment of Operatives [1] (continued)

3rd Quarter Each Year: Thousands

	1989	1990	1991	1992 [2]	1993	1994	1995	1996 [3]	1997	1998	1999
(c) By Region of Registration											
North	33.8	31.9	28.2	25.1	22.7	23.2	23.6	21.9	24.7	24.7	28.2
Yorkshire & Humberside	54.1	51.2	45.3	44.6	39.6	38.7	38.4	38.3	45.4	49.9	52.5
East Midlands	33.5	32.2	27.3	25.3	20.9	21.0	21.1	19.9	26.7	28.8	31.3
East Anglia	19.2	18.7	14.7	14.8	12.5	11.6	11.6	10.8	14.1	15.3	15.4
South East:											
Beds, Essex, Herts	27.2	27.1	21.3	19.8	16.4	17.9	17.0	17.1	22.3	25.3	28.8
Greater London	84.6	78.9	64.7	59.5	52.3	46.2	45.0	43.4	49.2	50.7	49.5
Kent, Surrey, Sussex	39.3	37.3	31.2	29.6	24.5	23.9	23.1	23.0	28.9	31.9	38.7
Berks, Bucks, Hants, Oxon	34.5	34.7	25.9	26.9	21.5	20.5	19.0	19.6	24.3	26.3	33.5
South East: All	185.6	178.0	143.1	135.8	114.6	108.5	104.2	103.1	125.7	134.2	150.5
South West	39.5	38.4	30.9	27.6	23.7	23.1	24.1	22.9	30.5	31.5	37.6
West Midlands	46.9	44.1	36.3	33.1	29.9	32.4	31.9	34.7	38.4	40.1	45.5
North West	54.0	49.9	41.7	41.6	38.6	38.3	35.1	36.2	43.4	46.8	54.8
England	466.6	444.4	367.3	348.0	302.7	296.9	290.1	287.7	348.9	371.4	415.8
Wales	19.6	19.3	16.0	15.4	13.6	13.9	14.1	13.0	16.3	17.7	18.8
Scotland	64.4	62.0	54.2	55.9	53.1	53.0	50.6	47.7	53.1	53.2	58.2
Great Britain	**550.5**	**525.6**	**437.5**	**419.2**	**369.4**	**363.7**	**354.9**	**348.3**	**418.2**	**442.3**	**492.8**

Notes

. . = not available due to technical problems with the trade type breakdown

1. Information relates to employment by firms on the Department's register.

2. There is a discontinuity between 1991 and 1992 when the Department's register was enhanced by integration with the ONS register.

3. There is a discontinuity in the series between 1995 and 1996 as improved survey techniques, resulting in better coverage and classification of businesses during the previous 24 months, have been incorporated in a single update. This affects the trade classification of all businesses and the number of businesses with 7 or fewer employees.

4. Excluding open-cast coal specialists.

5. Excluding firms whose major activity is plant hire without operatives.

Source of Data: Construction Market Intelligence, Department of the Environment, Transport and the Regions
Contact: Neville Price 020 7944 5587

Table 3.6 Private Contractors: Number of Firms [1,2] by Size and Trade of Firm

3rd Quarter 1998

Size of Firm (by Number Employed)	Main Trades				Specialist Trades			
	General Builders	Building and Civil Engineering Contractors	Civil Engineers[3]	All Main Trades	Plumbers	Carpenters and Joiners	Painters	Roofers
1	..	..	..	32,549	7,759	6,164	3,912	2,754
2-3	..	..	..	19,531	3,382	2,797	3,308	1,709
4-7	..	..	..	6,796	966	703	1,144	686
8-13	..	..	..	1,478	205	164	228	194
14-24	..	..	..	1,297	106	103	190	142
25-34	..	..	..	503	40	25	73	48
35-59	..	..	..	591	32	35	68	47
60-79	..	..	..	224	8	14	19	13
80-114	..	..	..	173	9	} 11	12	} 6
115-299	..	..	..	262	12			
300-599	..	..	..	85	-	-	} 15	
600-1,199	..	..	..	37	-			-
1,200 and Over	..	..	..	24	-	-	-	-
All Firms	..	..	..	63,550	12,519	10,016	8,969	5,599

Specialist Trades (continued)

Size of Firm	Plasterers	Glaziers Contractors	Demolition Specialists	Scaffolding Specialists	Reinforced Concrete Specialists	Heating and Ventilating Engineers	Electrical Contractors	Asphalt and Tar Sprayers
1	1,293	1,342	358	211	109	2,590	10,901	221
2-3	785	1,128	200	445	105	1,531	5,231	218
4-7	319	415	118	173	71	713	1,786	130
8-13	57	115	42	55	10	243	556	42
14-24	48	74	37	64	11	211	449	43
25-34	15	21	10	19	5	66	166	20
35-59	14	18	20	20		77	148	20
60-79		3		} 9	} 10	21	51	9
80-114	} 7	4	} 8			20	41	} 8
115-299				8	-	17	32	
300-599		} 8	-		-	5	12	
600-1,199	-		-	} 5	-	} 6	6	
1,200 and Over	-		-		-		6	-
All Firms	2,538	3,128	793	1,009	321	5,500	19,385	711

Specialist Trades (continued)

Size of Firm	Plant Hirers[4]	Flooring Contractors	Constructional Engineers	Insulating Specialists	Suspended Ceiling Specialists	Floor and Wall Tiling Specialists	Miscellaneous	All Trades
1	2,454	1,524	393	373	..	412	9,490	87,837
2-3	880	725	238	248	..	240	4,085	47,918
4-7	295	273	109	143	..	80	1,170	16,391
8-13	77	81	41	57	..	15	293	3,988
14-24	79	48	33	54	..	12	253	3,274
25-34	37	16	15	31	..		82	1,201
35-59	32	9	13	14	..		89	1,263
60-79	7				..	} 11	26	419
80-114	5	} 8	} 9	} 14	..		16	319
115-299					..		16	405
300-599	} 16	-	} 13		..	-	10	125
600-1,199		-			..	-		56
1,200 and Over	-	-	-	-	..	-	} 5	40
All Firms	3,882	2,684	864	934	..	770	15,535	163,236

Table 3.6 Private Contractors: Number of Firms [1,2] by Size and Trade of Firm (continued)

3rd Quarter 1999

Size of Firm (by Number Employed)	Main Trades				Specialist Trades			
	General Builders	Building and Civil Engineering Contractors	Civil Engineers[3]	All Main Trades	Plumbers	Carpenters and Joiners	Painters	Roofers
1	..	..	..	30,770	8,511	5,892	3,564	2,546
2-3	..	..	..	18,938	3,685	2,825	3,455	1,947
4-7	..	..	..	6,642	960	640	1,280	708
8-13	..	..	..	1,508	240	158	261	192
14-24	..	..	..	1,235	105	111	176	149
25-34	..	..	..	541	41	29	74	38
35-59	..	..	..	500	36	45	65	39
60-79	..	..	..	209	6	10	19	12
80-114	..	..	..	164	9	10	11	} 5
115-299	..	..	..	225	9	5		
300-599	..	..	..	64	-	-	} 16	-
600-1,199	..	..	..	34	-	-		-
1,200 and Over	..	..	..	27	-	-	-	-
All Firms	..	..	..	60,858	13,600	9,725	8,921	5,636

Specialist Trades (continued)

Size of Firm	Plasterers	Glaziers Contractors	Demolition Specialists	Scaffolding Specialists	Reinforced Concrete Specialists	Heating and Ventilating Engineers	Electrical Contractors	Asphalt and Tar Sprayers
1	1,455	2,000	509	436	109	3,346	10,366	109
2-3	821	1,145	191	382	134	1,642	5,155	191
4-7	310	465	184	262	68	562	2,027	126
8-13	65	127	45	62	21	226	531	41
14-24	51	64	44	57	} 12	176	469	51
25-34	21	17	14	29		79	198	31
35-59		10	27	17	} 7	62	155	18
60-79	} 18	} 6		} 7		20	46	} 11
80-114					-	22	32	
115-299			} 7	5	-	16	31	
300-599	-	} 6			-	5	12	} 6
600-1,199	-			5	-	} 5	7	
1,200 and Over	-	-			-		8	-
All Firms	2,741	3,841	1,021	1,262	351	6,161	19,036	584

Specialist Trades (continued)

Size of Firm	Plant Hirers[4]	Flooring Contractors	Constructional Engineers	Insulating Specialists	Suspended Ceiling Specialists	Floor and Wall Tiling Specialists	Miscellaneous	All Trades
1	2,182	1,637	473	291	..	400	10,038	88,018
2-3	764	821	325	229	..	248	4,964	49,350
4-7	349	184	126	126	..	116	1,532	16,969
8-13	86	75	55	62	..	14	360	4,148
14-24	78	47	27	68	..	14	324	3,271
25-34	29	19	7	29	..	5	124	1,332
35-59	31	11	17	17	..		103	1,188
60-79	7				..	} 6	39	397
80-114	6	} 5	} 5	} 11	..		24	304
115-299	13	5			..		31	379
300-599		-		-	..	-	16	105
600-1,199	} 5	-	} 8	-	..	-	} 5	58
1,200 and Over	-	-	-	-	..	-		42
All Firms	3,549	2,805	1,042	832	..	803	17,560	165,561

Notes

. . = not available due to technical problems with the trade type breakdown
- = nil or less than half the digit shown.
1. Information relates to the number of firms, on the Department's register.
2. The number of firms include some which were temporarily inactive.
3. Excluding open-cast coal specialists.
4. Excluding firms whose major activity is plant-hire without operatives.

Source of Data: Construction Market Intelligence, Department of the Environment, Transport and the Regions
Contact: Neville Price 020 7944 5587

Table 3.7 Private Contractors: Number of Firms[1,2,3] by Size of Firm and Region of Registration

Size of Firm (by Number Employed)	North	Yorks & Humber	East Midlands	East Anglia	South East — Beds, Essex, Herts	Greater London	Kent, Surrey, Sussex	Oxon, Berks, Bucks, Hants	All	South West	West Midlands	North West	England	Wales	Scotland	Great Britain
3rd Quarter 1998																
1	2,678	6,258	6,430	4,287	7,696	9,808	8,724	7,489	33,717	9,189	7,947	7,128	77,621	4,521	5,695	87,837
2-3	1,757	3,747	3,532	2,299	3,913	4,751	4,386	3,985	17,035	5,138	4,349	4,261	42,110	2,447	3,361	47,918
4-7	682	1,465	1,195	664	1,050	1,554	1,426	1,277	5,307	1,525	1,489	1,564	13,889	823	1,679	16,391
8-13	201	391	278	158	219	384	313	267	1,183	312	375	412	3,309	177	502	3,988
14-24	171	274	246	125	204	286	249	211	950	266	316	358	2,705	154	415	3,274
25-34	76	120	90	40	59	98	72	68	297	98	128	124	972	54	175	1,201
35-59	86	128	80	48	62	123	83	82	350	96	125	141	1,054	53	156	1,263
60-79	17	51	38	20	28	31	29	26	114	26	33	48	347	16	56	419
80-114	16	36	26	7	14	26	24	12	76	26	32	53	271	9	39	319
115-299	26	50	25	⎱ 20	24	46	28	26	124	16	31	48	337	13	55	405
300-599	9	12	11	(with above)	⎱ 12	12	7	⎱ 15	37	⎱ 13	11	16	103	⎱ 6	18	125
600-1,199	⎱ 5	⎱ 11	⎱ 5		(with above)	12	⎱ 7	(with above)	21	(with above)	⎱ 8	⎱ 9	51		⎱ 8	56
1,200 and Over	(with above)	(with above)	(with above)		(with above)	11	(with above)	(with above)	18	-	(with above)	(with above)	35	-	(with above)	40
All Firms	5,724	12,543	11,956	7,668	13,281	17,142	15,348	13,458	59,229	16,705	14,844	14,162	142,804	8,273	12,159	163,236
3rd Quarter 1999																
1	2,510	6,147	5,783	3,855	8,293	9,747	8,838	8,547	35,425	9,275	7,274	7,456	77,725	4,437	5,856	88,018
2-3	2,157	3,742	3,589	2,253	3,436	4,906	4,620	4,467	17,430	5,231	4,162	4,124	42,687	2,596	4,066	49,350
4-7	805	1,435	1,241	669	1,076	1,551	1,590	1,154	5,372	1,687	1,454	1,745	14,409	853	1,707	16,969
8-13	240	394	315	213	192	374	312	233	1,111	281	394	429	3,377	206	566	4,148
14-24	223	300	186	108	176	304	267	165	911	230	351	371	2,680	145	446	3,271
25-34	93	157	98	26	55	114	93	81	343	105	107	150	1,079	50	203	1,332
35-59	78	126	83	46	69	102	77	71	319	88	112	132	984	52	152	1,188
60-79	18	40	29	20	30	28	24	26	108	30	42	45	332	12	53	397
80-114	19	37	26	9	15	25	19	20	79	20	19	44	253	8	43	304
115-299	28	45	30	⎱ 17	16	36	29	21	102	18	29	47	313	⎱ 14	55	379
300-599	⎱ 12	8	8	(with above)	8	11	6	11	36	6	8	13	90	(with above)	14	105
600-1,199	(with above)	1	5	-	⎱ 7	7	⎱ 10	⎱ 7	21	⎱ 8	7	⎱ 10	52	⎱ 9	⎱ 9	58
1,200 and Over	(with above)	7	-	-	(with above)	11	(with above)	(with above)	21	(with above)	5	(with above)	37	-	(with above)	42
All Firms	6,183	12,442	11,396	7,220	13,388	17,222	15,897	14,810	61,317	16,985	13,974	14,576	144,093	8,382	13,179	165,561

Notes

- = nil or less than half the digit shown.

1. Information relates to the number of firms on the Department's register.
2. The number of firms include some which were temporarily inactive.
3. Based on the definition of the construction industry as given in the Revised 1992 Standard Industrial Classification.

Source of Data: Construction Market Intelligence, Department of the Environment, Transport and the Regions
Contact: Neville Price 020 7944 5587

Table 3.8 Private Contractors: Value of Work Done[1,2] by Size of Firm and Type of Work

Great Britain   £ Million

Size of Firm (by Number Employed)	3rd Quarter 1998[3]						3rd Quarter 1999[3]					
	All New Work	Housing: Public & Private Sectors	Repair & Maintenance				All New Work	Housing: Public & Private Sectors	Repair & Maintenance			
			Other		All Repair & Maintenance	All Work			Other		All Repair & Maintenance	All Work
			Public	Private					Public	Private		
1	392.6	533.5	61.5	172.5	767.6	1,160.2	519.1	603.4	113.0	235.5	952.0	1,471.1
2-3	405.9	449.4	61.6	152.4	663.4	1,069.4	483.1	569.8	138.5	173.4	881.7	1,364.8
4-7	345.9	320.1	81.8	142.0	543.9	889.8	495.7	397.6	91.6	188.8	678.0	1,173.7
8-13	270.3	131.4	58.5	114.2	304.1	574.4	280.3	121.3	44.4	108.3	274.1	554.4
14-24	458.7	150.0	93.0	163.2	406.2	865.0	430.1	120.7	83.8	138.6	343.1	773.2
25-34	306.7	83.2	52.4	90.7	226.3	533.0	343.5	77.4	39.8	97.2	214.4	557.9
35-59	686.7	98.0	70.1	149.4	317.4	1,004.1	742.1	99.4	63.4	154.4	317.2	1,059.3
60-79	351.1	44.3	37.1	71.3	152.6	503.8	372.7	55.1	45.6	100.7	201.3	574.1
80-114	463.2	68.5	40.4	58.7	167.6	630.9	474.2	58.1	34.1	60.9	153.1	627.4
115-299	1,070.3	107.1	71.4	139.8	318.3	1,388.5	1,085.2	101.6	88.2	130.1	320.0	1,405.2
300-599	719.1	27.6	63.4	124.9	215.8	934.9	700.9	36.2	36.1	95.5	167.8	868.6
600-1,199	759.7	40.1	220.6	100.1	360.7	1,120.4	827.2	46.4	166.6	59.2	272.1	1,099.3
1,200 and Over	1,307.3	25.6	19.8	135.7	181.1	1,488.4	1,412.8	80.2	38.7	198.3	317.2	1,730.0
All Firms	7,537.6	2,078.7	931.7	1,614.9	4,625.3	12,163.0	8,167.1	2,367.2	983.8	1,740.9	5,091.9	13,259.0

Notes

1. Information relates to work done by firms on the Department's register.
2. The numbers of firms include some of which were temporarily inactive.
3. Based on the definition of the construction industry as given in the Revised 1992 Standard Industrial Classification.

Source of Data: Construction Market Intelligence, Department of the Environment, Transport and the Regions
Contact: Neville Price 020 7944 5587

Table 3.9 Private Contractors: Value of Work Done[1] by Trade of Firm and Type of Work

Great Britain £ Million

| Size of Firm (by Number Employed) | 3rd Quarter 1998[2] | | | | | | 3rd Quarter 1999[3] | | | | | |
| | | Repair & Maintenance | | | | | | Repair & Maintenance | | | | |
	All New Work	Housing: Public & Private Sectors	Other Public	Other Private	All Repair & Maintenance	All Work	All New Work	Housing: Public & Private Sectors	Other Public	Other Private	All Repair & Maintenance	All Work
Main Trades:												
General Builders	..	..	..	..	..	..	..	..	..	..	..	..
Building and Civil Engineering Contractors	..	..	..	..	..	..	..	..	..	..	..	..
Civil Engineers[3]	..	..	..	..	..	..	..	..	..	..	..	..
All Main Trades	4,787.0	917.4	438.2	858.0	2,013.6	6,800.6	4,389.3	1,272.3	528.7	935.6	2,736.6	7,125.9
Specialist Trades:												
Plumbers	164.7	137.1	22.7	49.7	209.4	374.1	234.4	67.9	28.2	50.0	146.1	380.5
Carpenters and Joiners	128.5	93.2	16.9	54.1	164.3	292.7	207.2	60.1	25.0	44.2	129.2	336.4
Painters	110.9	112.6	34.5	71.8	218.9	329.8	230.9	66.9	27.8	49.2	143.9	374.8
Roofers	137.5	74.6	31.7	58.3	164.5	302.0	179.2	51.9	21.6	38.2	111.7	290.9
Plasterers	53.4	25.4	4.6	9.6	39.6	93.0	65.1	18.9	7.8	13.9	40.6	105.7
Glaziers	100.8	85.3	10.8	31.7	127.9	228.6	94.7	27.4	11.4	20.2	59.0	153.7
Demolition Contractors	45.6	2.7	6.0	5.0	13.7	59.3	77.7	22.5	9.4	16.6	48.4	126.2
Scaffolding Specialists	115.3	12.6	12.8	44.2	69.7	184.9	130.9	37.9	15.8	27.9	81.6	212.5
Reinforced Concrete Specialists	22.5	5.2	2.2	4.9	12.3	34.8	27.9	8.1	3.4	6.0	17.4	45.3
Heating and Ventilating Engineers	282.6	114.8	43.3	135.0	293.1	575.7	394.0	114.2	47.5	84.0	245.6	639.6
Electrical Contractors	684.5	217.9	92.1	227.4	537.4	1,221.8	886.5	257.0	106.8	189.0	552.7	1,439.2
Asphalt and Tar Sprayers	61.1	10.3	69.2	36.4	115.8	177.0	114.3	33.1	13.8	24.4	71.3	185.5
Plant Hirers[4]	195.3	6.8	12.0	12.3	31.0	226.3	147.5	42.8	17.8	31.5	92.0	239.5
Flooring contractors	51.5	29.2	12.6	24.7	66.5	118.0	71.3	20.7	8.6	15.2	44.5	115.8
Constructional Engineers	122.6	13.6	12.7	15.1	41.4	164.0	75.1	21.8	9.0	16.0	46.8	121.9
Insulating Specialists	38.7	26.5	6.6	27.0	60.0	98.7	62.2	18.0	7.5	13.3	38.8	101.0
Suspended Ceiling Specialists	..	..	..	..	..	..	..	..	..	..	..	..
Floor and Wall Tiling Specialists	14.2	8.1	3.4	7.1	18.6	32.8	22.4	6.5	2.7	4.8	13.9	36.3
Miscellaneous	369.8	155.5	93.2	126.0	374.7	744.5	642.1	186.1	77.3	136.9	400.3	1,042.4
All Trades	7,537.6	2,078.8	931.7	1,614.9	4,625.4	12,163.0	8,167.1	2,367.2	983.8	1,740.9	5,091.9	13,259.0

Notes

.. = not available due to technical problems with the trade type breakdown

1. Information relates to the value of work done by firms on the Department's register.
2. Based on the definition of the construction industry as given in the Revised 1992 Standard Industrial Classification.
3. Excluding work done by open-cast coal specialists.
4. Excluding firms whose major activity is work done by plant hirers without operatives.

Source of Data: Construction Market Intelligence, Department of the Environment, Transport and the Regions
Contact: Neville Price 020 7944 5587

Table 3.10 Private Contractors: Value of Work Done[1,2] by Size and Trade of Firm

3rd Quarter 1998: £ Million

Size of Firm (by Number Employed)	Main Trades				Specialist Trades			
	General Builders	Building and Civil Engineering Contractors	Civil Engineers[3]	All Main Trades	Plumbers	Carpenters and Joiners	Painters	Roofers
1	..	..	..	436.1	103.7	81.1	50.6	36.7
2-3	..	..	..	448.1	74.0	59.7	70.2	38.1
4-7	..	..	..	373.2	50.1	35.7	58.2	37.1
8-13	..	..	..	210.7	29.4	18.0	19.4	37.8
14-24	..	..	..	341.7	26.1	21.9	30.5	45.7
25-34	..	..	..	233.1	17.6	11.6	17.3	23.5
35-59	..	..	..	531.7	21.9	23.8	25.6	35.2
60-79	..	..	..	293.9	5.6	18.4	10.3	23.3
80-114	..	..	..	390.0	13.2	} 22.4	10.9	} 24.7
115-299	..	..	..	977.1	32.4			
300-599	..	..	..	705.3	-	-	} 36.9	-
600-1,199	..	..	..	790.2	-	-		
1,200 and Over	..	..	..	1,069.5	-	-	-	-
All Firms	..	..	..	6,800.6	374.1	292.7	329.8	302.0

Specialist Trades (continued)

Size of Firm	Plasterers	Glaziers Contractors	Demolition Specialists	Scaffolding Specialists	Reinforced Concrete Specialists	Heating and Ventilating Engineers	Electrical Contractors	Asphalt and Tar Sprayers
1	16.9	19.8	4.4	3.0	1.5	38.0	144.8	3.3
2-3	17.9	27.7	4.3	10.5	2.2	35.7	114.6	5.7
4-7	16.4	26.6	6.1	9.9	3.9	43.2	99.9	8.1
8-13	5.4	18.4	6.8	6.1	1.2	42.8	74.1	9.2
14-24	10.8	23.3	8.5	12.4	3.4	70.3	103.8	16.5
25-34	11.1	14.1	3.7	5.2	4.3	35.8	59.4	11.5
35-59	4.4	11.4	16.4	9.1	} 19.3	64.8	101.7	18.9
60-79	} 10.2	4.0	} 9.0	} 5.8		26.6	48.5	14.3
80-114		13.4				34.2	56.3	} 89.4
115-299		} 70.0		10.7	-	57.3	101.1	
300-599			-		-	30.3	73.4	
600-1,199	-		-	} 112.3	-	} 96.5	88.1	
1,200 and Over	-		-		-		156.1	-
All Firms	93.0	228.6	59.3	184.9	34.8	575.7	1,221.8	177.0

Specialist Trades (continued)

Size of Firm	Plant Hirers[4]	Flooring Contractors	Constructional Engineers	Insulating Specialists	Suspended Ceiling Specialists	Floor and Wall Tiling Specialists	Miscellaneous	All Trades
1	33.9	19.1	5.6	4.8	..	5.8	117.2	1,160.2
2-3	19.3	15.6	5.7	5.7	..	5.6	86.2	1,069.4
4-7	15.9	14.5	6.5	7.9	..	4.6	59.4	889.8
8-13	7.9	13.6	6.4	13.2	..	2.4	43.9	574.4
14-24	18.3	16.3	10.6	11.8	..	2.5	80.9	865.0
25-34	12.8	9.4	6.9	11.1	..		39.6	533.0
35-59	14.5	7.0	9.9	8.7	..		74.9	1,004.1
60-79	5.7	} 22.5	} 27.0		..	} 11.8	30.3	503.8
80-114	6.4			} 35.5	..		28.0	630.9
115-299					..		32.4	1,388.5
300-599	} 91.7	-	} 85.4		..	-	50.0	934.9
600-1,199		-			..	-	} 101.7	1,120.4
1,200 and Over	-	-	-	-	..	-		1,488.4
All Firms	226.3	118.0	164.0	98.7	..	32.8	744.5	12,163.0

Table 3.10 Private Contractors: Value of Work Done[1,2] by Size and Trade of Firm (continued)

3rd Quarter 1999: £ Million

Size of Firm (by Number Employed)	Main Trades				Specialist Trades			
	General Builders	Building and Civil Engineering Contractors	Civil Engineers[3]	All Main Trades	Plumbers	Carpenters and Joiners	Painters	Roofers
1	..	..	..	520.3	109.3	86.4	46.7	35.8
2-3	..	..	..	648.7	72.6	67.8	58.3	45.4
4-7	..	..	..	520.6	62.1	32.4	64.9	45.6
8-13	..	..	..	206.0	29.4	16.7	20.5	28.5
14-24	..	..	..	292.2	17.7	32.0	25.2	46.7
25-34	..	..	..	248.3	14.5	14.7	18.8	14.0
35-59	..	..	..	512.7	27.8	34.0	35.9	30.5
60-79	..	..	..	322.2	6.5	14.3	32.2	17.8
80-114	..	..	..	353.4	15.6	23.2	13.7	} 26.7
115-299	..	..	..	942.7	25.1	14.8	} 58.7	
300-599	..	..	..	600.9	-	-		-
600-1,199	..	..	..	726.0	-	-		-
1,200 and Over	..	..	..	1,232.2	-	-	-	-
All Firms	..	..	..	7,125.9	380.5	336.4	374.8	290.9

Specialist Trades (continued)

Size of Firm	Plasterers	Glaziers Contractors	Demolition Specialists	Scaffolding Specialists	Reinforced Concrete Specialists	Heating and Ventilating Engineers	Electrical Contractors	Asphalt and Tar Sprayers
1	30.8	19.0	16.6	16.6	5.9	87.8	181.1	1.1
2-3	14.5	34.2	9.7	9.7	4.6	40.2	112.1	5.0
4-7	14.3	27.6	12.6	11.4	15.7	49.3	114.7	9.0
8-13	11.4	15.5	6.0	8.1	1.6	35.6	60.4	11.3
14-24	14.4	15.1	7.4	10.0	} 3.3	40.8	123.8	16.3
25-34	5.7	6.5	4.6	9.6		34.8	75.9	15.3
35-59	}	7.5	23.1	10.9		56.3	120.4	21.7
60-79	} 14.6	} 8.3		} 6.8	} 14.1	32.7	50.5	} 19.8
80-114	}	}	} 46.1			49.7	53.4	
115-299	}	} 19.9		9.7	-	52.7	82.5	} 86.0
300-599	-			} 119.5	-	46.1	82.6	
600-1,199	-				-	} 113.6	109.6	
1,200 and Over	-	-			-		272.1	-
All Firms	105.7	153.7	126.2	212.5	45.3	639.6	1,439.2	185.5

Specialist Trades (continued)

Size of Firm	Plant Hirers[4]	Flooring Contractors	Constructional Engineers	Insulating Specialists	Suspended Ceiling Specialists	Floor and Wall Tiling Specialists	Miscellaneous	All Trades
1	41.7	20.2	4.3	5.5	..	6.2	169.3	1,471.1
2-3	13.0	15.2	16.5	7.7	..	9.0	140.3	1,364.8
4-7	16.9	13.0	5.8	10.5	..	7.6	100.9	1,173.7
8-13	11.6	14.3	7.9	13.4	..	1.6	51.1	554.4
14-24	17.0	10.8	4.9	13.1	..	2.2	76.9	773.2
25-34	9.2	8.6	2.2	9.9	..	1.7	56.8	557.9
35-59	22.8	7.7	14.7	20.2	..		84.9	1,059.3
60-79	6.9	} 12.4	} 9.4	} 111.7	..	} 8.1	49.1	574.1
80-114	12.9				..		53.8	627.4
115-299	45.6	13.6			..		76.0	1,405.2
300-599	} 42.0	-	} 56.1	-	..	-	87.7	868.6
600-1,199		-		-	..	-	} 95.5	1,099.3
1,200 and Over	-	-	-	-	..	-		1,730.3
All Firms	239.5	115.8	121.9	101.0	..	36.3	1,042.4	13,259.0

Notes to table 3.10

. . = not available due to technical problems with the trade type breakdown

- = nil or less than half the final digit shown.

1. Information relates to work done by firms on the Department's register.

2. Based on the definition of the construction industry as given in the Revised 1992 Standard Industrial Classification.

3. Excluding work done by open-cast coal specialists.

4. Excluding work done by plant-hirers without operatives.

Source of Data: Construction Market Intelligence, Department of the Environment, Transport and the Regions

Contact: Neville Price 020 7944 5587

Table 3.11 Private Contractors: Employment of Operatives (including Trainees)[1,2] by Size and Trade of Firm

3rd Quarter 1998: Thousands

Size of Firm (by Number Employed)	Main Trades				Specialist Trades			
	General Builders	Building and Civil Engineering Contractors	Civil Engineers[3]	All Main Trades	Plumbers	Carpenters and Joiners	Painters	Roofers
2-3	..	..	..	27.0	5.4	4.4	4.0	2.3
4-7	..	..	..	19.3	2.8	2.0	3.3	2.0
8-13	..	..	..	10.3	1.4	1.1	1.7	1.3
14-24	..	..	..	16.5	1.4	1.4	2.6	1.7
25-34	..	..	..	10.4	0.9	0.6	1.7	1.0
35-59	..	..	..	19.0	1.1	1.1	2.5	1.5
60-79	..	..	..	10.2	0.4	0.7	1.1	0.6
80-114	..	..	..	10.5	0.6	} 1.1	0.9	} 0.4
115-299	..	..	..	28.2	1.2			
300-599	..	..	..	21.0	-	-	} 3.0	-
600-1,199	..	..	..	16.7	-	-		-
1,200 and Over	..	..	..	23.4	-	-	-	-
All Firms	..	..	..	212.4	15.3	12.3	20.9	10.8

Specialist Trades (continued)

Size of Firm	Plasterers	Glaziers Contractors	Demolition Specialists	Scaffolding Specialists	Reinforced Concrete Specialists	Heating and Ventilating Engineers	Electrical Contractors	Asphalt and Tar Sprayers
2-3	1.1	1.4	0.3	0.5	0.1	2.3	8.0	0.3
4-7	0.9	1.2	0.3	0.5	0.2	2.1	5.1	0.4
8-13	0.4	0.8	0.3	0.4	0.1	1.6	3.7	0.3
14-24	0.6	0.9	0.5	0.9	0.1	2.5	5.8	0.6
25-34	0.3	0.4	0.2	0.4	0.1	1.3	3.5	0.4
35-59	0.5	0.5	0.7	0.7	} 0.4	2.2	4.8	0.7
60-79	} 0.7	0.1	} 0.6	} 0.6		1.0	2.7	0.5
80-114		0.2				1.2	2.9	} 2.3
115-299		} 2.1		1.2	-	2.1	4.2	
300-599			-	} 6.7	-	1.2	4.3	
600-1,199	-		-		-	} 5.9	4.0	
1,200 and Over	-		-		-		8.3	-
All Firms	4.4	7.5	3.0	11.9	1.1	23.2	57.3	5.4

Specialist Trades (continued)

Size of Firm	Plant Hirers[4]	Flooring Contractors	Constructional Engineers	Insulating Specialists	Suspended Ceiling Specialists	Floor and Wall Tiling Specialists	Miscellaneous	All Trades
2-3	1.6	1.1	0.3	0.3	..	0.4	6.2	68.7
4-7	0.9	0.8	0.3	0.4	..	0.2	3.2	46.6
8-13	0.6	0.5	0.3	0.4	..	0.1	1.8	27.2
14-24	1.1	0.6	0.4	0.7	..	0.1	3.0	41.6
25-34	0.9	0.3	0.3	0.6	..	} 0.4	1.6	25.2
35-59	1.2	0.3	0.4	0.4	..		2.7	40.8
60-79	0.4	} 0.5	} 0.5		..		1.3	20.0
80-114	0.4			} 2.3	..		0.9	20.4
115-299	} 3.4		} 2.2		..	-	1.5	45.7
300-599		-			..	-	2.8	33.2
600-1,199		-			..	-	} 2.8	28.3
1,200 and Over	-	-	-	-	..	-		44.8
All Firms	10.5	4.0	4.7	5.2	..	1.3	27.8	442.4

Table 3.11 Private Contractors: Employment of Operatives (including Trainees)[1,2] by Size and Trade of Firm (continued)

3rd Quarter 1999: Thousands

Size of Firm (by Number Employed)	Main Trades				Specialist Trades			
	General Builders	Building and Civil Engineering Contractors	Civil Engineers[3]	All Main Trades	Plumbers	Carpenters and Joiners	Painters	Roofers
2-3	..	..	..	24.0	2.9	3.5	3.4	2.8
4-7	..	..	..	33.3	4.4	2.8	6.4	3.3
8-13	..	..	..	12.1	1.7	1.2	2.0	1.5
14-24	..	..	..	17.6	1.2	1.4	2.5	1.8
25-34	..	..	..	10.6	0.7	0.6	1.4	0.7
35-59	..	..	..	17.8	1.4	1.6	2.5	1.3
60-79	..	..	..	10.5	0.3	0.5	1.0	0.6
80-114	..	..	..	11.0	0.7	0.8	0.9	} 0.4
115-299	..	..	..	29.4	1.0	0.5	} 3.2	
300-599	..	..	..	16.5	-	-		-
600-1,199	..	..	..	16.9	-	-		-
1,200 and Over	..	..	..	32.2	-	-	-	-
All Firms	..	..	..	232.0	14.4	12.8	23.4	12.3

Specialist Trades (continued)

Size of Firm	Plasterers	Glaziers Contractors	Demolition Specialists	Scaffolding Specialists	Reinforced Concrete Specialists	Heating and Ventilating Engineers	Electrical Contractors	Asphalt and Tar Sprayers
2-3	0.7	0.9	0.4	0.8	0.2	1.7	7.2	0.2
4-7	1.2	1.9	1.2	1.5	2.0	2.5	9.5	0.9
8-13	0.6	0.8	0.4	0.5	0.1	1.7	3.9	0.5
14-24	0.7	0.8	0.6	0.9	} 0.1	2.2	6.7	0.7
25-34	0.4	0.3	0.3	0.8		1.8	4.0	0.6
35-59	} 0.9	0.3	1.0	0.7		2.1	5.5	0.7
60-79		} 0.4		} 0.5	} 0.5	0.9	2.3	} 0.7
80-114						1.7	2.5	
115-299			} 1.8	0.7	-	2.0	4.0	} 2.0
300-599	-	} 0.7		} 6.4	-	1.7	4.1	
600-1,199	-				-	} 4.2	3.7	
1,200 and Over	-	-			-		12.5	-
All Firms	4.5	6.2	5.7	12.7	3.0	22.3	65.7	6.2

Specialist Trades (continued)

Size of Firm	Plant Hirers[4]	Flooring Contractors	Constructional Engineers	Insulating Specialists	Suspended Ceiling Specialists	Floor and Wall Tiling Specialists	Miscellaneous	All Trades
2-3	1.2	0.6	0.8	0.3	..	0.4	6.2	59.6
4-7	1.6	0.9	0.4	0.7	..	0.5	7.0	83.2
8-13	0.6	0.6	0.4	0.7	..	0.1	2.6	32.1
14-24	1.3	0.6	0.3	0.9	..	0.2	3.9	44.6
25-34	0.6	0.4	0.1	0.5	..	0.1	2.2	26.0
35-59	1.1	0.4	0.5	1.0	..		3.7	42.7
60-79	0.3	} 0.2	} 0.2	} 1.0	..	} 0.3	1.7	19.5
80-114	0.6				..		1.5	21.7
115-299	1.9	0.5			..		4.0	49.2
300-599	} 2.0	-	} 2.0	-	..	-	4.9	29.5
600-1,199		-		-	..	-	} 3.8	30.3
1,200 and Over	-	-	-	-	..	-		54.7
All Firms	11.1	4.1	4.9	5.0	..	1.6	41.4	493.0

Notes to table 3.11

. . = not available due to technical problems with the trade type breakdown

- = nil or less than half the final digit shown.

1. Information relates to operatives employed by firms on the Department's register.
2. Based on the definition of the construction industry as given in the Revised 1992 Standard Industrial Classification.
3. Excluding operatives employed by open-cast coal specialists.
4. Excluding firms whose major activity is plant-hire without operatives.

Source of Data: Construction Market Intelligence, Department of the Environment, Transport and the Regions

Contact: Neville Price 020 7944 5587

Table 3.12 Private Contractors: Employment of Administrative, Professional, Technical and Clerical Staff by Size and Trade of Firm[1,2]

3rd Quarter 1998

Size of Firm (by Number Employed)	Main Trades				Specialist Trades			
	General Builders	Building and Civil Engineering Contractors	Civil Engineers[3]	All Main Trades	Plumbers	Carpenters and Joiners	Painters	Roofers
2-3	..	..	..	8,379	1,645	1,328	1,220	729
4-7	..	..	..	6,498	924	657	1,029	659
8-13	..	..	..	3,769	499	325	402	553
14-24	..	..	..	6,034	451	373	549	788
25-34	..	..	..	3,863	267	165	390	374
35-59	..	..	..	7,271	337	385	461	585
60-79	..	..	..	5,249	135	221	163	290
80-114	..	..	..	6,276	205	} 299	187	} 280
115-299	..	..	..	18,147	624		} 621	
300-599	..	..	..	14,344	-	-		-
600-1,199	..	..	..	15,412	-	-		-
1,200 and Over	..	..	..	24,753	-	-	-	-
All Firms	..	..	..	119,993	5,087	3,752	5,021	4,258

Specialist Trades (continued)

Size of Firm	Plasterers	Glaziers Contractors	Demolition Specialists	Scaffolding Specialists	Reinforced Concrete Specialists	Heating and Ventilating Engineers	Electrical Contractors	Asphalt and Tar Sprayers
2-3	325	449	88	148	39	710	2,454	87
4-7	293	445	110	152	63	765	1,714	119
8-13	151	347	126	128	25	778	1,452	103
14-24	241	399	143	201	44	1,217	2,016	158
25-34	83	233	67	91	62	597	1,180	100
35-59	86	291	192	164	} 145	1,146	1,712	189
60-79	} 185	133	} 122	} 118		487	771	123
80-114		185			-	733	1,008	} 864
115-299				177	-	909	1,488	
300-599		} 3,289	-	} 2,232	-	894	1,263	
600-1,199	-		-		-	} 2,584	1,747	
1,200 and Over	-		-		-		4,611	-
All Firms	1,364	5,770	846	3,411	378	10,821	21,416	1,742

Specialist Trades (continued)

Size of Firm	Plant Hirers[4]	Flooring Contractors	Constructional Engineers	Insulating Specialists	Suspended Ceiling Specialists	Floor and Wall Tiling Specialists	Miscellaneous	All Trades
2-3	482	333	110	106	..	109	1,923	21,190
4-7	275	288	107	144	..	79	1,145	15,746
8-13	137	282	104	164	..	47	886	10,420
14-24	276	249	160	227	..	70	1,459	15,219
25-34	178	151	155	262	..		763	9,058
35-59	227	102	171	194	..	} 161	1,201	14,940
60-79	72	} 346	} 395		..		514	8,643
80-114	55			} 526	..		621	10,466
115-299					..		1,082	25,266
300-599	} 1,002	-	} 1,414		..	-	1,335	18,898
600-1,199		-			..	-	} 2,483	20,862
1,200 and Over	-	-	-	-	..	-		37,192
All Firms	2,705	1,751	2,616	1,622	..	466	13,412	207,899

Table 3.12 Private Contractors: Employment of Administrative, Professional, Technical and Clerical Staff by Size and Trade of Firm[1,2] (continued)

3rd Quarter 1999

Size of Firm (by Number Employed)	Main Trades				Specialist Trades			
	General Builders	Building and Civil Engineering Contractors	Civil Engineers[3]	All Main Trades	Plumbers	Carpenters and Joiners	Painters	Roofers
2-3	..	..	..	2,572	436	336	293	394
4-7	..	..	..	6,915	1,047	433	1,158	828
8-13	..	..	..	4,223	592	363	508	542
14-24	..	..	..	5,593	457	686	633	933
25-34	..	..	..	4,012	258	196	405	211
35-59	..	..	..	7,047	401	473	555	472
60-79	..	..	..	5,561	105	209	247	328
80-114	..	..	..	5,832	248	245	165	} 406
115-299	..	..	..	18,841	510	557		
300-599	..	..	..	11,394	-	-	} 1,290	-
600-1,199	..	..	..	14,471	-	-		-
1,200 and Over	..	..	..	25,437	-	-	-	-
All Firms	..	..	..	111,897	4,054	3,498	5,255	4,113

Specialist Trades (continued)

	Plasterers	Glaziers Contractors	Demolition Specialists	Scaffolding Specialists	Reinforced Concrete Specialists	Heating and Ventilating Engineers	Electrical Contractors	Asphalt and Tar Sprayers
2-3	25	184	59	92	17	829	805	8
4-7	231	516	253	261	261	841	2,291	82
8-13	229	408	128	178	61	736	1,494	129
14-24	211	402	162	254	} 52	802	2,198	231
25-34	95	159	82	183		618	1,436	170
35-59		157	279	158		893	1,835	184
60-79	} 250	} 209		} 119	} 201	453	976	} 246
80-114			} 1,107			720	866	
115-299				195	-	1,056	1,967	
300-599	-	} 534			-	838	1,760	} 1106
600-1,199	-			} 2,826	-		1,946	
1,200 and Over	-	-			-	} 2,060	7,067	-
All Firms	1,041	2,569	2,068	4,267	591	9,847	24,641	2,155

Specialist Trades (continued)

	Plant Hirers[4]	Flooring Contractors	Constructional Engineers	Insulating Specialists	Suspended Ceiling Specialists	Floor and Wall Tiling Specialists	Miscellaneous	All Trades
2-3	67	60	101	75	..	50	1,285	7,849
4-7	269	360	114	163	..	163	1,812	18,647
8-13	238	296	134	232	..	34	1,220	11,815
14-24	316	236	115	274	..	59	1,843	15,561
25-34	183	168	45	186	..	36	1,212	9,737
35-59	289	121	234	283	..		1,262	15,021
60-79	139	} 177	} 210	} 337	..	} 116	1,043	9,749
80-114	126				..		876	9,974
115-299	760	307			..		2,400	29,210
300-599	} 887	-	} 889	-	..	-	2,477	17,755
600-1,199		-		-	..		} 2,377	21,273
1,200 and Over	-	-	-	-	..	-		37,593
All Firms	3,275	1,725	1,843	1,550	..	459	17,807	204,184

Notes to table 3.12

. . = not available due to technical problems with the trade type breakdown

- = nil or less than half the final digit shown.

1. Information relates to employment by firms, on the Department's register.
2. Based on the definition of the construction industry as given in the Revised 1992 Standard Industrial Classification.
3. Excluding employment by open-cast coal specialists.
4. Excluding employment by firms whose major activity is plant-hire without operatives.

Source of Data: Construction Market Intelligence, Department of the Environment, Transport and the Regions

Table 3.13 Private Contractors: Total Employment[1] by Size of Firm

Great Britain

Size of Firm (by Number Employed)	3rd Quarter 1998[2]			3rd Quarter 1999[2]		
	Firms[3]	Working Proprietors	All Employment[4]	Firms[3]	Working Proprietors	All Employment[4]
1	87,837	77,350	77,400	88,018	86,774	183,200
2-3	47,918	58,425	148,300	49,350	53,070	120,500
4-7	16,391	18,800	81,200	16,969	17,412	119,300
8-13	3,988	4,015	41,600	4,148	4,017	48,000
14-24	3,274	2,745	59,500	3,271	3,001	63,200
25-34	1,201	900	35,100	1,332	290	36,000
35-59	1,263	835	56,600	1,188	412	58,100
60-79	419	170	28,800	397	83	29,300
80-114	319	125	30,900	304	53	31,700
115-299	405	80	71,100	379	61	78,500
300-599	125	10	52,100	105	29	47,300
600-1,199	56	0	49,100	58	-	51,600
1,200 and Over	40	5	82,000	42	5	92,300
All Firms	163,236	163,463	813,600	165,561	165,208	958,800

Notes

- = nil or less than half the final digit shown.
1. Information relates to firms on the Department's register.
2. Based on the definition of the construction industry as given in the Revised 1992 Standard Industrial Classification.
3. The numbers of firms include some which were temporarily inactive.
4. Working proprietors and employees (including trainees).

Source of Data: Construction Market Intelligence, Department of the Environment, Transport and the Regions
Contact: Neville Price 020 7944 5587

Table 3.14 Private Contractors: Total Employment by Trade of Firm[1]

Trade of Firm	3rd Quarter 1998[2]			3rd Quarter 1999[2]		
	Firms[3]	Working Proprietors	All Employment[4]	Firms[3]	Working Proprietors	All Employment[4]
Main Trades:						
General Builders	..	..	..	..	..	..
Building and Civil Engineering Contractors	..	..	..	..	..	..
Civil Engineers	..	..	..	..	..	..
All Main Trades	63,550	63,825	396,200	60,858	60,554	439,996
Specialist Trades:						
Plumbers	12,519	12,460	32,800	13,600	13,968	39,478
Carpenters and Joiners	10,016	10,175	26,200	9,725	9,621	31,758
Painters	8,969	9,380	35,300	8,921	9,082	41,000
Roofers	5,599	5,630	20,700	5,636	5,757	24,424
Plasterers	2,538	2,665	8,500	2,741	2,923	10,491
Glaziers	3,128	3,340	16,600	3,841	3,998	13,930
Demolition Contractors	793	765	4,600	1,021	829	9,607
Scaffolding Specialists	1,009	1,080	16,400	1,262	1,358	19,880
Reinforced Concrete Specialists	321	325	1,800	351	401	4,496
Heating and Ventilating Engineers	5,500	5,425	39,500	6,161	6,164	44,690
Electrical Contractors	19,385	19,240	98,000	19,036	18,085	118,741
Asphalt and Tar Sprayers	711	730	7,900	584	582	9,022
Plant Hirers[5]	3,882	3,800	17,000	3,549	3,419	20,056
Flooring Contractors	2,684	2,630	8,400	2,805	2,761	10,424
Constructional Engineers	864	860	8,200	1,042	1,153	8,140
Insulating Specialists	934	905	7,700	832	764	7,625
Suspended Ceiling Specialists	..	..	..	..	..	..
Floor and Wall Tiling Specialists	770	785	2,500	803	805	3,042
Miscellaneous	15,535	15,090	56,300	17,560	17,865	87,759
All Trades	163,236	163,465	813,700	165,561	165,208	958,798

Notes

.. = not available due to technical problems with the trade type breakdown

1. Information relates to firms on the Department's register.
2. Based on the definition of the construction industry as given in the Revised 1992 Standard Industrial Classification.
3. The numbers of firms include some which were temporarily inactive.
4. Working proprietors and employees (including trainees).
5. Excluding firms whose major activity is plant-hire without operatives.

Source of Data: Construction Market Intelligence, Department of the Environment, Transport and the Regions
Contact: Neville Price 020 7944 5587

Table 3.15 Private Contractors: Total Employment[1,2] by Region of Registration

Region of Registration	Firms[3]	Working Proprietors	Administrative, Professional, Technical and Clerical Staff	Operatives[4]	All Employment
					3rd Quarter 1998
North	5,724	5,730	7,060	24,735	37,525
Yorkshire & Humberside	12,543	12,530	19,455	49,880	81,870
East Midlands	11,956	12,025	14,240	28,750	55,015
East Anglia	7,668	7,695	8,370	15,295	31,355
South East:					
Beds, Essex, Herts	13,281	13,305	13,900	25,325	52,530
Greater London	17,142	16,870	36,170	50,740	103,780
Kent, Surrey, Sussex	15,348	15,260	16,425	31,900	63,585
Berks, Bucks, Hants, Oxon	13,458	13,530	14,420	26,260	54,205
South East: All	59,229	58,960	80,920	134,215	274,100
South West	16,705	16,835	13,480	31,505	61,820
West Midlands	14,844	14,780	19,475	40,145	74,400
North West	14,162	14,050	20,565	46,810	81,420
England	142,831	142,720	183,560	371,340	697,620
Wales	8,273	8,340	5,980	17,720	32,045
Scotland	12,159	12,385	18,355	53,205	83,950
Great Britain	**163,263**	**163,450**	**207,900**	**442,265**	**813,615**
					3rd Quarter 1999
North	6,183	6,286	7,842	30,930	45,058
Yorkshire & Humberside	12,442	11,947	118,238	59,800	89,985
East Midlands	11,396	11,657	12,767	37,118	61,543
East Anglia	7,220	6,807	5,403	20,386	32,596
South East:					
Beds, Essex, Herts	13,388	13,578	15,892	35,565	65,035
Greater London	17,222	16,527	31,452	58,419	106,397
Kent, Surrey, Sussex	15,897	15,227	17,567	48,005	80,799
Berks, Bucks, Hants, Oxon	14,810	15,267	17,031	45,330	77,627
South East: All	61,317	60,599	81,942	187,318	329,858
South West	16,985	17,366	12,986	46,241	76,593
West Midlands	13,974	13,748	21,818	52,372	87,937
North West	14,576	14,551	22,031	60,825	97,406
England	144,093	142,962	183,026	494,989	820,977
Wales	8,382	8,764	5,715	23,040	37,519
Scotland	13,179	13,482	21,513	65,307	100,302
Great Britain	**165,561**	**165,208**	**210,254**	**583,336**	**958,798**

Notes

1. Information relates to firms on the Department's register.
2. Based on the definition of the construction industry as given in the Revised 1992 Standard Industrial Classification.
3. The numbers of firms include some which were temporarily inactive.
4. Operative trainees are included in the figures for 'Operatives' and 'All employment'.

Source of Data: Construction Market Intelligence, Department of the Environment, Transport and the Regions
Contact: Neville Price 020 7944 5587

CHAPTER 4
Price Indices

Table 4.1 Tender Price Indices

Index 1995 = 100

Year & Quarter		Social Housing[1]		Public Sector Building[2] (Non-Housing)	Road Construction[3]
		New Build	Rehab		
1989		99	98	106	82
1990	Q1	100	99	103	82
	Q2	101	100	101	82
	Q3	99	99	97	79
	Q4	96	97	95	79
1991	Q1	94	95	94	79
	Q2	92	92	91	80
	Q3	91	91	89	74
	Q4	91	91	87	72
1992	Q1	90	91	86	71
	Q2	89	89	83	70
	Q3	87	88	82	70
	Q4	85	86	82	68
1993	Q1	86	87	83	67
	Q2	89	91	85	73
	Q3	91	93	86	76
	Q4	94	95	88	80
1994	Q1	97	100	90	83
	Q2	101	103	92	87
	Q3	103	105	95	92
	Q4	104	105	97	93
1995	Q1	102	102	99	99
	Q2	100	99	101	100
	Q3	99	99	101	101
	Q4	99	100	99	100
1996	Q1	99	101	100	96
	Q2	101	103	100	94
	Q3	103	104	101	100
	Q4	105	105	102	101
1997	Q1	103	103	103	99
	Q2	105	105	105	97
	Q3	107	106	106	101
	Q4	110	111	107	102
1998	Q1	112	110	108	99
	Q2	111	109	109	102
	Q3	111	109	110	97
	Q4	115	113	110	96
1999	Q1	117	115	111	93
	Q2	119	118	114	103
	Q3	122	121	115	98
	Q4	125(p)	124(p)	114	105

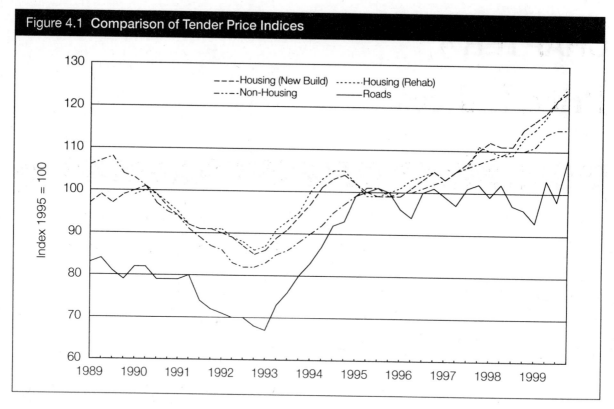

Figure 4.1 Comparison of Tender Price Indices

Notes

p = provisional

1. The Tender Price Index for Social Housing measures tender price levels in social housing new build commissioned by local authorities and housing associations in England and Wales. The data for the index comes from local authorities and housing associations in England and Wales. Multiply the index for any given quarter by the corresponding regional factor (table 4.2) to obtain a project index in that location. The Rehab Tender Price Index is derived from the New Build Tender   Price Index.

2. The Public Sector Building (Non-Housing) Index measures tender price levels for new build, non-housing building in the public sector for Great Britain. The data come from Government departments and local authorities. Multiply the index for for any given quarter by the corresponding regional factor (table 4.3) and function factor (table 4.4) to obtain a projectindex in that location and for that building type.

3. The Roads Index measures tender price levels for new build and repair & maintenance road contracts in Great Britain over £250,000 in value. The data for the index come from the Highways Agency and local authorities. Multiply the index for any given quarter by the corresponding regional factor (table 4.5), road type factor (table 4.6) and value factor (table 4.7) to obtain a project index in that location, for that road type and for that project value.

Source of Data: Construction Market Intelligence, Department of the Environment, Transport and the Regions
Contact: Marcella Douglas 020 7944 5594

Table 4.2 Regional Factors for the Social Housing Tender Price Index[1]

Year & Quarter		North Location Factor	SS	Wales Location Factor	SS	Midlands Location Factor	SS	East Location Factor	SS	South West Location Factor	SS	South East Location Factor	SS	London Location Factor	SS
1989		0.93		0.97		0.94		1.09		0.96		1.08		1.28	
1990	Q1	0.93	365	0.96	122	0.94	232	1.09	181	0.97	142	1.08	228	1.28	53
	Q2	0.93	349	0.96	113	0.93	233	1.09	170	0.97	126	1.09	220	1.30	51
	Q3	0.93	313	0.96	107	0.93	219	1.08	163	0.97	117	1.09	205	1.28	48
	Q4	0.94	279	0.96	104	0.93	216	1.07	159	0.97	114	1.08	191	1.27	46
1991	Q1	0.95	237	0.96	89	0.92	195	1.06	137	0.98	102	1.09	172	1.25	40
	Q2	0.95	218	0.97	81	0.93	187	1.05	131	0.97	98	1.09	154	1.26	33
	Q3	0.97	208	0.96	68	0.93	179	1.04	135	0.98	92	1.09	151	1.23	31
	Q4	0.96	183	0.96	47	0.94	175	1.03	141	0.98	87	1.09	139	1.23	22
1992	Q1	0.97	170	0.96	37	0.94	173	1.02	141	0.99	75	1.08	132	1.21	19
	Q2	0.98	167	0.97	33	0.94	170	1.02	131	0.99	73	1.08	118	1.22	19
	Q3	0.98	154	0.98	34	0.94	150	1.01	134	0.99	71	1.08	106	1.21	18
	Q4	0.99	149	0.98	29	0.95	139	1.00	123	0.99	58	1.08	95	1.20	14
1993	Q1	0.99	136	0.98	30	0.94	131	1.00	119	1.00	59	1.08	84	1.19	13
	Q2	0.99	135	0.99	29	0.95	122	0.99	117	1.01	59	1.07	76	1.16	10
	Q3	0.99	135	0.99	27	0.95	115	0.99	111	1.01	56	1.06	74	1.16	10
	Q4	0.99	137	0.98	20	0.96	110	0.99	105	1.01	49	1.07	69	..	9
1994	Q1	0.99	145	1.00	20	0.96	109	1.00	105	1.00	48	1.06	66	1.15	10
	Q2	0.98	144	0.99	19	0.96	103	1.01	106	1.01	51	1.06	64	1.17	14
	Q3	0.97	145	1.01	16	0.96	94	1.01	102	0.99	49	1.06	60	1.27	14
	Q4	0.96	139	1.02	14	0.96	94	1.01	93	0.99	49	1.06	54	1.29	16
1995	Q1	0.97	135	1.03	13	0.96	87	1.02	91	0.98	51	1.05	46	1.31	14
	Q2	0.96	134	1.03	10	0.97	81	1.02	86	0.98	41	1.06	37	1.30	15
	Q3	0.95	127	1.03	7	0.98	76	1.02	80	0.98	42	1.05	39	1.32	14
	Q4	0.96	138	..	6	0.98	74	1.02	81	0.97	44	1.05	37	1.29	16
1996	Q1	0.96	144	..	5	0.98	72	1.02	78	0.97	41	1.05	34	1.27	16
	Q2	0.96	144	..	5	0.98	68	1.03	77	0.96	40	1.06	35	1.27	16
	Q3	0.96	149	..	3	0.98	63	1.02	74	0.97	41	1.05	32	1.27	18
	Q4	0.96	151	..	3	0.97	61	1.03	73	0.96	46	1.06	32	1.25	19
1997	Q1	0.96	155	..	2	0.98	61	1.03	72	0.96	48	1.06	31	1.23	19
	Q2	0.97	156	..	2	0.98	61	1.01	66	0.96	45	1.07	29	1.20	17
	Q3	0.98	142	..	2	0.98	59	1.02	61	0.96	47	1.07	26	1.18	17
	Q4	0.98	147	..	2	1.00	56	1.03	62	0.96	48	1.07	23	1.19	18
1998	Q1	0.97	140	..	2	0.99	51	1.04	61	0.95	51	1.09	27	1.20	19
	Q2	0.96	135	..	2	0.96	43	1.05	61	0.97	50	1.07	28	1.22	17
	Q3	0.95	133	..	3	0.96	43	1.07	53	0.97	45	1.06	26	1.22	17
	Q4	0.95	122	..	2	0.95	45	1.06	54	0.97	42	1.07	29	1.24	18
1999	Q1	0.94	118	..	1	0.95	45	1.05	56	0.96	47	1.08	37	1.25	18
	Q2	0.94	115	..	1	0.95	44	1.05	56	0.96	49	1.09	37	1.25	18
	Q3	0.93	113	..	1	0.95	45	1.06	53	0.95	48	1.10	38	1.29	16
	Q4(p)	0.94	115	..	1	0.97	42	1.02	49	0.97	44	1.11	39	1.29	15

Notes

p = provisional

.. = not available due to small number of contracts

SS = sample size

1. In any given quarter, the social housing tender price index represents a regional factor of 1.00.

Source of Data: Construction Market Intelligence, Department of the Environment, Transport and the Regions
Contact: Marcella Douglas 020 7944 5594

Table 4.3 Regional Factors for the Public Sector Building (Non-Housing) Tender Price Index[1]

Year & Quarter	Scotland Location Factor	SS	North Location Factor	SS	Wales Location Factor	SS	Midlands Location Factor	SS	East Location Factor	SS	South West Location Factor	SS	South East Location Factor	SS	London Location Factor	SS
1989	0.90		0.88		0.90		0.91		1.08		0.97		1.10		1.18	
1990 Q1	0.89	175	0.90	219	0.91	50	0.91	168	1.08	232	0.97	129	1.08	208	1.14	44
Q2	0.89	168	0.90	219	0.90	47	0.91	172	1.08	221	0.95	132	1.10	203	1.16	43
Q3	0.90	168	0.91	207	0.86	44	0.92	172	1.10	201	0.96	133	1.09	194	1.20	42
Q4	0.91	165	0.91	208	0.90	43	0.91	171	1.09	195	0.96	129	1.10	186	1.20	39
1991 Q1	0.91	163	0.91	190	0.88	43	0.93	173	1.08	175	0.97	128	1.08	180	1.19	41
Q2	0.90	160	0.93	186	0.89	37	0.91	166	1.06	163	0.99	127	1.07	166	1.18	43
Q3	0.93	158	0.94	189	0.90	40	0.93	169	1.03	149	0.98	121	1.09	151	1.18	42
Q4	0.94	156	0.96	179	0.91	36	0.93	170	1.03	135	0.98	119	1.05	144	1.16	47
1992 Q1	0.96	144	0.97	176	0.95	39	0.94	161	1.02	121	0.97	112	1.01	134	1.16	45
Q2	0.98	139	0.97	174	0.92	43	0.96	157	0.98	112	0.97	107	1.02	122	1.13	42
Q3	1.00	137	0.99	171	0.91	45	0.95	149	0.99	107	0.98	99	0.98	112	1.13	42
Q4	1.01	137	0.99	169	0.91	46	0.96	149	1.00	100	0.98	91	0.98	107	1.09	40
1993 Q1	1.03	136	1.00	163	0.92	48	0.95	148	0.98	92	0.96	90	0.98	99	1.01	39
Q2	1.05	139	1.00	158	0.93	45	0.95	142	1.00	89	0.95	86	0.96	96	1.03	37
Q3	1.04	135	1.00	161	0.94	44	0.96	149	1.00	87	0.95	81	0.98	101	1.01	38
Q4	1.03	140	0.99	156	0.94	45	0.94	146	0.99	92	0.95	80	0.97	98	1.02	43
1994 Q1	1.03	130	0.99	153	0.96	45	0.96	144	0.98	88	0.93	74	0.98	87	1.03	47
Q2	1.02	126	1.00	155	0.95	44	0.97	143	0.97	79	0.94	76	0.98	89	1.04	51
Q3	1.02	126	0.99	153	0.96	41	0.97	151	1.00	75	0.94	69	0.99	84	1.04	52
Q4	1.01	118	1.00	151	0.96	40	0.96	144	0.99	73	0.94	68	0.99	90	1.03	49
1995 Q1	1.01	118	0.99	149	0.96	35	0.96	138	1.00	81	0.94	65	0.99	98	1.04	48
Q2	1.02	124	0.98	140	0.96	31	0.96	138	1.00	78	0.95	67	1.00	108	1.04	50
Q3	1.00	123	0.98	139	0.96	28	0.97	135	1.02	75	0.95	62	1.00	104	1.05	51
Q4	1.02	119	0.97	142	0.95	28	0.97	151	1.00	78	0.96	61	1.01	109	1.06	51
1996 Q1	1.00	119	0.97	148	0.92	26	0.96	151	1.02	77	0.97	59	1.01	120	1.05	54
Q2	0.99	115	0.99	150	0.94	27	0.95	158	1.02	85	0.96	62	1.01	119	1.04	60
Q3	0.98	112	0.97	144	0.95	27	0.96	157	1.01	91	0.95	66	1.01	128	1.06	66
Q4	0.98	104	0.97	149	0.96	25	0.96	164	1.01	93	0.95	74	1.01	138	1.08	69
1997 Q1	0.97	109	0.97	162	0.94	25	0.96	159	1.00	96	0.95	81	1.01	146	1.08	64
Q2	0.97	113	0.97	164	0.94	31	0.96	159	1.01	104	0.96	80	1.02	147	1.08	61
Q3	0.96	117	0.97	158	0.95	36	0.95	148	1.01	106	0.96	82	1.02	148	1.09	62
Q4	0.96	121	0.96	166	0.95	35	0.95	158	1.01	109	0.97	83	1.02	146	1.13	64
1998 Q1	0.96	120	0.97	169	0.94	38	0.95	156	1.01	104	0.96	86	1.03	138	1.14	65
Q2	0.97	121	0.97	178	0.94	38	0.95	156	1.01	106	0.97	87	1.03	130	1.14	63
Q3	0.97	124	0.97	189	0.94	43	0.96	155	1.00	120	0.98	94	1.02	134	1.14	63
Q4	0.96	135	0.97	185	0.93	40	0.95	140	1.02	127	0.98	91	1.04	135	1.14	63
1999 Q1	0.97	137	0.97	191	0.94	40	0.95	138	1.01	132	0.98	102	1.04	126	1.14	54
Q2	0.98	145	0.97	190	0.94	41	0.96	132	1.03	128	0.99	100	1.04	130	1.16	52
Q3	0.98	153	0.96	195	0.94	40	0.96	133	1.02	122	0.99	99	1.06	122	1.16	47
Q4	0.98	175	0.97	204	0.92	41	0.96	122	1.03	111	0.98	95	1.07	115	1.13	39

Notes

SS = sample size

1. In any given quarter, the public sector building (non-housing) tender price index represents a regional factor of 1.00

Source of Data: Construction Market Intelligence, Department of the Environment, Transport and the Regions
Contact: Marcella Douglas 020 7944 5594

Table 4.4 Function Factors for the Public Sector Building (Non-Housing) Tender Price Index[1]

Year & Quarter	Courts, Police Stations & Prisons Factor	SS	Hospitals, Clinics & Day Centres Factor	SS	Schools, Colleges & Training Centres Factor	SS	Homes, Hostels & Barracks Factor	SS	Offices, Tech Buildings & Factories Factor	SS	Miscellaneous Factor	SS
1989	0.98		0.98		0.99		0.98		0.96		1.02	
1990 Q1	0.97	60	0.98	321	0.99	358	0.97	111	0.96	236	1.01	139
Q2	0.98	59	0.98	307	0.99	363	0.97	119	0.97	227	1.01	130
Q3	1.02	57	0.98	294	1.00	370	0.97	110	0.97	208	1.01	122
Q4	1.01	54	0.99	278	0.98	383	0.98	102	0.97	202	1.00	117
1991 Q1	1.01	48	0.99	277	0.97	373	1.00	89	0.99	198	0.99	108
Q2	1.02	47	1.00	269	0.97	369	0.99	85	0.99	177	0.98	101
Q3	1.01	43	1.00	249	0.97	371	0.98	82	0.98	179	0.99	95
Q4	1.01	39	0.99	241	0.97	366	0.98	84	0.96	166	1.01	90
1992 Q1	1.02	37	0.99	241	0.98	346	0.99	81	0.98	153	1.00	74
Q2	1.01	36	0.99	243	0.98	337	0.98	78	0.99	139	1.00	63
Q3	1.01	29	0.99	225	0.99	347	0.97	75	0.96	128	0.99	58
Q4	0.99	27	1.00	215	0.99	338	0.96	79	0.97	125	0.98	55
1993 Q1	0.98	27	0.99	205	1.00	337	0.96	79	0.98	118	0.99	49
Q2	0.99	26	0.98	201	1.00	334	0.94	72	0.99	111	0.98	48
Q3	0.96	26	0.98	202	1.00	340	0.95	74	1.00	103	0.99	51
Q4	0.96	28	0.99	203	1.00	333	0.95	83	0.99	101	0.99	52
1994 Q1	0.93	32	0.99	198	1.00	313	0.96	89	0.97	90	0.99	46
Q2	0.97	33	0.99	202	1.00	306	0.96	90	0.98	85	0.99	47
Q3	0.96	41	0.98	212	1.00	294	0.97	92	0.97	67	1.00	45
Q4	0.95	44	0.99	209	1.01	289	0.98	88	0.95	58	0.99	45
1995 Q1	0.97	45	0.99	192	1.00	300	0.98	91	0.98	61	0.98	43
Q2	0.97	43	0.99	191	1.00	304	0.97	82	0.98	62	0.98	54
Q3	0.98	47	0.98	188	1.01	287	0.99	77	0.98	59	0.98	59
Q4	0.97	52	0.97	196	1.01	299	0.98	69	0.97	56	1.00	67
1996 Q1	1.00	53	0.98	203	1.00	316	1.00	62	0.97	50	0.97	70
Q2	0.98	53	0.98	202	1.00	333	1.01	61	0.99	49	0.97	78
Q3	0.97	52	0.98	212	1.00	325	0.99	63	0.98	51	0.99	88
Q4	0.98	54	0.98	223	1.00	345	0.99	54	0.98	45	1.00	95
1997 Q1	0.98	52	0.98	219	1.00	370	0.99	54	0.98	46	1.00	101
Q2	0.98	52	0.98	208	1.00	387	0.99	54	0.99	51	1.00	107
Q3	0.98	47	0.99	202	1.00	389	0.99	48	1.00	56	1.01	115
Q4	0.98	44	0.99	200	1.00	412	0.98	44	1.01	61	1.00	121
1998 Q1	0.98	39	1.00	200	0.99	407	0.99	40	1.00	56	1.00	134
Q2	0.97	43	1.00	183	0.98	418	0.99	42	1.01	54	1.01	137
Q3	0.97	40	0.99	183	0.98	450	1.01	44	1.02	59	0.99	146
Q4	0.96	38	0.99	182	0.98	446	1.02	41	1.01	60	1.00	156
1999 Q1	0.96	35	0.99	172	0.98	446	1.02	42	1.01	60	1.01	165
Q2	0.98	33	0.99	171	0.99	444	1.03	41	1.02	57	1.01	172
Q3	0.98	31	1.00	159	0.99	459	1.03	37	1.00	56	1.00	169
Q4	0.98	24	1.00	158	0.99	451	1.03	38	0.99	57	1.01	174

Notes
SS = sample size
1. In any given quarter, the public sector building (non-housing) tender price index represents a function factor of 1.00.

Source of Data: Construction Market Intelligence, Department of the Environment, Transport and the Regions
Contact: Marcella Douglas 020 7944 5594

Table 4.5 Regional Factors for the Road Construction Tender Price Index [1,2]

Year & Quarter	Scotland Location Factor	SS	North Location Factor	SS	Wales Location Factor	SS	Midlands Location Factor	SS	East Location Factor	SS	South West Location Factor	SS	South East Location Factor	SS	London Location Factor	SS
1989	0.83		0.94		1.00		0.99		1.05		1.05		1.22		1.13	
1990 Q1	0.83	32	0.90	75	0.98	29	0.99	76	1.08	50	1.01	38	1.22	48	1.17	14
Q2	0.82	27	0.91	78	0.99	31	0.99	72	1.09	49	1.00	35	1.24	49	1.20	17
Q3	0.84	28	0.93	69	0.97	31	0.98	74	1.07	49	0.99	37	1.23	45	1.23	16
Q4	0.84	27	0.94	76	0.93	34	0.99	79	1.07	49	0.99	38	1.17	53	1.23	17
1991 Q1	0.85	24	0.93	71	0.92	37	0.98	81	1.05	52	0.98	32	1.17	58	1.28	22
Q2	0.84	24	0.94	75	0.92	39	0.99	87	1.06	57	0.98	32	1.18	67	1.29	22
Q3	0.85	21	0.96	83	0.93	43	0.99	88	1.02	62	0.97	37	1.18	76	1.3	25
Q4	0.84	19	0.98	90	0.94	49	0.99	92	1.01	63	0.98	36	1.18	79	1.24	26
1992 Q1	0.86	16	0.98	95	0.93	48	0.99	89	1.01	65	0.97	40	1.16	82	1.25	27
Q2	0.85	15	0.98	103	0.94	46	1.00	92	1.02	66	0.97	41	1.19	78	1.26	24
Q3	0.84	13	0.99	100	0.94	41	0.99	86	0.99	58	0.96	40	1.17	80	1.29	22
Q4	0.82	13	0.99	102	0.92	32	0.99	87	1.00	55	0.98	41	1.17	80	1.32	21
1993 Q1	0.83	13	1.00	101	0.93	31	0.99	84	0.98	45	0.98	39	1.16	76	1.34	17
Q2	0.85	13	1.00	102	0.94	29	1.00	82	0.99	44	0.97	39	1.16	74	1.33	15
Q3	0.83	10	1.00	106	0.93	28	0.99	77	1.00	43	0.97	38	1.15	74	1.29	14
Q4	0.85	9	0.99	100	0.94	25	0.99	72	1.03	42	0.98	41	1.18	68	1.26	16
1994 Q1	0.86	9	0.99	101	0.94	23	1.00	64	1.03	37	0.96	42	1.19	65	1.15	15
Q2	0.87	6	1.00	98	0.98	20	1.00	56	1.01	29	0.96	42	1.17	55	1.14	16
Q3	0.87	6	0.99	92	1.01	16	0.99	54	1.07	24	0.94	36	1.17	47	1.14	13
Q4	0.87	6	0.99	87	0.97	12	0.97	43	1.09	21	0.95	35	1.2	47	1.22	12
1995 Q1	0.85	4	0.99	83	0.94	11	0.97	40	1.09	21	0.97	34	1.15	40	1.15	12
Q2	0.90	3	1.00	73	0.97	12	0.96	37	1.06	18	0.97	31	1.10	35	1.18	12
Q3	0.90	3	1.01	69	0.97	14	0.96	34	1.06	18	0.96	30	1.10	33	1.18	12
Q4	..	1	1.00	63	0.98	13	0.95	32	0.98	18	0.96	27	1.10	30	1.18	13
1996 Q1	..	2	1.00	61	1.04	13	0.95	34	0.95	17	0.95	30	1.09	34	1.18	13
Q2	..	1	1.01	52	1.00	14	0.96	33	0.92	16	0.95	28	1.09	34	1.19	12
Q3	..	1	1.00	47	1.00	14	0.98	31	0.95	17	0.95	26	1.09	36	1.16	14
Q4	..	3	1.01	48	1.00	14	0.95	29	0.98	20	1.01	24	1.09	38	1.20	15
1997 Q1	0.85	4	0.99	47	0.99	14	0.95	27	1.01	21	1.01	22	1.09	37	1.23	11
Q2	0.83	4	1.00	46	1.00	16	0.94	27	1.04	22	1.04	20	1.11	36	1.17	10
Q3	0.82	4	0.99	55	0.89	19	0.92	28	1.08	28	1.04	21	1.10	33	1.15	11
Q4	0.79	5	0.98	63	0.91	22	0.93	29	1.10	31	0.98	21	1.10	32	1.14	15
1998 Q1	0.80	5	1.00	72	0.90	22	0.91	32	1.12	31	0.96	20	1.09	30	1.15	14
Q2	0.80	5	1.01	80	0.92	21	0.91	33	1.11	32	0.95	21	1.10	31	1.15	17
Q3	0.80	5	1.00	88	0.93	20	0.94	37	1.11	38	0.94	22	1.12	29	1.15	19
Q4	0.78	5	1.00	87	0.95	21	0.94	39	1.1	47	0.93	21	1.13	29	1.11	21
1999 Q1	0.85	4	1.00	101	0.94	27	0.95	39	1.08	51	0.89	17	1.13	23	1.14	26
Q2	0.85	4	1.00	107	0.93	27	0.94	41	1.07	50	0.90	21	1.15	22	1.14	28
Q3	0.85	4	0.99	110	0.93	30	0.94	43	1.06	50	0.92	21	1.17	20	1.17	27
Q4	..	2	1.00	121	0.92	31	0.95	44	1.05	50	0.88	20	1.19	16	1.15	26

Notes

.. = not available due to small number of contracts

ss = sample size

1. Contracts of £250,000 or more.
2. In any given quarter, the road construction tender price index represents a regional factor of 1.00.

Source of Data: Construction Market Intelligence, Department of the Environment, Transport and the Regions
Contact: Marcella Douglas 020 7944 5594

Table 4.6 Road Type Factors for the Road Construction Tender Price Index [1,2]

Year & Quarter		New Construction Factor	SS	Motorway Widening Factor	SS	Major Maintenance Factor	SS
1989		1.00		1.23		..	
1990	Q1	1.00	358	1.22	4	..	0
	Q2	1.00	353	1.13	5	..	0
	Q3	1.00	344	1.14	5	..	0
	Q4	1.00	360	1.15	3	1.04	10
1991	Q1	1.00	357	1.20	3	1.02	17
	Q2	1.00	363	1.14	4	1.02	36
	Q3	0.99	371	1.20	5	1.02	59
	Q4	0.99	381	1.20	5	1.04	68
1992	Q1	0.99	373	1.21	5	1.03	84
	Q2	0.99	363	1.23	5	1.03	97
	Q3	0.98	326	1.24	5	1.03	109
	Q4	0.98	308	1.19	7	1.03	116
1993	Q1	0.98	276	1.09	10	1.03	120
	Q2	0.98	262	1.08	12	1.03	124
	Q3	0.98	249	1.08	12	1.03	130
	Q4	0.97	234	1.10	12	1.04	128
1994	Q1	0.97	218	1.09	12	1.04	126
	Q2	0.96	198	1.07	10	1.07	114
	Q3	0.95	183	1.06	9	1.09	96
	Q4	0.94	154	1.03	9	1.08	100
1995	Q1	0.94	145	1.03	9	1.09	91
	Q2	0.93	131	0.99	9	1.12	81
	Q3	0.93	131	0.98	9	1.12	73
	Q4	0.93	121	0.96	7	1.13	69
1996	Q1	0.94	135	0.95	3	1.13	66
	Q2	0.94	127	..	0	1.12	63
	Q3	0.93	120	..	0	1.12	66
	Q4	0.93	122	..	0	1.13	69
1997	Q1	0.94	113	..	0	1.10	70
	Q2	0.96	113	..	0	1.06	78
	Q3	0.98	118	..	0	1.03	82
	Q4	0.98	128	..	0	1.04	90
1998	Q1	0.99	126	..	0	1.02	100
	Q2	0.98	135	..	0	1.03	105
	Q3	0.98	138	..	1	1.03	119
	Q4	0.97	136	..	1	1.03	132
1999	Q1	0.96	132	..	1	1.04	152
	Q2	0.95	138	..	1	1.04	158
	Q3	0.95	138	..	1	1.04	163
	Q4	0.94	129	..	1	1.04	189

Notes

.. = not available due to small number of contracts

SS = sample size

1. Contracts of £250,000 or more.
2. In any given quarter, the road construction tender price index represents a road type factor of 1.00.

Source of Data: Construction Market Intelligence, Department of the Environment, Transport and the Regions

Contact: Marcella Douglas 020 7944 5594

Table 4.7 Value Factors for the Road Construction Tender Price Index [1,2]

Year & Quarter		£1m Factor	£2m Factor	£4m Factor	£7m Factor	£12m Factor	£20m Factor	£30m Factor	£50m Factor
1989		1.00	1.00	1.00	1.00	1.00	1.01	1.01	1.00
1990	Q1	1.00	1.00	1.00	0.99	0.99	0.98	0.98	0.99
	Q2	1.01	1.01	1.00	1.00	1.00	0.99	0.99	0.98
	Q3	1.00	1.00	1.00	0.99	0.99	0.99	0.99	0.98
	Q4	1.01	1.01	1.00	1.00	0.99	0.99	0.99	0.95
1991	Q1	1.01	1.00	0.99	0.98	0.97	0.96	0.96	0.95
	Q2	1.02	1.00	0.99	0.98	0.97	0.96	0.96	0.95
	Q3	1.01	1.00	0.99	0.98	0.97	0.95	0.95	0.94
	Q4	1.01	1.00	0.98	0.96	0.95	0.93	0.93	0.91
1992	Q1	1.02	1.00	0.98	0.97	0.95	0.94	0.93	0.92
	Q2	1.02	1.00	0.98	0.96	0.94	0.93	0.92	0.90
	Q3	1.03	1.00	0.98	0.96	0.93	0.91	0.90	0.88
	Q4	1.02	1.01	0.99	0.97	0.96	0.94	0.93	0.92
1993	Q1	1.03	1.01	0.99	0.97	0.96	0.94	0.93	0.92
	Q2	1.04	1.01	0.99	0.97	0.94	0.93	0.91	0.89
	Q3	1.04	1.01	0.98	0.96	0.94	0.92	0.90	0.88
	Q4	1.03	1.00	0.97	0.94	0.92	0.90	0.88	0.86
1994	Q1	1.03	1.00	0.98	0.95	0.93	0.91	0.89	0.87
	Q2	1.03	1.00	0.98	0.96	0.94	0.92	0.90	0.89
	Q3	1.03	1.00	0.97	0.95	0.93	0.91	0.89	0.97
	Q4	1.04	1.02	0.99	0.97	0.95	0.94	0.92	0.90
1995	Q1	1.04	1.02	1.00	0.99	0.97	0.96	0.95	0.93
	Q2	1.00	0.99	0.99	0.99	0.98	0.98	0.98	0.98
	Q3	1.00	1.00	0.99	0.99	0.99	0.99	0.98	0.98
	Q4	1.01	1.01	1.00	1.00	1.00	1.00	1.00	0.99
1996	Q1	1.02	1.02	1.01	1.01	1.01	1.01	1.00	1.00
	Q2	1.02	1.01	1.01	1.01	1.01	1.01	1.01	1.00
	Q3	1.01	1.01	1.01	1.01	1.00	1.00	1.00	1.00
	Q4	1.01	1.01	1.01	1.00	1.00	1.00	1.00	1.00
1997	Q1	1.01	1.01	1.00	1.00	1.00	1.00	1.00	0.99
	Q2	1.01	1.01	1.01	1.00	1.00	1.00	1.00	0.99
	Q3	1.01	1.00	0.99	0.98	0.97	0.96	0.95	0.94
	Q4	1.02	1.00	0.99	0.97	0.96	0.95	0.94	0.92
1998	Q1	1.02	1.00	0.99	0.96	0.95	0.92	0.91	0.89
	Q2	1.01	0.99	0.98	0.95	0.93	0.91	0.89	0.88
	Q3	1.00	0.98	0.96	0.94	0.92	0.90	0.89	0.87
	Q4	1.01	0.99	0.97	0.96	0.94	0.93	0.90	0.89
1999	Q1	1.01	0.98	0.97	0.95	0.93	0.92	0.90	0.89
	Q2	1.02	0.99	0.97	0.95	0.93	0.92	0.90	0.89
	Q3	1.02	1.00	0.98	0.96	0.95	0.93	0.92	0.91
	Q4	1.01	0.99	0.97	0.96	0.95	0.93	0.92	0.91

Notes

1. Contracts of £250,000 or more.
2. In any given quarter, the road construction tender price index represents a value factor of 1.00.

Source of Data: Construction Market Intelligence, Department of the Environment, Transport and the Regions
Contact: Marcella Douglas 020 7944 5594

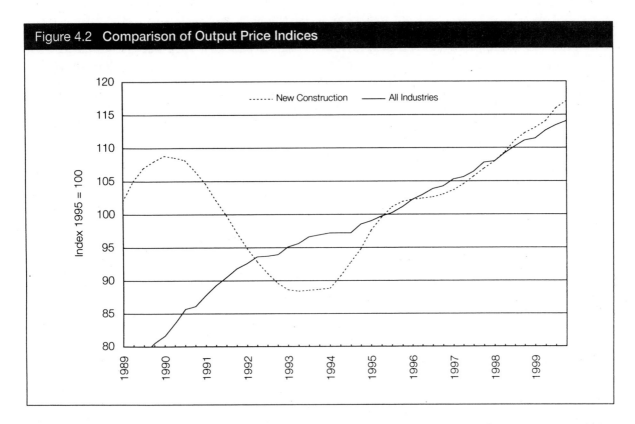

Figure 4.2 Comparison of Output Price Indices

Source of Data: Table 4.8

Table 4.8 Comparison of Output Price Indices

Year & Quarter		All New Construction[1]	All Industries[2]
			Index 1995 = 100
1989		105	78
1990	Q1	109	82
	Q2	109	84
	Q3	108	86
	Q4	107	86
1991	Q1	105	88
	Q2	102	89
	Q3	100	91
	Q4	97	92
1992	Q1	95	93
	Q2	93	94
	Q3	91	94
	Q4	90	94
1993	Q1	89	95
	Q2	88	96
	Q3	89	97
	Q4	89	97
1994	Q1	89	97
	Q2	91	97
	Q3	93	97
	Q4	95	99
1995	Q1	98	99
	Q2	100	100
	Q3	101	100
	Q4	102	101
1996	Q1	102	102
	Q2	102	103
	Q3	103	104
	Q4	103	104
1997	Q1	104	105
	Q2	105	106
	Q3	106	106
	Q4	107	108
1998	Q1	108	108
	Q2	109	109
	Q3	111	110
	Q4	112	111
1999	Q1	113	111
	Q2	114	113
	Q3	116	113
	Q4	117	114

Notes

1. The All New Construction Output Price Index measures price levels incurred in all payments by clients of construction on all new build construction underway at the time of each index value. It is based on lagged values of tender price indices, cost indices of labour and materials, and on values of construction New Orders for Great Britain.
2. The All Industries GDP Deflator measures inflation in goods and services throughout the economy at market prices for the UK

Source of Data: Construction Market Intelligence, Department of the Environment, Transport and the Regions
Contact: Marcella Douglas 020 7944 5594

Table 4.9 Output Price Indices[1]

Index 1995 = 100

Year & Quarter		Public Housing	Private Housing	Infra-structure	Public Building	Private Industrial	Private Commercial	All New Construction
1989		96	91	105	112	106	113	105
1990	Q1	99	96	106	117	111	115	109
	Q2	100	98	106	117	109	114	109
	Q3	100	99	105	117	108	113	108
	Q4	100	99	104	115	105	111	107
1991	Q1	99	99	102	112	102	109	105
	Q2	97	99	100	108	98	105	102
	Q3	95	98	98	105	95	103	100
	Q4	94	97	94	101	93	100	97
1992	Q1	92	96	91	98	90	97	95
	Q2	92	95	88	95	88	95	93
	Q3	91	94	86	93	87	93	91
	Q4	89	93	84	91	88	91	90
1993	Q1	88	93	83	90	89	89	89
	Q2	88	93	82	89	90	88	88
	Q3	89	93	82	88	92	88	89
	Q4	90	94	82	88	91	88	89
1994	Q1	91	94	83	89	89	88	89
	Q2	92	95	85	91	91	90	91
	Q3	94	97	88	93	94	93	93
	Q4	96	98	90	95	96	95	95
1995	Q1	99	99	96	97	99	98	98
	Q2	100	100	99	99	100	99	100
	Q3	101	100	102	101	101	101	101
	Q4	101	101	103	103	100	102	102
1996	Q1	100	101	104	103	100	103	102
	Q2	100	102	103	104	100	103	102
	Q3	101	103	103	104	99	103	103
	Q4	102	104	104	104	98	103	103
1997	Q1	103	105	104	105	99	104	104
	Q2	103	106	104	106	101	105	105
	Q3	103	107	105	107	103	106	106
	Q4	104	109	106	108	104	108	107
1998	Q1	105	110	106	109	105	109	108
	Q2	107	112	107	111	106	111	109
	Q3	109	114	108	112	109	112	111
	Q4	110	116	108	113	110	114	112
1999	Q1	112	119	107	114	111	115	113
	Q2	113	121	106	114	111	117	114
	Q3	115	124	107	115	112	119	116
	Q4	116	127	108	116	111	120	117

Notes

1. The Output Price Indices measure price levels incurred in payments by construction clients on construction of the appropriate type underway at the time of each index value. They are based on lagged values of tender price indices (Public Housing, Infrastructure, Public Building, Private Industrial and Private Commercial), of costs of labour and materials and mortgage levels (Private Housing) and the values of Construction New Orders. The All New Construction Output Price Index is a weighted derivation from the other new Construction Output Price Indices. The principal area of application of the Output Price Indices is the conversion of output expressed in current prices to output in volume terms

Source of Data: Construction Market Intelligence, Department of the Environment, Transport and the Regions
Contact: Marcella Douglas 020 7944 5594

Table 4.10 Public Works Output Price Indices[1]

Index 1995 = 100

Year & Quarter		Building	Civil Engineering	Non-Roads	Roads	All Pubic Works
1989		112	109	111	104	109
1990	Q1	117	111	116	105	113
	Q2	117	111	116	104	113
	Q3	117	111	115	104	112
	Q4	115	108	113	103	110
1991	Q1	112	106	110	101	108
	Q2	108	104	107	100	105
	Q3	105	102	104	98	102
	Q4	101	91	100	93	98
1992	Q1	98	96	97	90	95
	Q2	95	93	95	87	92
	Q3	93	90	92	85	90
	Q4	91	87	90	83	88
1993	Q1	90	86	89	81	87
	Q2	89	85	88	81	86
	Q3	88	84	87	80	86
	Q4	88	84	87	80	86
1994	Q1	89	85	88	82	87
	Q2	91	88	90	84	88
	Q3	93	90	92	86	91
	Q4	95	91	94	89	93
1995	Q1	97	95	96	96	96
	Q2	99	98	99	99	99
	Q3	101	102	102	102	102
	Q4	103	105	103	103	103
1996	Q1	103	106	104	104	103
	Q2	104	106	104	104	104
	Q3	104	106	104	104	104
	Q4	104	106	105	105	104
1997	Q1	105	106	105	105	105
	Q2	106	106	106	105	105
	Q3	107	106	107	105	106
	Q4	108	108	108	106	107
1998	Q1	109	108	110	106	108
	Q2	111	109	111	106	109
	Q3	112	110	112	107	110
	Q4	113	110	113	106	111
1999	Q1	114	109	113	105	111
	Q2	114	109	113	105	111
	Q3	115	109	114	106	112
	Q4	116	109	115	106	112

Notes

1. For a description of Output Price Indices see footnote 1 of Table 4.9

Source of Data: Construction Market Intelligence, Department of the Environment, Transport and the Regions
Contact: Marcella Douglas 020 7944 5594

Table 4.11 Direct Labour Output Deflators and Contractors' Output Deflators[1]

Year & Quarter		Direct Labour Output Deflators				Contractors' Output Deflators
		Public Housing	Public Works	Public Housing R&M	Public Works R&M	Repair & Maintenance (R&M)
1989		56	77	75	76	76
1990	Q1	57	80	80	80	80
	Q2	58	81	80	80	81
	Q3	60	82	80	81	84
	Q4	60	86	86	86	85
1991	Q1	60	86	87	87	86
	Q2	61	87	87	87	87
	Q3	63	86	87	87	89
	Q4	63	89	90	90	89
1992	Q1	89	89	91	90	89
	Q2	89	89	91	90	90
	Q3	89	89	91	90	92
	Q4	92	92	95	94	92
1993	Q1	92	92	95	95	93
	Q2	93	93	95	95	93
	Q3	93	93	95	95	94
	Q4	94	93	95	95	94
1994	Q1	95	94	96	95	94
	Q2	95	95	96	96	95
	Q3	96	96	96	96	97
	Q4	98	97	98	98	97
1995	Q1	99	99	99	99	98
	Q2	100	100	99	99	99
	Q3	100	100	99	99	101
	Q4	102	102	103	102	101
1996	Q1	102	102	103	102	101
	Q2	102	102	103	103	102
	Q3	102	102	103	103	103
	Q4	104	104	106	105	103
1997	Q1	104	104	106	105	104
	Q2	105	104	106	106	104
	Q3	105	104	106	106	105
	Q4	106	105	107	107	107
1998	Q1	107	107	110	109	107
	Q2	108	107	110	109	108
	Q3	108	107	110	109	112
	Q4	113	112	118	117	112
1999	Q1	113	112	118	117	112
	Q2	113	112	118	116	111
	Q3	112	111	117	116	116
	Q4	117	117	125	122	116

Notes

1. The deflators are Output Price Indices, see footnote 1 of Table 4.9. All except Public Housing New Build and Public Works New Build use costs of labour and materials in place of Tender Price Indices. Their principal application is the conversion of output expressed in current prices to output in volume terms.

Source of Data: Construction Market Intelligence, Department of the Environment, Transport and the Regions
Contact: Marcella Douglas 020 7944 5594

CHAPTER 5

Cost Indices

Table 5.1 Resource Cost Index of Building Non-Housing[1]

Year	Quarter	Combined Index (100%)	Mechanical Index (15%)	Electrical Index (10%)	Building Index (75%)	Derived Building Indices Labour & Plant	Materials
1989		75	73	73	76	76	77
1990	Q1	78	76	77	79	79	80
	Q2	80	80	77	80	79	82
	Q3	82	80	78	84	85	83
	Q4	83	81	80	84	85	83
1991	Q1	84	85	85	84	85	83
	Q2	85	86	85	85	85	84
	Q3	86	86	85	86	89	84
	Q4	86	88	86	86	90	83
1992	Q1	87	89	89	87	90	84
	Q2	88	89	90	88	90	86
	Q3	90	90	91	90	94	86
	Q4	90	91	91	90	95	86
1993	Q1	90	92	91	91	95	87
	Q2	92	92	92	92	95	89
	Q3	93	93	93	92	95	91
	Q4	93	93	93	93	95	91
1994	Q1	94	94	93	94	95	93
	Q2	94	95	94	94	95	94
	Q3	96	96	95	96	97	96
	Q4	97	97	96	97	97	96
1995	Q1	98	98	98	97	97	98
	Q2	99	100	100	99	98	100
	Q3	101	100	101	102	102	101
	Q4	102	102	101	102	102	101
1996	Q1	102	103	104	102	103	101
	Q2	102	103	104	102	103	101
	Q3	103	103	104	103	106	101
	Q4	103	104	104	103	106	100
1997	Q1	104	106	106	103	106	101
	Q2	104	106	106	104	106	102
	Q3	105	106	106	104	107	102
	Q4	106	107	106	106	110	102
1998	Q1	106	108	107	105	110	102
	Q2	107	109	110	106	110	102
	Q3	110	110	110	110	120	102
	Q4	110	113	110	109	121	101
1999	Q1	110	111	111	109	121	100
	Q2	109	109	111	109	120	100
	Q3	111	110	110	112	129	100
	Q4	113	113	111	113	130	100

Notes

1. The Resource Cost Index of Building Non-Housing (NOCOS) gives a measure of the notional trend of costs to a contractor of changes in the cost of labour, materials and plant by application of the Price Adjustment Formulae for Building and Specialist Engineering Works (Series 3) to a cost model for a Non-Housing Building Project.

Source of Data: Construction Market Intelligence, Department of the Environment, Transport and the Regions
Contact: Marcella Douglas 020 7944 5594

Table 5.2 Resource Cost Index of House Building[1]

Year	Quarter	Combined Index (100%)	Mechanical Index (10%)	Electrical Index (10%)	Building Index (80%)	Derived Building Indices	
						Labour & Plant	Materials
1989		76	75	74	77	77	79
1990	Q1	80	78	78	80	80	81
	Q2	81	81	78	81	80	83
	Q3	84	81	79	85	85	84
	Q4	84	82	81	85	85	84
1991	Q1	85	86	85	85	85	85
	Q2	86	87	85	86	86	86
	Q3	87	87	86	87	90	85
	Q4	87	88	87	87	90	85
1992	Q1	88	89	90	87	90	86
	Q2	89	91	90	88	90	86
	Q3	90	91	92	90	95	86
	Q4	90	92	92	90	95	86
1993	Q1	91	93	92	90	95	87
	Q2	92	93	93	92	95	89
	Q3	93	94	94	93	95	90
	Q4	93	94	94	93	95	91
1994	Q1	94	94	94	94	95	92
	Q2	94	96	95	94	95	94
	Q3	96	96	95	96	97	95
	Q4	97	97	96	97	97	97
1995	Q1	98	98	99	98	97	98
	Q2	99	100	99	99	98	101
	Q3	102	100	101	102	102	101
	Q4	101	102	101	101	102	100
1996	Q1	102	103	104	102	103	101
	Q2	102	103	104	102	103	101
	Q3	103	103	104	103	106	101
	Q4	103	105	104	103	106	101
1997	Q1	105	107	106	104	106	102
	Q2	106	109	106	105	106	103
	Q3	105	109	106	105	107	103
	Q4	107	110	106	106	110	104
1998	Q1	107	111	107	106	110	103
	Q2	108	112	111	107	110	103
	Q3	111	113	110	111	120	104
	Q4	111	116	110	111	121	104
1999	Q1	111	114	111	111	121	103
	Q2	111	112	111	111	120	103
	Q3	114	113	111	115	129	103
	Q4	115	115	111	115	130	103

Notes

1. The Resource Cost Index of House Building (HOCOS) gives a measure of the notional trend of costs to a contractor of changes in the cost of labour, materials and plant by application of the Price Adjustment Formulae for Building and Specialist Engineering Works (Series 3) to a cost model for a House Building Project.

Source of Data: Construction Market Intelligence, Department of the Environment, Transport and the Regions
Contact: Marcella Douglas 020 7944 5594

Table 5.3 Resource Cost Index of Road Construction[1]

Year	Quarter	Combined Index	Derived Indices	
			Labour & Plant	Materials
1989		77	78	75
1990	**Q1**	80	81	78
	Q2	81	81	80
	Q3	84	86	81
	Q4	84	86	81
1991	**Q1**	85	87	84
	Q2	87	87	86
	Q3	88	90	86
	Q4	88	90	86
1992	**Q1**	84	90	80
	Q2	85	90	81
	Q3	86	94	80
	Q4	87	94	82
1993	**Q1**	88	94	83
	Q2	90	95	97
	Q3	91	95	87
	Q4	91	95	88
1994	**Q1**	92	96	89
	Q2	93	96	90
	Q3	95	97	93
	Q4	96	98	95
1995	**Q1**	98	99	98
	Q2	100	99	101
	Q3	101	101	101
	Q4	101	101	101
1996	**Q1**	101	102	101
	Q2	103	103	103
	Q3	104	105	103
	Q4	104	106	103
1997	**Q1**	105	105	105
	Q2	106	105	107
	Q3	107	107	107
	Q4	108	109	107
1998	**Q1**	108	109	108
	Q2	110	109	110
	Q3	111	114	109
	Q4	110	113	108
1999	**Q1**	110	113	107
	Q2	111	115	108
	Q3	114	120	109
	Q4	114	120	108

Notes

1. The Resource Cost Index of Road Construction (ROCOS) gives a measure of the notional trend of costs to a contractor of changes in the cost of labour, materials and plant by application of the Price Adjustment Formulae for Civil Engineering Works (1990 Series) to a cost model for a Road Construction Project.

Source of Data: Construction Market Intelligence, Department of the Environment, Transport and the Regions
Contact: Marcella Douglas 020 7944 5594

Table 5.4 Resource Cost Index of Infrastructure[1]

Year	Quarter	Combined Index	Derived Indices	
			Labour & Plant	Materials
1989		76	77	75
1990	Q1	79	80	78
	Q2	79	80	78
	Q3	85	86	84
	Q4	85	86	84
1991	Q1	85	86	84
	Q2	86	87	85
	Q3	87	89	85
	Q4	87	90	85
1992	Q1	85	89	82
	Q2	86	90	81
	Q3	87	93	82
	Q4	88	94	82
1993	Q1	88	94	83
	Q2	90	95	87
	Q3	91	95	88
	Q4	92	95	89
1994	Q1	93	96	91
	Q2	94	96	92
	Q3	96	97	94
	Q4	96	97	96
1995	Q1	98	99	98
	Q2	100	99	100
	Q3	101	101	101
	Q4	101	101	100
1996	Q1	101	102	100
	Q2	102	103	102
	Q3	103	105	101
	Q4	103	106	101
1997	Q1	104	106	102
	Q2	104	106	103
	Q3	106	108	104
	Q4	106	109	104
1998	Q1	107	109	105
	Q2	108	110	106
	Q3	109	113	105
	Q4	108	113	104
1999	Q1	108	113	103
	Q2	109	115	103
	Q3	111	120	104
	Q4	111	120	106

Notes
1. The Resource Cost Index of Infrastructure (FOCOS) gives a measure of the notional trend of costs to a contractor of changes in the cost of labour, materials and plant by application of the Price Adjustment Formulae for Civil Engineering Works (1990 Series) to a cost model for an Infrastructure Project.

Source of Data: Construction Market Intelligence, Department of the Environment, Transport and the Regions
Contact: Marcella Douglas 020 7944 5594

Table 5.5 Resource Cost Index of Maintenance for Building Non-Housing[1]

Year	Quarter	Combined Index (100%)	Mechanical Index (15%)	Electrical Index (13%)	Building Index (72%)	Derived Building Indices Labour & Plant	Materials
1989		76	74	75	77	76	78
1990	Q1	79	77	79	80	79	81
	Q2	81	81	79	81	79	83
	Q3	83	81	80	84	85	84
	Q4	84	82	82	84	85	84
1991	Q1	85	85	85	85	85	84
	Q2	85	87	86	85	85	85
	Q3	87	87	86	87	89	85
	Q4	87	88	87	87	90	85
1992	Q1	88	89	90	87	90	85
	Q2	89	90	91	88	90	86
	Q3	90	91	92	90	95	85
	Q4	90	91	92	90	95	86
1993	Q1	91	93	93	91	95	87
	Q2	92	93	94	92	95	89
	Q3	93	94	94	93	95	90
	Q4	93	94	94	93	95	91
1994	Q1	94	95	95	94	95	93
	Q2	94	96	95	94	95	94
	Q3	96	96	96	96	97	95
	Q4	97	98	97	97	97	96
1995	Q1	98	99	99	98	97	98
	Q2	99	100	100	99	98	101
	Q3	101	100	100	102	102	101
	Q4	101	101	101	102	103	101
1996	Q1	102	103	104	102	103	101
	Q2	102	103	104	102	103	101
	Q3	103	103	104	103	106	101
	Q4	103	104	104	103	106	101
1997	Q1	105	106	106	104	106	102
	Q2	105	106	106	104	106	103
	Q3	105	106	106	105	107	103
	Q4	107	107	107	107	110	103
1998	Q1	107	109	107	106	110	102
	Q2	108	109	111	107	111	102
	Q3	111	111	111	112	121	102
	Q4	112	115	111	112	121	102
1999	Q1	112	114	113	112	121	102
	Q2	111	112	113	111	121	101
	Q3	115	113	112	116	130	101
	Q4	116	117	113	116	130	101

Notes

1. The Resource Cost Index of Maintenance for Building Non-Housing (NOMACOS) gives a measure of the notional trend of costs to a contractor of changes in the cost of labour, materials and plant by application of the Price Adjustment Formulae for Building and Specialist Engineering Works (Series 3) to a cost model for a Non-Housing Building Maintenance Project.

Source of Data: Construction Market Intelligence, Department of the Environment, Transport and the Regions
Contact: Marcella Douglas 020 7944 5594

Table 5.6 Resource Cost Index of Maintenance for House Building[1]

Year	Quarter	Combined Index (100%)	Mechanical Index (17%)	Electrical Index (15%)	Building Index (68%)	Derived Building Indices Labour & Plant	Materials
1989		78	78	76	78	76	78
1990	Q1	80	80	79	80	79	81
	Q2	81	84	79	81	79	83
	Q3	84	84	80	85	85	84
	Q4	85	85	82	85	85	84
1991	Q1	85	87	86	85	85	85
	Q2	86	89	86	85	85	85
	Q3	87	89	86	87	89	85
	Q4	88	90	87	88	90	85
1992	Q1	89	91	91	88	90	85
	Q2	89	92	92	88	90	86
	Q3	91	92	92	90	95	85
	Q4	91	93	92	91	95	85
1993	Q1	92	94	93	91	95	87
	Q2	93	95	94	92	95	89
	Q3	94	95	95	93	95	90
	Q4	94	95	95	94	95	91
1994	Q1	94	95	95	94	95	93
	Q2	95	97	96	95	95	94
	Q3	96	97	96	96	97	96
	Q4	97	98	97	97	97	97
1995	Q1	98	99	99	98	97	99
	Q2	99	100	100	99	98	101
	Q3	101	100	100	102	102	100
	Q4	101	102	101	101	102	100
1996	Q1	102	103	104	102	103	100
	Q2	102	103	104	102	103	100
	Q3	103	103	104	103	106	101
	Q4	104	105	104	103	106	101
1997	Q1	105	107	106	104	106	102
	Q2	106	108	106	105	106	103
	Q3	106	108	106	105	107	103
	Q4	108	109	107	107	110	103
1998	Q1	107	111	107	106	110	102
	Q2	109	112	112	107	111	102
	Q3	113	114	111	113	121	102
	Q4	113	117	111	113	121	102
1999	Q1	113	116	113	112	121	101
	Q2	113	115	113	111	121	101
	Q3	116	116	113	117	130	101
	Q4	117	120	114	117	130	101

Notes
1. The Resource Cost Index of Maintenance for House Building (HOMACOS) gives a measure of the notional trend of costs to a contractor of changes in the cost of labour, materials and plant by application of the Price Adjustment Formulae for Building and Specialist Engineering Works (Series 3) to a cost model for a Housing Building Maintenance Project.

Source of Data: Construction Market Intelligence, Department of the Environment, Transport and the Regions
Contact: Marcella Douglas 020 7944 5594

CHAPTER 6

Investment

Table 6.1 Gross Fixed Capital Formation at Current Purchaser's Prices: Analysis by Type of Asset and Sector

United Kingdom | | | | | | | | | | £ Million

	1989	1990	1991	1992	1993	1994	1995	1996	1997	1998	1999
New Dwellings, excluding Land											
Public Non-Financial Corporations	223	201	153	172	150	139	162	151	123	114	114
Private Non-Financial Corporations	203	189	181	192	206	211	217	236	253	277	305
Financial Corporations	0	0	0	0	0	0	0	0	0	0	0
Central Government	125	252	210	218	368	320	221	314	287	273	240
Local Government	3,465	3,728	2,399	2,189	2,250	2,489	2,421	1,834	1,499	1,443	1,249
Households and NPISH	18,755	16,678	15,396	16,054	16,918	18,074	18,567	20,231	21,765	23,724	25,691
Total	**22,771**	**21,048**	**18,339**	**18,825**	**19,892**	**21,233**	**21,588**	**22,766**	**23,927**	**25,831**	**27,599**
Other Buildings and Structures											
Public Non-Financial Corporations	3,551	2,659	3,015	3,332	3,416	3,492	3,781	3,397	2,675	2,691	2,489
Private Non-Financial Corporations	16,212	19,716	17,977	15,922	13,721	13,906	15,106	17,180	20,196	22,016	26,097
Financial Corporations	2,953	3,306	2,214	1,499	1,007	1,263	1,461	1,474	1,823	2,543	2,387
Central Government	4,152	5,101	5,601	5,246	4,733	4,745	4,484	4,059	3,318	2,866	2,452
Local Government	4,276	4,947	4,409	4,682	5,207	5,265	5,513	5,134	4,602	5,164	5,655
Households and NPISH	1,187	1,301	1,156	957	1,015	1,044	1,274	1,754	2,333	2,414	2,616
Total	**32,331**	**37,030**	**34,372**	**31,638**	**29,099**	**29,715**	**31,619**	**32,998**	**34,947**	**37,694**	**41,696**
Other											
Public Non-Financial Corporations	2,378	2,639	1,245	2,165	1,891	1,996	1,833	1,708	1,856	1,495	1,755
Private Non-Financial Corporations	42,232	43,445	40,181	38,304	41,296	42,336	48,698	55,485	61,168	66,978	70,753
Financial Corporations	5,375	3,585	4,124	3,512	2,789	4,304	3,590	3,843	3,324	6,775	5,352
Central Government	1,751	2,226	2,367	2,575	2,506	2,181	2,176	1,030	834	1,102	1,379
Local Government	-2,152	-1,691	-924	-666	-1,342	-942	-810	-1,055	-668	-370	-789
Households and NPISH	6,347	6,032	4,976	3,925	5,099	6,567	7,666	8,900	8,765	8,566	10,627
Total	**55,931**	**56,236**	**51,969**	**49,815**	**52,239**	**56,442**	**63,153**	**69,911**	**75,279**	**84,546**	**89,077**
Gross Fixed Capital Formation											
Public Non-Financial Corporations	6,152	5,499	4,413	5,669	5,457	5,627	5,776	5,256	4,654	4,300	4,358
Private Non-Financial Corporations	58,647	63,350	58,339	54,418	55,223	56,453	64,021	72,901	81,616	89,269	97,154
Financial Corporations	8,328	6,891	6,338	5,011	3,796	5,567	5,051	5,317	5,147	9,318	7,739
Central Government	6,028	7,579	8,178	8,039	7,607	7,246	6,881	5,403	4,439	4,241	4,071
Local Government	5,589	6,984	5,884	6,205	6,115	6,812	7,124	5,913	5,433	6,237	6,115
Households and NPISH	26,289	24,011	21,528	20,936	23,032	25,685	27,507	30,885	32,864	34,706	38,935
UK Total Economy	**111,033**	**114,314**	**104,680**	**100,278**	**101,230**	**107,390**	**116,360**	**125,675**	**134,153**	**148,071**	**158,372**

Source of Data: Construction Market Intelligence, Department of the Environment, Transport and the Regions
Contact: Neville Price 020 7944 5587

Table 6.2 Gross Domestic Fixed Capital Formation at Current Purchaser's Prices: Analysis by Broad Sector and Type of Asset

United Kingdom   £ Million

	1989	1990	1991	1992	1993	1994	1995	1996	1997	1998	1999
Private Sector											
New Dwellings, excluding Land	18,958	16,867	15,577	16,246	17,124	18,285	18,784	20,467	22,018	24,001	25,996
Other Buildings and Structures	20,352	24,323	21,347	18,378	15,743	16,213	17,841	20,408	24,352	26,973	31,100
Total	**39,310**	**41,190**	**36,924**	**34,624**	**32,867**	**34,498**	**36,625**	**40,875**	**46,370**	**50,974**	**57,096**
Public Non-Financial Corporations											
New Dwellings, excluding Land	223	201	153	172	150	139	162	151	123	114	114
Other Buildings and Structures	3,551	2,659	3,015	3,332	3,416	3,492	3,781	3,397	2,675	2,691	2,489
Total	**3,774**	**2,860**	**3,168**	**3,504**	**3,566**	**3,631**	**3,943**	**3,548**	**2,798**	**2,805**	**2,603**
General government											
New Dwellings, excluding Land	3,590	3,980	2,609	2,407	2,618	2,809	2,642	2,148	1,786	1,716	1,489
Other New Buildings and Works	8,428	10,048	10,010	9,928	9,940	10,010	9,997	9,193	7,920	8,030	8,107
Total	**12,018**	**14,028**	**12,619**	**12,335**	**12,558**	**12,819**	**12,639**	**11,341**	**9,706**	**9,746**	**9,596**
Other											
Transport Equipment	10,699	10,593	8,668	8,397	9,594	11,329	11,055	12,163	13,163	15,928	17,297
Other Machinery and Equipment and Cultivated Assets	38,018	37,817	35,075	34,690	35,566	37,693	44,464	49,216	52,229	57,987	59,564
Intangible Fixed Assets	2,823	3,571	4,063	3,782	3,648	3,613	3,939	4,136	4,249	4,547	4,470
Costs associated with the transfer of ownership of non-produced assets	4,391	4,255	4,163	2,946	3,431	3,807	3,695	4,396	5,638	6,084	7,746
Total	**55,931**	**56,236**	**51,969**	**49,815**	**52,239**	**56,442**	**63,153**	**69,911**	**75,279**	**84,546**	**89,077**
UK Total Gross Fixed Capital Formation	**111,033**	**114,314**	**104,680**	**100,278**	**101,230**	**107,390**	**116,360**	**125,675**	**134,153**	**148,071**	**158,372**

Source of Data: Construction Market Intelligence, Department of the Environment, Transport and the Regions
Contact: Neville Price 020 7944 5587

Table 6.3 Gross Domestic Fixed Capital Formation at 1995 Purchaser's Prices: Analysis by Broad Sector and Type of Asset Total Economy

United Kingdom

£ Million at 1995 Prices

	1989	1990	1991	1992	1993	1994	1995	1996	1997	1998	1999
Private Sector											
New Dwellings, excluding Land	22,775	18,746	16,603	17,311	18,353	19,076	18,784	19,903	20,824	21,521	22,109
Other Buildings and Structures	19,018	21,821	20,273	19,280	17,449	17,914	17,841	19,099	22,860	24,315	26,777
Total	**41,793**	**40,567**	**36,876**	**36,591**	**35,802**	**36,990**	**36,625**	**39,002**	**43,684**	**45,836**	**48,886**
Public Non-Financial Corporations											
New Dwellings, excluding Land[2]	..	..	..	..	..	150	162	147	119	106	100
Other Buildings and Structures[2]	..	..	..	..	..	3,847	3,781	3,154	2,417	2,424	2,168
Total	..	..	..	..	..	**3,997**	**3,943**	**3,301**	**2,536**	**2,530**	**2,268**
General Government											
New Dwellings, excluding Land	3,793	3,958	2,633	2,570	2,932	3,041	2,642	2,104	1,726	1,604	1,313
Other Buildings and Structures	7,825	8,759	9,211	10,296	11,174	11,162	9,997	8,511	7,313	7,213	7,191
Total	**11,618**	**12,717**	**11,844**	**12,866**	**14,106**	**14,203**	**12,639**	**10,615**	**9,039**	**8,817**	**8,504**
Other											
Transport Equipment	13,108	11,907	9,202	8,482	10,027	11,927	11,055	11,777	12,982	14,913	16,256
Other Machinery and Equipment and Cultivated Assets	42,165	40,736	38,590	37,042	36,185	38,171	44,464	49,124	54,241	63,833	68,701
Intangible Fixed Assets	3,645	4,230	4,473	3,915	3,678	3,631	3,939	4,162	4,103	4,156	3,988
Costs associated with the transfer of ownership of non-produced assets	4,806	4,334	4,174	3,641	3,846	4,123	3,695	4,061	4,661	4,352	4,669
Total	**63,724**	**61,207**	**56,439**	**53,080**	**53,736**	**57,852**	**63,153**	**69,124**	**75,987**	**87,254**	**93,614**
UK Total Gross Fixed Capital Formation	**117,135**	**114,491**	**105,159**	**102,537**	**103,644**	**113,042**	**116,360**	**122,042**	**131,246**	**144,437**	**153,272**

Notes

1. For the years before 1994, totals differ from the sum of their components.
2. Data not available for years before 1994.

Source of Data: Construction Market Intelligence, Department of the Environment, Transport and the Regions
Contact: Neville Price 020 7944 5587

CHAPTER 7

Commercial Floorspace

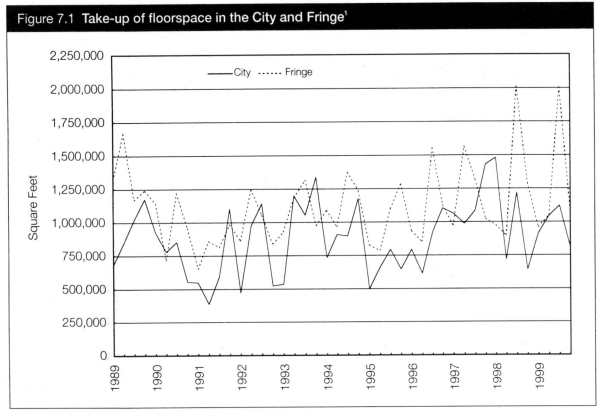

Figure 7.1 Take-up of floorspace in the City and Fringe[1]

Source of Data: Table 7.1

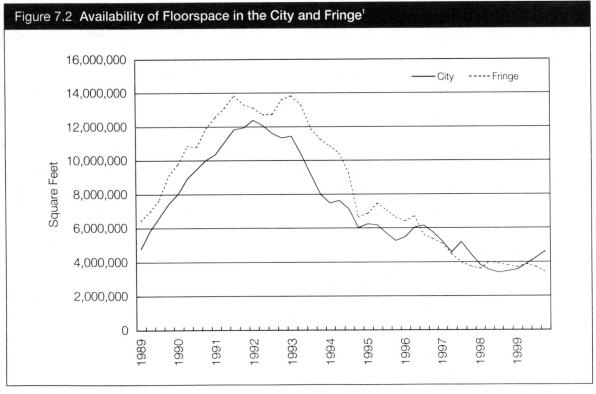

Figure 7.2 Availability of Floorspace in the City and Fringe[1]

Source of Data: Table 7.1

Table 7.1 Take-up and Availability of Floorspace in the City of London and its Surrounding Area[1]

Square Feet

Year		Take-Up				Average Availability			
		City		Fringe		City		Fringe	
		New	Secondary	New	Secondary	New	Secondary	New	Secondary
1989		3,051,117	667,205	2,078,945	3,319,575	20,861,351	3,836,262	15,360,214	14,851,763
1990	Q1	763,478	161,588	435,251	709,990	6,485,459	1,528,596	5,008,555	4,763,041
	Q2	636,954	141,725	220,485	495,411	7,200,716	1,730,560	5,568,463	5,276,306
	Q3	617,873	234,932	518,721	703,514	5,124,311	4,337,489	4,491,746	6,277,959
	Q4	205,125	349,028	209,795	750,333	4,109,102	5,901,620	4,303,817	7,576,231
1991	Q1	156,575	391,140	68,712	582,590	4,990,603	5,346,714	4,506,279	8,047,683
	Q2	121,518	265,429	298,718	560,160	5,519,116	5,580,176	4,440,869	8,661,711
	Q3	314,944	278,828	123,736	688,860	5,642,552	6,182,298	4,518,727	9,310,587
	Q4	820,938	280,535	302,267	682,308	5,442,209	6,485,906	4,239,255	9,059,785
1992	Q1	188,386	283,112	143,249	718,354	5,958,868	6,425,422	4,129,556	8,974,797
	Q2	510,914	463,366	517,935	731,954	5,650,790	6,429,031	3,718,183	8,988,363
	Q3	561,696	578,973	244,315	809,947	5,187,836	6,406,233	3,830,028	8,893,070
	Q4	230,501	293,085	179,006	657,401	4,738,194	6,595,258	4,165,037	9,431,435
1993	Q1	172,026	362,720	194,895	733,489	4,900,651	6,535,829	4,374,312	9,450,273
	Q2	535,241	663,771	260,396	923,072	4,372,167	5,963,231	4,147,484	9,123,544
	Q3	529,811	523,015	662,638	653,558	3,768,404	5,341,059	3,309,132	8,554,630
	Q4	944,080	391,970	255,534	718,703	3,088,937	4,903,544	2,850,544	8,395,618
1994	Q1	258,608	475,196	374,039	716,640	2,704,926	4,774,941	2,477,226	8,375,803
	Q2	256,755	651,899	146,787	817,217	2,756,443	4,868,201	2,331,140	8,080,914
	Q3[2]	896,965		1,375,196		7,139,044		9,268,786	
	Q4[2]	1,177,227		1,234,446		6,011,623		6,632,412	
1995	Q1	267,587	229,239	186,626	640,038	3,188,367	3,045,511	1,499,935	5,364,532
	Q2	269,160	391,515	102,962	682,868	3,182,931	2,990,467	1,678,405	5,773,370
	Q3	480,162	312,679	393,005	706,220	2,819,734	2,860,787	1,398,120	5,662,811
	Q4	361,644	284,484	80,102	1,204,588	2,725,569	2,519,895	1,609,624	5,011,675
1996	Q1	467,739	326,148	116,196	816,982	3,008,509	2,459,361	1,620,870	4,776,697
	Q2	368,038	244,767	184,071	667,642	3,646,616	2,388,569	1,909,611	4,798,046
	Q3	596,419	327,220	465,342	1,091,797	3,745,735	2,407,081	1,246,947	4,344,798
	Q4	808,878	294,138	223,224	903,085	3,164,933	2,573,988	1,041,744	4,313,493
1997	Q1	722,192	341,623	130,698	840,895	2,982,444	2,250,183	1,059,425	4,033,096
	Q2	392,620	594,769	659,934	904,094	3,391,091	1,175,751	628,119	3,840,269
	Q3	632,807	451,921	275,560	1,039,936	3,551,408	1,626,030	553,545	3,421,190
	Q4	1,196,386	234,718	206,617	815,765	2,900,978	1,542,946	492,269	3,237,916
1998	Q1	970,626	508,989	104,925	871,671	2,483,232	1,337,159	528,986	3,083,675
	Q2	551,976	164,847	157,309	740,142	2,282,208	1,252,271	654,153	3,355,521
	Q3	636,471	581,625	273,410	1,741,385	2,168,786	1,204,649	908,249	3,005,773
	Q4	332,123	309,625	83,156	1,206,371	2,245,601	1,218,279	1,012,233	2,791,186
1999	Q1	594,252	319,489	374,080	585,439	2,324,993	1,234,378	1,020,298	2,668,221
	Q2	641,585	402,706	293,729	753,782	2,649,507	1,235,848	1,135,764	2,738,391
	Q3	737,914	381,908	470,884	1,535,627	2,957,373	1,291,960	1,038,848	2,658,280
	Q4	361,371	451,714	170,038	909,815	3,128,916	1,489,160	1,106,871	2,298,008

Notes

1. Description of Data: The "City" is defined as the area within the City of London Corporation boundaries. The "Fringe" is defined as the area covered by the EC1-4, WC1-2, SE1 and E1 postal districts outside the City of London Corporation boundaries. "Take-Up" for a quarter is the amount of floorspace, in use classes B1 and A2, that is let, pre-let prior to practical completion, sold with vacant possession to an owner-occupier, or under offer in the quarter. "Availability" is the amount of floorspace, in use classes B1 and A2, available at the end of each month for occupation within the next 12 months. The quarterly figures are an average of the three months. For data prior to August 1990, "New" floorspace refers to floorspace in refurbished buildings or buildings constructed since 1950, while "Secondary" floorspace refers to all other floorspace. For data after August 1990, "New" floorspace refers to floorspace in refurbished buildings or buildings constructed since January 1988, while "Secondary" floorspace refers to all other floorspace.

2. Only totals collected for the last half of 1994.

Source of Data: Ingleby Trice Kennard, Chartered Surveyors Valuers and Property Consultants
Contact: Malcolm Trice 020 7606 7461

Table 7.2 New Office Designs Awarded BREEAM¹ Certificates

Floor area of buildings awarded BREEAM certificates over the last 12 months include Government Agencies (e.g. MAFF) and private sector firms (e.g. BT).

Square metres

Year & Quarter		London	South East	South West	Eastern	West Mids	East Mids	North West	Yorks. & Humb.	Northern	Wales	Scotland
						DETR Regions²						
1992	Q4	14,886	0	0	0	0	0	0	0	0	0	0
1993	Q1	0	0	0	0	0	0	0	0	12,980	0	0
	Q2	6,358	0	0	0	0	0	0	0	0	0	0
	Q3	68,476	0	0	12,600	0	0	0	0	0	0	0
	Q4	0	0	0	0	0	0	1,700	840	2,400	1,350	0
1994	Q1	6,700	5,000	3,800	0	4,900	0	920	7,705	720	0	0
	Q2	23,750	0	570	0	11,360	7,652	2,069	7,860	0	0	3,772
	Q3	0	1,400	0	1,634	0	0	0	0	0	0	5,970
	Q4	6,800	2,798	0	0	0	0	53,050	8,000	0	0	0
1995	Q1	108,443	4,208	0	0	5,900	990	1,100	0	12,643	6,900	0
	Q2	16,241	29,450	11,810	1,463	7,425	7,000	34,100	10,557	0	0	38,237
	Q3	58,729	10,144	5,322	7,641	12,000	0	23,688	10,157	0	8,500	0
	Q4	38,035	21,038	1,850	2,178	0	56,706	13,324	6,968	2,920	0	6,503
1996	Q1	21,293	5,500	0	0	2,019	3,000	0	9,293	0	6,500	0
	Q2	9,642	5,570	0	6,250	11,879	1,500	0	0	0	0	0
	Q3	6,724	9,360	1,600	4,000	11,160	0	0	0	5,130	0	0
	Q4	24,528	19,344	0	0	0	0	947	10,400	20,254	0	0
1997	Q1	11,887	6,611	0	0	5,480	0	6,877	3,070	7,313	0	7,200
	Q2	24,519	0	1,740	350	2,787	2,508	3,483	0	2,880	0	0
	Q3	2,745	8,602	2,580	0	1,200	0	2,364	0	0	0	6,316
	Q4	15,799	350	0	0	540	0	0	5,000	0	0	0
1998	Q1	8,800	0	1,954	0	0	8,700	1,900	0	3,502	0	0
	Q2	20,702	8,820	0	0	0	0	3,637	0	0	7,927	0
	Q3	5,289	7,626	1,858	0	10,460	0	0	4,988	76,200	0	0
	Q4	73,382	1,301	0	0	0	0	3,600	7,486	4,127	0	1,977

Year & Quarter		London	South East	South West	Eastern	West Mids	East Mids	North West	Mersey-side	Yorks. & Humb.	North East	Wales	Scotland
						DETR Regions²							
1999	Q1	0	0	0	0	8,500	0	0	0	0	0	0	0
	Q2	26,599	2,162	0	1,974	11,810	0	0	0	3,400	12,260	0	15,000
	Q3	16,530	0	0	0	0	0	0	7,004	9,632	0	0	0
	Q4	21,829	0	9,104	0	0	0	0	0	4,800	0	0	0

Notes

1. BREEAM is the Building Research Establishment Environmental Assessment Method, under which new office schemes are awarded BREEAM Certificates provided they meet a number of environmental quality criteria.
2. The DETR Regions are defined in Appendix 1. Note the change in regional classification from 1999 onwards.

Source of Data: Centre for Sustainable Construction, BRE
Contact: Hilary John 01923 664462

CHAPTER 8

Government Departments' Construction Plans

Public private partnerships (PPPs), including the Private Finance Initiative, are playing an increasingly important role in the delivery of new public infrastructure, with a significant increase in the number of projects awarded. This chapter provides an introduction to PPPs and PFI and some of the areas where they are delivering projects. It also covers recent changes to the PFI, information about the Gershon review of civil procurement, and local government PFI.

PUBLIC/PRIVATE PARTNERSHIPS (PPPS) AND THE PRIVATE FINANCE INITIATIVE (PFI)

The Government is committed to making partnerships between the public and private sectors work. It wants to see improvements in public services and infrastructure and public/private partnerships are seen as playing an increasingly important role in delivering this. Public/private partnerships (PPPs) demand a transformation of the roles and responsibilities of both the public and private sectors. Government bodies are moving from being owners and operators of assets into becoming intelligent clients purchasing long term services.

The Private Finance Initiative (PFI) is one form of public/private partnership and involves the public sector purchasing services rather than capital assets. The public sector should provide an output specification detailing the core service requirements, and the appropriate quality standards, without defining how they are to be delivered. The private sector can no longer simply build things but needs to take a longer term involvement in the maintenance and operation of assets required to deliver a service. Therefore, to achieve the best commercial result, it must design and build with a view to the whole life costs of an asset.

PFI under review - Bates I and Bates II
In 1997 the newly elected Government commissioned Sir Malcolm Bates to conduct a rapid review of how the PFI was working. The Review concluded that the Private Finance Initiative provides a mechanism through which the public sector can secure improved value for money in partnership with the private sector. The Review favoured the simplification of public sector structures and the clarification of their responsibilities.

A new Treasury Taskforce was established in late 1997 to develop PFI policy and to provide a central input on 'significant' projects, helping to streamline the PFI procurement process. It has worked with Departments and the private sector to issue a series of statements (to set the policy context) and technical notes which provide practical advice to those involved with PFI procurements. These include a comprehensive guide on the standardisation of PFI contracts.

Two years on from his original review, Sir Malcolm Bates was invited to look again at the PFI process and the implementation of his proposals, which included the creation of the Treasury private finance Taskforce. Sir Malcolm concluded that there was a need to consolidate and strengthen central co-ordination in PFI procurement and he identified a need for greater standardisation of contractual terms and improvements to the skills base within government.

The main recommendation of the second Bates review was the creation of a new body which will help increase and improve investment in the UK's public services from private sources. Sir Malcolm felt that in many ways the Treasury Taskforce had served its purpose but that a new entity would be needed to assist novel and complicated projects which would not ordinarily be favoured by the PFI process, and to help ensure that financially sound deals offering best value are procured.

In July 1999 the then Chief Secretary to the Treasury, Alan Milburn, announced that the Government had accepted the recommendations of the second Bates review and that it would include provision within legislation for the creation of Partnerships UK to succeed the treasury taskforce.

The Gershon Review of Government Civil Procurement

At the same time the Government accepted the recommendations of the second Bates Review it also accepted the recommendations put forward by Peter Gershon following his review of government's approach to civil procurement.

In order to ensure best practice and put in place a framework for obtaining significant cost savings, he concluded that a new organisation – the Office of Government Commerce (OGC) - should be established. OGC became operational in April 2000 with a brief to establish an integrated strategic framework and process for procurement. It includes those parts of government with policy responsibilities for procurement as well as government's own major procuring agencies - Property Advisor to the Civil Estate (PACE), Central Computer and Telecommunications Agency (CCTA), and the Buying Agency (TBA).

The Department of the Environment, Transport and the Regions

The Department of the Environment, Transport and the Regions was created in 1997 by the merger of the former Departments of the Environment and Transport. The Department has responsibility for a wide range of policy programmes where public/private partnerships and the PFI are playing an increasingly central role in their development, including:

• Local Government

• Integrated Transport

• Construction

• Housing and Regeneration

• The Environment

Although it began as a central government initiative, the PFI approach to procurement is now being pursued by a range of organisations helping to deliver public services for which the Department has policy responsibility:

Sponsored bodies. The Department sponsors a range of Executive Agencies and public bodies from the Highways Agency and the Civil Aviation Authority to the Environment Agency and the Health and Safety Commission. A number of projects are in procurement, including the estate rationalisation of the Health and Safety Laboratory and the public/private partnership to gain investment in London Underground's infrastructure.

Government Offices for the Regions (GORs). In order to better co-ordinate the central government response to local issues the Departments of Environment, Transport & the Regions, Trade & Industry and for Education & Employment combined their separate regional structures to form ten integrated Government Offices. In addition, the Home Office has formal links to the GORs. The Offices provide a unified, comprehensive and accessible service for businesses, local authorities and local people.

Regional Development. The Department is actively developing the Government's regional policies. Nine Regional Development Agencies (RDAs) have been established to encourage the partnership approach to improving regional co-operation and economic development. A tenth RDA for London will be effective from July 2000.

PFI Projects

The transport sector has been at the forefront of the PFI since its inception and accounts for the majority (by value) of PFI projects to date. These represent the greater part of the

£7.5 billion of private investment to be delivered through PFI deals signed by the Department. They range in size from the Channel Tunnel Rail Link (£4 billion) to the first design, build, finance and operate (DBFO) road scheme with a capital value of £9m. PFI has been successfully applied in the case of river crossings (Dartford and Severn), roads construction and maintenance (DBFOs), light rail (Manchester Metro and Croydon Tramlink), and several significant projects for London Underground delivering power, new ticketing (prestige), communications (connect) and rolling stock (Northern Line train services). PFI transport schemes in prospect include major new DBFO road projects and extensions to light rail networks in Greater Manchester and London Docklands.

The Department's flagship urban regeneration public/private partnerships project is the Docklands Light Railway extension to Lewisham (total capital investment of £200m). In September 1996 DLR Ltd awarded a concession to City Greenwich Lewisham Rail Link plc to design, build, finance and maintain the Lewisham extension for 25 years, following a PFI competition between four prequalified private sector consortia. The 4.2km extension will take the railway underneath the Thames in two new tunnels, adding 5 new stations south of the river and 2 reconstructed stations on the Isle of Dogs. The extension is expected to open in time for the Millennium celebrations at Greenwich peninsular. In June 1998 the Deputy Prime Minister announced approval in principle for a further extension of the railway to provide a connection to the London City Airport.

Local Authorities

To permit local authorities to make maximum use of the opportunities offered by partnerships with the private sector, the Government has acted to clarify its position to enable local authorities to enter into such projects. To this end a succession of measures have been introduced.

Working with the Treasury, other Departments and the Public Private Partnerships Programme (4Ps) the Department has clarified and streamlined the process by which authorities can apply to the Government for additional revenue to support PFI projects. At

April 2000, a total of 126 local government projects had been endorsed by a Treasury-chaired Projects Review Group, 38 of which have reached contract signature. Endorsement provides an assurance to the private sector that the project has been "road-tested" - it is commercially viable - and provides the local authority with an assurance that government support will be forthcoming - the project is affordable.

Local authority PFI projects in procurement cover a wide range of services: schools, residential homes, waste services, leisure centres, magistrates' courts, local transport, and more recently the first housing pathfinder projects.

The new guidance for Local Transport Plans includes a checklist to determine whether projects are suitable for a PFI route. Local authorities are undertaking many transport projects in partnership with the private sector such as the vehicle and depot service in Islington and the A130 in Essex. Signed local authority projects which are not yet in construction include the Doncaster Transport Interchange and the Nottingham Express Transit light rail scheme

Guidance

In March 2000, the Treasury issued "Public Private Partnerships: The Government's Approach". This sets out the Government's commitment to PPPs in delivering its policies and the underlying principles and themes which apply to the various forms of PPPs. This publication is available the Stationery Office and can viewed on the Treasury Taskforce website at: http://www.treasury-projects-taskforce.gov.uk/index.html

Table 8.1 Department for Education and Employment PFI/PPP Projects

The following includes projects at different stages of development in the education sectors.

Project Name	Description	Capital Value (£m)/ Comment
DfEE and Employment Service PFI/PPP Projects including Pathfinders		
– Signed Projects		
ESCOM IT Partnership	Computerised Replacement	4
Human Resources Partnerships	IT systems	116
(formerly NewPay)	Payroll IT project	6
Further Education PFI/PPP Projects including Pathfinders		
– Institutions Testing PPP/PFI		
Canterbury College	Relocation	30
Newbury College	Relocation	9
North Hertfordshire College, Stevenage	Disposal and Development	8
Uxbridge College	New-build and Refurbishment	3
West Hertfordshire College	Rationalisation on to 2 sites	10
Lowestoft College	IT Strategy and Procurement	n/a
Gloucestershire College of Arts & Technology	Relocation	16
– Signed Projects		
New College, Nottingham Wyggeston &	New Centre	17
Queen Elizabeth I College, Leicester	New Sports Laboratory and Fitness Room	Over 2
Tynemouth College, North Shields	IT Partnership Deal	n/a
Runshaw College, Leyland	New-build Adult Education Centre	2
Hammersmith & West London College	Conversion	2
South Thames College	New-build	3
City of Bristol College	Relocation	12
City of Liverpool Community College	Relocation and Redevelopment	18
Darlington College of Technology	New Telematics Centre	n/a
Manchester College of Arts & Technology	Replacement	16
Weymouth College	Relocation	13
Schools/LEA PFI/PPP Projects including Pathfinders		
– Signed Projects Receiving PFI Credits		
Dorset LEA	DBFO Rebuild School Project	Contract signed in Nov-97
Kingston upon Hull LEA	DBFO New-build School Project	Contract signed in Jul-98
Barnhill School, Hillingdon	Procurement of School	Contract signed in Nov-98
Dudley LEA	DBFO IT Schools Project	Contract signed Jan-99
Enfield LEA	DBFO New-build School Project	Contract signed in Feb-99
Lewisham LEA	Replacement School Kitchens	Contract signed Feb-99
Portsmouth LEA	DBFO New-build School Project	Contract signed Mar-99
Leeds Cardinal Heenan	DBFO Rebuild School Project	Signed Jun-99
Birmingham LEA	Refurbishment of Schools	

Table 8.1 Department for Education and Employment PFI/PPP Projects (continued)

The following includes projects at different stages of development in the education sectors.

Project Name	Description	Capital Value (£m)/ Comment
Projects Endorsed by the Project Review Group (PRG) **– Not Yet Signed**		
Manchester LEA	New School and Old School Redevelopment	
Staffordshire LEA	Rebuild/Refurbishment - Schools	Endorsed at Feb-98 PRG
Stoke on Trent LEA	Repair & Maintenance and Energy Project - Schools	
Westminster LEA	DBFO School Project	
Sheffield LEA	Rebuild - Schools	Endorsed at Jun-98 PRG
Tower Hamlets LEA	Energy Conservation and Management Scheme	
East Sussex LEA	New-build/Rebuild/Refurbishment - Schools	
East Riding LEA	New-build/Rebuild/Refurbishment - Schools	
Lambeth LEA - Lilian Baylis School	Rebuild - Schools	Endorsed at Aug-98
Lancashire LEA - Fleetwood High School	Rebuild/Refurbishment - Schools	
Torbay LEA	Rebuild - Schools	
Tower Hamlets LEA - Mulberry School	Rebuild - Schools	
Wirral LEA	Replacement/Refurbishment/Replacement - Schools	
Cornwall LEA	Repair/Rationalise - Schools	
Essex LEA	New School	Endorsed at Aug-98
Haringey LEA	New-build/Expansion/Refurbishment - Schools	
Kent LEA	New Schools	
Leeds LEA	New-build/Replacement/Refurbishment - Schools	
Liverpool LEA	Replacement - School	
North Yorkshire LEA	Refurbishment - Schools	
Nottinghamshire LEA	Refurbishment or Replacement - Schools	Endorsed at Oct-98/Nov-98 PRG
Tameside LEA	Rationalisation/Replacement - Schools	
Waltham Forest LEA	New School	
Wiltshire LEA	New-build/Extension/Refurbishment - Schools	
Projects Endorsed by the DfEE and Treasury TaskForce (TTF)		
Brent LEA	Refurbishment/Maintenance - Schools	
Kirklees LEA	DBFO Repair/Refurbishment Schools Project	
Grant Maintained Schools Projects Currently Being Developed		
Greenford High School, Ealing	New Catering and Sports Facilities	
St. Thomas More, Willenhall	Redevelopment of School	
St. Wilfred's CE High School, Blackburn	Relocation of Site	Not yet endorsed by the PRG
Frances Bardsley School, Romford	Consolidate onto one Site	
Jews Free School	Relocation of Site	
Hinchley Wood School, Surrey	New Sports and Dining Facilities	
Projects Supported by New Deal for Schools (NDS) **– Signed Projects**		
Kent LEA	Library Extension and New Technology Room	Signed in Jun-98
Waltham Forest LEA	Music Centre	Signed Nov-97
Projects Approved by DfEE **– Not Yet Signed**		
Blackpool LEA	Sports Centre	
Cambridgeshire LEA	Rebuild School	
Essex LEA	Schools Energy Project	
Hertfordshire LEA	Nursery Classroom	
Swindon LEA	Schools Energy Project	
Tower Hamlets LEA	Rebuild School	
Warrington LEA	Sports Project	
West Sussex LEA	New Dual Classroom for School	
	Replace 2 Classrooms in School	
	Information & Learning Centre for School	

Source of Data: Department for Further Education & Employment
Contract: Sue Meehan 020 7925 6348

Table 8.2 Department of the Environment, Transport & the Regions PFI Projects[1]

Project Name	Description	Contract Signed	Contractor/Consortium	Capital Value (£m)
Signed Projects:				
Dartford - Thurrock Crossing	River Crossing	Apr-87	Dartford River Crossing Ltd	180
Second Severn Crossing	River Crossing	Apr-90	John Laing/GTM	331
Birmingham Northern Relief Road	Tolled Road	Feb-92	Midland Expressway Ltd	450
London Underground - Jubilee Line Extension[1]	Light Rail	Oct-93	Bechtel	3,200
London Underground - Northern Line Trains	Light Rail	Apr-95	GEC Alsthom	409
Midland Metro Line One	Metro	Jul-95	Centro/Altram/Travel West Midlands Ltd	145
8 DBFO Roads	Roads	Jan-96 to Oct-96	Various	590
Channel Tunnel Rail Link	Rail	Feb-96	London & Continental Railways (LCR)	4,178
Docklands Light Railway Lewisham Extension	Light Rail	Sep-96	City Greenwich Lewisham Rail Link plc	202
Croydon Tramlink	Light Rail	Nov-96	Tramtrack Croydon Ltd	205
Salford Quays and Eccles	Metro	Dec-96	Altram	160
QEII Conference Centre	Catering	Jan-97	Leiths	3
Waltham Forest Housing Action Trust	Housing	Jan-97	Peabody Trust	15
New Oceanic Flight Data Processing System	Air Traffic System	Jun-97	EDS Ltd	18
Tower Hamlets Housing Action Trust	Housing	Apr-98	Circle 33/Old Ford Housing Association	23
London Underground - Power	Power Supply	Aug-98	Powerlink	108
London Underground - Prestige	Ticketing System	Aug-98	Transys	137
London Underground - British Transport Police	Accommodation	Mar-99	AP Services (Amey plc)	13
DBFO Road - A13 Thames Gateway	Roads	Apr-99	Road Management Services	146
London Underground - Connect	Communications	Nov-99	CityLink Telecommunications Consortium	355
Vehicle Inspectorate MOT Computerisation	IT	Mar-00	Siemans Business Systems	n/a
Thameslink 2000[1]	Rail		Railtrack/London & Continental Railways (LCR)	680
Luton Airport Parkway	Railway Station		Railtrack/London Luton Airport Ltd	20
Heathrow Express[1]	Rail		BAA plc	440
Total				**11,988**
Projects Yet to Reach Signature:				
Manchester Metrolink - Extensions	Light Rail			500[2]
Piccadilly Line Extension to Heathrow Terminal 5	Light Rail			70[2]
Docklands Light Railway - City Airport Extension	Light Rail			35[2]
Health & Safety Laboratory Estate Rationalisation	Accommodation			n/k
Highways Agency - Traffic Control Centre	IT System			n/k

Notes

n/a = not available
n/k = not known
DBFO = Design, Build, Finance and Operate
1. Other significant construction projects.
2. Broad estimate of capital value.

Source of Data: Private Finance Unit, Department of the Environment, Transport and the Regions
Contact: Rachel Edwards 020 7944 5015

Table 8.3 Department of Health PFI Projects

Project Name	Project Description	Capital Value (£m)[1,2,3]	FBC Approval Status Yes/No	Financial Close Status Yes/No	Operation Status Yes/No
Northern & Yorkshire					
Leeds Teaching Hospitals NHS Trust	Implementation of Leeds Acute Services Reconfiguration Strategy	125.00	No	No	No
Newcastle upon Tyne Hospitals NHS Trust[4]	Completion of Reconfiguration of Acute Hospital Services	123.80	No	No	No
South Tees Acute Hospitals NHS Trust	Single Site Development/ Centralisation of Acute Hospital Services	122.00	Yes	Yes	No
Carlisle Hospitals NHS Trust	Centralisation to Cumberland Infirmary Site	64.70	Yes	Yes	No
Calderdale Health Care NHS Trust	Centralisation of Acute Hospital Services	64.60	Yes	Yes	No
North Durham Health Care NHS Trust	New DGH	61.00	Yes	Yes	No
Leeds Community & Mental Health Services NHS Trust	Reprovision of Mental Health Services	47.00	Yes	Yes	No
South Durham Health Care NHS Trust	Redevelopment of Hospital	41.20	Yes	Yes	No
Hull and East Yorkshire Hospitals NHS Trust	Maternity and Acute Development	21.70	No	No	No
Northumbria Health Care NHS Trust	Redevelopment of Hospital	21.20	No	No	No
Northumbria Health Care NHS Trust	Phase II Development of Hospital	14.10	No	No	No
Hull and East Yorkshire Hospitals NHS Trust	Phase 5 - Reprovision of Outpatient, Radiology and Urology Services	7.90	No	No	No
South Durham Health Care NHS Trust	Hospital	6.24	No	No	No
North Tees & Hartlepool NHS Trust	Rehabilitation Unit and Site Rationalisation	5.70	No	No	No
Leeds Community & Mental Health Services NHS Trust	Community Units for the Elderly	5.00	Yes	Yes	Yes
Bradford Community Health NHS Trust	Range of Primary and Community Services	4.00	Yes	Yes	No
Hull and East Riding Community Health NHS Trust	Hospital and Resource Centre	3.30	Yes	Yes	Yes
Northallerton Health Services NHS Trust	Conversion	3.00	Yes	Yes	Yes
Hull and East Yorkshire Hospitals NHS Trust	Joint Procurement of an Integrated Patient Management System	3.00	Yes	Yes	No
Leeds Community & Mental Health Services NHS Trust	Community and Mental Health Information System	1.80	Yes	Yes	No
Leeds Teaching Hospitals NHS Trust	Replacement Linear Accelerators and Simulators	1.39	Yes	Yes	No
Northumbria Health Care NHS Trust	CHP Energy Scheme	1.35	Yes	Yes	Yes
York Health Services NHS Trust	Provision of MRI Service	1.00	Yes	Yes	Yes
Regional Total		**749.98**			
Trent					
Southern Derbyshire Acute Hospitals NHS Trust	Acute Reconfiguration	177.00	No	No	No
Queens Medical Centre, Nottingham University Hospital	ENT/Ophthalmology	16.60	Yes	Yes	No
Nottingham Health Care NHS Trust	Elderly/Mental Health	12.10	No	No	No
Doncaster and South Humber NHS Trust	Elderly Mental Health and Mental Health Rehabilitation	12.00	No	No	No
East Midlands Ambulance Service NHS Trust	Radio Control System	3.26	No	No	No
Nottingham City Hospital NHS Trust	Headquarters	3.25	Yes	Yes	Yes
Rotherham General Hospitals NHS Trust	Catering Reprovision	3.20	Yes	Yes	Yes
Queens Medical Centre, Nottingham University Hospital	CHP Generator	2.80	Yes	Yes	Yes
Rotherham Priority Health Service NHS Trust	Elderly Mental Health	2.10	Yes	Yes	Yes
Barnsley District General Hospital NHS Trust	Catering Reprovision	1.80	No	No	No
Kings Mill Centre for Health Care Services NHS Trust	Residential Accommodation	1.60	Yes	Yes	No
Leicestershire & Rutland Health Care NHS Trust	Headquarters	1.60	Yes	Yes	Yes
Doncaster Royal Infirmary & Montagu Hospital NHS Trust	MRI Scanner	1.50	Yes	Yes	No
Bassetlaw Hospital & Community Health Services	Health Centre	1.30	No	No	No
Central Nottinghamshire Health Care NHS Trust	Learning Disability/Mental Health Team Base	1.30	No	No	No

Table 8.3 Department of Health PFI Projects (continued)

Project Name	Project Description	Capital Value (£m)[1,2,3]	FBC Approval Status Yes/No	Financial Close Status Yes/No	Operation Status Yes/No
Lincoln & Louth NHS Trust	Boiler Plant	1.20	No	No	No
East Midlands Ambulance Service NHS Trust	Radio Control System	1.10	Yes	Yes	Yes
Rotherham General Hospitals NHS Trust	Entrance Redevelopment	1.00	Yes	Yes	Yes
Queens Medical Centre, Nottingham University Hospital	Catering Reprovision	1.00	Yes	Yes	No
Regional Total		**245.71**			
Eastern					
Norfolk and Norwich Health Care NHS Trust	New DGH	158.00	Yes	Yes	No
Luton and Dunstable Hospital NHS Trust	St Mary's Wing	14.70	No	No	No
Essex & Herts Community NHS Trust	Hospital	11.40	No	No	No
East Hertfordshire NHS Trust	Ambulatory Care Centre	10.00	No	No	No
Fulbourn	School of Nursing/Admin.	9.90	Yes	Yes	No
Norfolk and Norwich Health Care NHS Trust	Staff Accommodation	9.80	No	No	No
Essex Rivers Health Care NHS Trust	Staff Accommodation	8.80	No	No	No
Local Health Partnerships NHS Trust	Mental Health Unit	3.60	No	No	No
Luton and Dunstable Hospital NHS Trust	PAS	2.30	No	No	No
Regional Total		**228.50**			
London					
Barts & The London NHS Trust	Acute Site Rationalisation	620.00	No	No	No
University College London Hospitals NHS Trust	Acute Rationalisation	274.00	No	No	No
Havering Hospitals NHS Trust	Redevelopment of Hospital Facilities	148.00	No	No	No
Bromley Hospitals NHS Trust	New DGH	117.90	Yes	Yes	No
Greenwich Health Care NHS Trust	Refurbishment of DGH	93.00	Yes	Yes	No
King's Health Care NHS Trust	New Block	64.00	Yes	Yes	No
Barnet & Chase Farm Hospitals NHS Trust	Barnet Phase 1B	54.00	Yes	Yes	No
St. George's Health Care NHS Trust	Neurol Cardiac Unit	49.00	No	Yes	No
West Middlesex University NHS Trust	Acute Rationalisation	33.00	No	No	No
The Whittington Hospital NHS Trust	Redevelopment of Acute Hospital Services	23.00	No	No	No
Newham Health Care NHS Trust	Rationalisation	20.00	No	No	No
Oxleas NHS Trust	Reprovision of Mental Health	15.00	Yes	Yes	No
Camden & Islington Community Health Service	Mental Health	14.00	No	No	No
Newham Community Health Services NHS Trust	Mental Health Reprovision	12.08	No	No	No
Parkside Health NHS Trust	Willesden	11.00	No	No	No
Redbridge Health Care NHS Trust	Mental Health Reprovision and Geriatric Day Centre	10.80	No	No	No
Mayday Health Care NHS Trust	Energy Centre	8.00	Yes	Yes	No
Ealing Hammersmith and Fulham Mental Health NHS Trust	Energy Centre	5.80	No	No	No
Enfield Community Care NHS Trust	Elderly Mentally Ill Home	3.54	Yes	No	No
North West London Hospitals NHS Trust	Vascular/Breast Unit	3.50	Yes	No	No
North West London Hospitals NHS Trust	ACAD Equipment	3.50	Yes	Yes	No
Mayday Health Care NHS Trust	Front Entrance	1.80	Yes	Yes	Yes
West Middlesex University NHS Trust	MRI	1.50	Yes	No	No
Royal National Orthopaedic Hospital NHS Trust	Scanner	1.00	Yes	Yes	Yes
Regional Total		**1587.42**			
South East					
Dartford and Gravesham NHS Trust	Acute General Hospital Development	94.00	Yes	Yes	No
Portsmouth Hospitals NHS Trust	Acute Rationalisation	75.00	No	No	No
Oxford Radcliffe Hospitals NHS Trust	Reprovide Radcliffe Infirmary	71.00	Yes	Yes	No

Table 8.3 Department of Health PFI Projects (continued)

Project Name	Project Description	Capital Value (£m)[1,2,3]	FBC Approval Status Yes/No	Financial Close Status Yes/No	Operation Status Yes/No
South Buckinghamshire NHS Trust	Site Rationalisation	45.10	Yes	Yes	No
West Berkshire Priority Care Services NHS Trust	Mental Health Reprovision	29.70	No	No	No
Stoke Mandeville Hospital NHS Trust	Partial Redevelopment	23.70	No	No	No
Nuffield Orthopaedic Centre NHS Trust	Orthopaedics/Medicine Redevelopment	23.60	No	No	No
Kent & Sussex Weald NHS Trust	Graylingwell Hospital Reprovision	22.00	Yes	Yes	No
Northampton Community Health Care NHS Trust	Mental Health Acute Hospital Reprovision	19.80	No	No	No
West Berkshire Priority Care Services NHS Trust	Consolidation of Sites	18.00	No	No	No
Southampton Community Health Service	Hospital	18.00	No	No	No
Surrey Hampshire Borders NHS Trust	Resource Centre	15.00	No	No	No
Portsmouth Health Care NHS Trust	Mental Health and Learning Disability	12.00	No	No	No
Thames Gateway NHS Trust	Hospital	10.00	No	No	No
Mid Kent Health Care NHS Trust	Ophthalmology	10.00	No	No	No
Oxfordshire Mental Health Care NHS Trust	Mental Health Medium Security Unit	9.00	Yes	Yes	No
Surrey Hampshire Borders NHS Trust	Health Centre	8.00	No	No	No
Rockingham Forest NHS Trust	Acute Mental Health Ward Reprovision	6.00	No	No	No
Thames Gateway NHS Trust	Hospital	5.50	Yes	Yes	No
Ashford & St Peter's Hospitals NHS Trust	Energy Management Scheme	5.00	No	No	No
Kettering General Hospital NHS Trust	IM&T	3.70	No	No	No
Worthing and Southlands Hospitals NHS Trust	Catering Service	3.30	No	No	No
Mid-Sussex NHS Trust	Low Secure Unit	3.30	No	No	No
Southampton University Hospitals NHS Trust	Energy Management Scheme	3.10	No	No	No
Surrey & Sussex Health Care NHS Trust	Hospital	3.00	No	No	No
Mid-Sussex NHS Trust	Residential Accommodation	3.00	No	No	No
North Hampshire Hospitals NHS Trust	Energy Management Scheme	2.70	Yes	Yes	Yes
Royal Surrey County & St Lukes Hospital NHS Trust	Residential Accommodation	2.20	No	No	No
Royal Berkshire & Battle Hospitals NHS Trust	RBBH Pas	2.00	No	No	No
North Hampshire Hospitals NHS Trust	MRI Scanner	1.90	Yes	Yes	No
Surrey & Sussex Health Care NHS Trust	Hospital Energy Management	1.80	Yes	Yes	No
Royal Berkshire & Battle Hospitals	RBBH Renal	1.70	No	No	No
North Hampshire Hospitals NHS Trust	Patient Management System	1.50	Yes	Yes	Yes
Queen Victoria Hospital NHS Trust	Energy Management	1.44	Yes	Yes	Yes
Regional Total		**555.04**			
South West					
Swindon & Marlborough NHS Trust	New DGH	96.00	Yes	Yes	No
Gloucestershire Royal NHS Trust	Site Redevelopment	37.00	No	No	No
Exeter & District Community Health Service NHS Trust	Hospital	10.50	No	No	No
Cornwall Health Care NHS Trust	Hospital	10.20	No	No	No
Cornwall Health Care NHS Trust	Hospital	7.40	No	No	No
Plymouth Hospitals NHS Trust	Patient Admin. System	7.00	Yes	Yes	No
North Bristol NHS Trust	Brain Injury Rehabilitation Unit	4.90	Yes	Yes	Yes
South Devon Health Care NHS Trust	Hospital	3.70	Yes	Yes	Yes
Poole Hospital NHS Trust	Residences	3.48	Yes	Yes	No
Plymouth Hospitals NHS Trust	Linear Accelerators	2.73	Yes	Yes	Yes
Gloucestershire Royal NHS Trust	Residences	2.65	No	No	No
Wiltshire Healthcare NHS Trust	Hospital	2.00	No	No	No
Regional Total		**187.56**			
West Midlands					
Walsgrave Hospitals NHS Trust	New DGH	178.00	No	No	No
Worcester Royal Infirmary NHS Trust	New DGH	86.60	Yes	Yes	No
Dudley Group of Hospitals NHS Trust	New DGH	67.70	No	No	No

Table 8.3 Department of Health PFI Projects (continued)

Project Name	Project Description	Capital Value (£m)[1,2,3]	FBC Approval Status Yes/No	Financial Close Status Yes/No	Operation Status Yes/No
Hereford Hospitals NHS Trust	New DGH	64.10	Yes	Yes	No
North Staffordshire Combined Health Care NHS Trust	Reprovision of Mental Health Facilities	19.20	Yes	Yes	No
City Hospital NHS Trust	Ambulatory Care Centre	19.20	No	No	No
Royal Wolverhampton Hospitals NHS Trust	Radiology Unit	12.50	No	No	No
Northern Birmingham Mental Health NHS Trust	Reprovision of Mental Health Facilities	12.30	No	No	No
Black Country Mental Health NHS Trust	Mental Health Unit	5.20	Yes	Yes	Yes
Dudley Priority Health NHS Trust	Health and Social Care Centre	3.80	Yes	Yes	No
South Birmingham Mental Health NHS Trust	Reprovision of Accommodation	2.20	Yes	Yes	No
Regional Total		**470.80**			
North West					
Central Manchester Health Care NHS Trust	Service Reconfiguration	250.00	No	No	No
South Manchester University Hospitals NHS Trust	Site Rationalisation	65.60	Yes	Yes	No
Blackburn, Hyndburn and Ribble Valley Health NHS Trust	Site Rationalisation	52.00	No	No	No
Blackpool Wyre and Flyde Community Health Services	New Elderly Unit	9.90	No	No	No
Bay Community NHS Trust	New Mentally Ill Unit	7.00	Yes	Yes	Yes
Stockport Health Care NHS Trust	New Mentally Ill Unit	4.70	Yes	Yes	Yes
Wigan & Leigh Health NHS Trust	New Primary Care Resource Centre	2.75	Yes	No	No
Mancunian Community Health NHS Trust	New Primary Care Resource Centre	2.00	Yes	Yes	No
Blackburn, Hyndburn and Ribble Valley Health Care	New Medical Staff Residences	1.30	Yes	No	No
Regional Total		**395.25**			
National Total		**4420.26**			

Notes

1. The capital value of PFI schemes are approximate and defined as: Total Capital Cost to the Private Sector includes the costs of land, construction, equipment and professional fees but excludes VAT, rolled up interest and financing costs such as bank arrangement fees, bank due diligence fees, banks' lawyers fees and third party equity costs. As PFI procures a service rather than the underlying asset, capital values shown are necessarily estimates.

2. Capital Costs have been adjusted so that they are in a consistent price base.

3. For schemes still in early stages of development, estimates are based on highest cost.

4. £123.80 - provisional figure for revised proposals.

Figures may not sum due to rounding.

Source of Data: NHS Executive, Private Finance Unit
Contact: Christian Richardson 01132 547385

Table 8.4 Capital Programme for Non-PFI Health Building Construction Projects over £2.5 Million[1]

Live schemes under construction[2,3] as at 31st March 2000

	Value Bands									
	Up to £10m		£10m - 25m		£25m - 50m		Over £50m		Total	
Regions	No.	Value £000s	No.	Value £000s	No.	Value £000s	No.	Value 000s	No.	Value 000s
Eastern	2	14	1	13	0	0	0	0	3	27
London	1	4	1	15	0	0	1	75	3	94
North West	5	35	2	48	1	31	0	0	8	114
Northern & Yorkshire	2	12	1	22	1	40	0	0	4	73
South & West	1	5	1	25	1	40	0	0	3	69
South East	5	22	1	21	0	0	1	62	7	105
Trent	1	8	1	12	1	26	0	0	3	46
West Midlands	2	6	0	0	0	0	0	0	2	6
TOTAL	**19**	**107**	**8**	**154**	**4**	**137**	**2**	**137**	**33**	**534**

Notes

1. This is not an exhaustive list of construction schemes over that value in the National Health Service, but as reported andrecorded on the Capital Investment Manual database, which is held by NHS Estates. Only projects which are publicly fundedfunded are included.

2. The Total Approved Cost of a scheme includes works costs, equipment, non-works cost, fees and contingencies. Any costs associated with VAT are excluded.
3. Live schemes under construction includes all schemes which are on site, and have not yet reached Practical Completion.

Source of Data: NHS Estates

Contact: Simon Wright 0113 2547185

Figure 8.1 Number of Homes for Elderly and for Younger Physically Disabled People in England

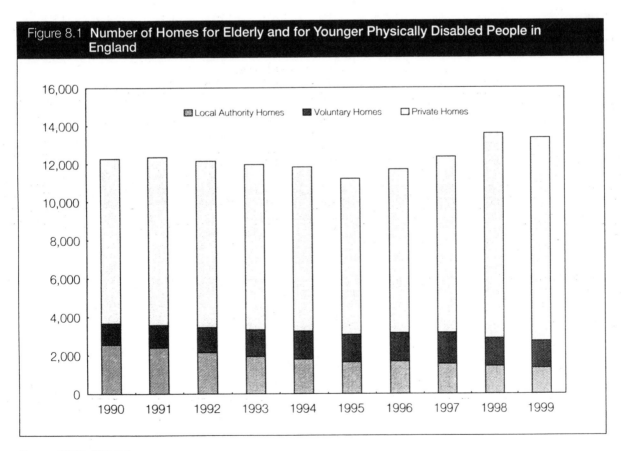

Source of Data: Table 8.5

Table 8.5 Number of Homes for Elderly and for Younger physically Disabled People by Type of Home

Local Authority Homes

SSI INSPECTION REGIONS[1]	1991	1992	1993	1994	1995	1996	1997	1998	1999
North West	388	290	260	239	210	214	199	186	184
North East	544	509	437	403	364	368	352	307	293
Central	535	511	459	435	415	416	372	398	356
Southern & South West	370	325	293	286	262	280	249	231	230
London East	307	279	264	244	219	220	204	189	177
London West	254	237	225	191	169	181	170	117	97
ENGLAND	2,398	2,151	1,938	1,798	1,639	1,679	1,546	1,428	1,337

Voluntary Homes

SSI INSPECTION REGIONS[1]	1991	1992	1993	1994	1995	1996	1997	1998	1999
North West	148	216	239	244	240	242	230	240	208
North East	111	120	136	146	147	156	176	137	132
Central	192	197	215	223	227	235	301	274	241
Southern & South West	273	301	327	330	321	332	350	317	326
London East	210	215	222	232	228	232	265	218	216
London West	232	238	247	277	265	273	293	265	264
ENGLAND	1,166	1,287	1,386	1,452	1,428	1,470	1,615	1,451	1,387

Private Homes

SSI INSPECTION REGIONS[1]	1991	1992	1993	1994	1995	1996	1997	1998	1999
North West	1,305	1,275	1,255	1,222	1,143	1,197	1,317	1,491	1,569
North East	1,269	1,246	1,239	1,229	1,199	1,296	1,322	1,529	1,475
Central	1,675	1,679	1,692	1,687	1,616	1,700	1,889	2,537	2,571
Southern & South West	2,456	2,474	2,429	2,413	2,271	2,331	2,496	2,864	2,771
London East	1,292	1,267	1,264	1,269	1,205	1,260	1,337	1,444	1,358
London West	796	774	768	759	720	764	837	843	873
ENGLAND	8,793	8,715	8,647	8,579	8,154	8,548	9,198	10,708	10,617

Total All Homes

SSI INSPECTION REGIONS[1]	1991	1992	1993	1994	1995	1996	1997	1998	1999
North West	1,841	1,781	1,754	1,705	1,593	1,653	1,746	1,917	1,961
North East	1,924	1,875	1,812	1,778	1,710	1,820	1,850	1,973	1,900
Central	2,402	2,387	2,366	2,345	2,258	2,351	2,562	3,209	3,168
Southern & South West	3,099	3,100	3,049	3,029	2,854	2,943	3,095	3,412	3,327
London East	1,809	1,761	1,750	1,745	1,652	1,712	1,806	1,851	1,751
London West	1,282	1,249	1,240	1,227	1,154	1,218	1,300	1,225	1,234
ENGLAND	12,357	12,153	11,971	11,829	11,221	11,697	12,359	13,587	13,341

Notes

1. SSI regions are defined in Appendix 2.

Source of Data: Statistics Division 3B, Department of Health
Contact: Laurent Ortmans 020 7972 5600

Table 8.6 Highway Agency PFI Projects

The following table gives the position on each of the DBFO road projects as at March 2000

Project Name	Length (km)	Schemes	Est. Cons. Cost (£m)	OJEC	Awarded/ Status	Contractor
A1 (M) Alconbury to Peterborough	20.8	A1 (M) Alconbury to Peterborough Improvement	128.0	Aug-94	8-Feb-96	Road Management Services (Peterborough) Ltd
M1-A1 Yorkshire Link	29.8	M1-A1 Link Road: Lofthouse to Bramham	214.0	Aug-94	26-Mar-96	Yorkshire Link Ltd
A419/A417 Swindon to Gloucester	54.7	A419/A417 Cirencester & Stratton Bypass A419 Latton Bypass A417 North of Stratton to Nettleton Improvement	49.0	Aug-94	8-Feb-96	Road Management Services (Gloucester) Ltd
A69 Carlisle to Newcastle	83.8	A69 Haltwhistle Bypass	9.4	Aug-94	12-Jan-96	Road Link Ltd
Dishforth to Tyne Tunnel	103.4	A19 Norton-Parkway Interchange Improvement	29.4	Feb-95	14-Oct-96	Autolink Concessionaires Ltd
A30/A35 Exeter to Bere Regis	104.3	A35 Honiton to Exeter Improvement A35 Tolpuddle to Puddletown Bypass	75.7	Feb-95	24-Jul-96	Connect (A50) Ltd
A50 Stoke to Derby Link	55.7	A564 Doveridge Bypass	20.6	Feb-95	20-May-96	Connect (A50) Ltd
M40 Junctions 1 - 15	121.3	M40 Junction 1A-3 Widening	65	Feb-95	8-Oct-96	UK Highways Ltd
A13 Thames Gateway DBFO Project	21.5	A13 Ironbridge to Cannington Improvement A13/A117 Woolwich Manor Way Junction Improvement A13 Movers Lane Junction Improvement Contingent Scheme: A13/A112 Prince Regent Lane Junction Improvement	146[1]	Jan-98	Provisional Prefered Bidder Selected Dec-99	
Doncaster to Durham DBFO Project	n/p	A1 (M) Wetherby to Walshford Widening A1 (M) Ferrybridge to Hook Moor widening	210	n/p	n/p	
Kent DBFO Project	n/p	A2 Bean to Cobham Widening Phase 2 A2/A282 Dartford Improvement A249 Iwade to Queenborough Improvement	152	n/p	n/p	
Traffic Control Centre Project				Dec-98	Tenders shortlisted to 2 consortia Feb-00	

Notes

n/p = not published yet

1. The estimated construction cost (£m) includes the contingent schemes.

Source of Data: Private Finance Team, Highways Agency

Contact: Ian Farrand 020 7921 4169

Table 8.7 Major (£1m and over) Non-PFI Road Contracts Awarded by the Highways Agency

Maintenance						£ Million
Regions [2,3]		South	Midlands	Northern	London	Total
1990	Q1	5.8 (3)	42.9 (7)	36.3 (6)	16.2 (1)	101.2 (17)
	Q2	21.7 (6)	21.6 (10)	2.6 (2)	1.5 (1)	47.4 (19)
	Q3	14.5 (9)	22.2 (13)	23.6 (6)	4.2 (1)	64.6 (29)
	Q4	16.6 (9)	17.4 (2)	25.5 (8)	-	59.6 (19)
1991	Q1	16.6 (6)	10.5 (7)	16.4 (7)	30.1 (4)	73.6 (24)
	Q2	10.6 (7)	23.5 (7)	15.7 (5)	-	49.9 (19)
	Q3	28.0 (11)	15.0 (9)	26.5 (7)	5.6 (2)	75.0 (29)
	Q4	7.3 (4)	6.1 (7)	9.1 (4)	2.9 (1)	25.4 (16)
1992	Q1	9.1 (6)	7.7 (4)	14.1 (6)	-	30.9 (16)
	Q2	16.7 (7)	12.2 (6)	21.0 (7)	2.1 (1)	52.0 (21)
	Q3	31.3 (8)	7.7 (3)	27.1 (10)	1.8 (1)	67.9 (22)
	Q4	16.5 (9)	14.2 (8)	14.0 (8)	-	44.7 (25)
1993	Q1	14.9 (7)	19.7 (8)	73.7 (9)	18.5 (2)	126.8 (26)
	Q2	11.9 (7)	14.9 (7)	20.9 (6)	2.1 (2)	49.9 (22)
	Q3	16.9 (8)	24.2 (10)	12.6 (5)	-	53.7 (23)
	Q4	46.1 (12)	16.7 (10)	14.1 (3)	11.4 (4)	88.3 (29)
1994	Q1	15.4 (8)	19.5 (7)	14.1 (5)	17.1 (2)	66.0 (22)
	Q2	37.5 (8)	6.0 (4)	29.7 (9)	13.6 (2)	86.8 (23)
	Q3	29.8 (8)	31.7 (12)	21.9 (4)	-	83.4 (24)
	Q4	31.3 (6)	1.6 (1)	20.5 (8)	-	53.5 (15)
1995	Q1	21.0 (8)	15.7 (3)	13.7 (5)	-	50.4 (16)
	Q2	10.0 (3)	13.5 (5)	11.2 (6)	1.1 (1)	35.8 (15)
	Q3	49.2 (8)	17.3 (9)	14.1 (6)	-	80.6 (23)
	Q4	3.6 (1)	2.8 (1)	2.2 (1)	6.2 (2)	14.7 (4)
1996	Q1	12.1 (2)	-	10.4 (1)	47.9 (2)	70.4 (5)
	Q2	-	7.3 (2)	-	-	7.3 (2)
	Q3	-	3.5 (2)	-	2.5 (2)	6.0 (4)
	Q4	-	1.4 (1)	-	5.6 (3)	7.0 (4)
1997	Q1	-	-	-	-	-
	Q2	-	-	-	-	-
	Q3	-	3.5 (2)	-	-	3.5 (2)
Regions [2,3]		Birmingham	Bedford	Leeds	Dorking	Total
1997	Q4	44.3 (9)	4.7 (3)	1.3 (1)	17.0 (7)	67.3 (20)
1998	Q1	53.7 (8)	30.4 (2)	4.3 (2)	32.3 (3)	120.7 (15)
	Q2	7.2 (3)	5.5 (3)	3.3 (2)	-	16.0 (8)
	Q3	22.0 (5)	12.8 (5)	21.3 (3)	16.9 (8)	73.0 (21)
	Q4	50.6 (7)	5.7 (2)	55.7 (6)	4.4 (2)	116.4 (17)
1999	Q1	6.4 (2)	16.5 (3)	2.3 (2)	22.2 (2)	47.4 (9)
	Q2	6.9 (3)	14.7 (5)	6.7 (3)	16.4 (5)	44.7 (16)
	Q3	28.8 (10)	26.7 (6)	5.7 (4)	10.9 (3)	72.1 (23)
	Q4	19.4 (6)	11.7 (4)	4.1 (3)	9.6 (4)	44.8 (17)

Table 8.7 Major (£1m and over) Non-PFI Road Contracts Awarded by the Highways Agency[1] (continued)

New Construction							£ Million
Regions [2,3]		South	Midlands	Northern	London	MWU [4,5]	Total
1990	Q1	14.2 (2)	78.3 (6)	30.5 (4)	-	-	123.1 (12)
	Q2	31.2 (2)	74.2 (4)	33.4 (5)	-	-	138.9 (11)
	Q3	25.9 (3)	84.1 (5)	-	-	-	110.1 (8)
	Q4	100.7 (5)	18.3 (4)	9.1 (4)	-	-	128.1 (13)
1991	Q1	-	31.7 (3)	2.7 (1)	26.6 (1)	1.1 (1)	62.1 (6)
	Q2	21.7 (1)	34.9 (4)	7.8 (3)	-	59.4 (1)	123.7 (9)
	Q3	99.5 (3)	7.0 (1)	6.7 (1)	-	-	113.2 (5)
	Q4	36.4 (4)	101.1 (6)	10.1 (2)	-	-	147.6 (12)
1992	Q1	22.8 (4)	9.3 (1)	5.6 (1)	-	-	37.6 (6)
	Q2	-	2.2 (1)	4.1 (1)	-	-	6.3 (2)
	Q3	48.8 (3)	-	-	-	-	48.8 (3)
	Q4	8.2 (1)	101.5 (5)	67.1 (2)	-	-	176.8 (8)
1993	Q1	65.1 (1)	26.0 (1)	49.3 (2)	-	14.7 (2)	155.0 (6)
	Q2	20.6 (2)	4.4 (1)	90.9 (4)	31.3 (2)	-	147.2 (9)
	Q3	1.2 (1)	31.6 (1)	5.5 (1)	-	-	38.3 (3)
	Q4	76.9 (3)	94.6 (8)	-	41.5 (2)	-	213.1 (13)
1994	Q1	149.8 (6)	51.9 (2)	68.8 (3)	83.1 (2)	-	353.6 (12)
	Q2	-	38.7 (1)	-	-	-	38.7 (1)
	Q3	-	1.2 (1)	48.9 (1)	-	-	50.2 (2)
	Q4	-	-	-	-	-	-
1995	Q1	143.4 (2)	209.7 (3)	122.7 (3)	36.6 (1)	14.4 (1)	526.9 (10)
	Q2	-	5.0 (1)	11.2 (1)	-	-	16.2 (2)
	Q3	-	-	-	-	-	-
	Q4	-	-	1.8 (1)	-	-	1.8 (1)
1996	Q1	-	-	98.7 (1)	-	-	98.7 (1)
	Q2	73.7 (1)	-	-	-	-	73.7 (1)
	Q3	30.8 (1)	-	-	50.9 (1)	-	81.7 (2)
	Q4	-	57.3 (2)	-	-	-	57.3 (2)

Table 8.7 Major (£1m and over) Non-PFI Road Contracts Awarded by the Highways Agency[1] (continued)

New Construction						£ Million
Regions [2,3]		Birmingham	Bedford	Leeds	Dorking	Total
1997	Q1	-	-	-	-	-
	Q2	-	-	-	-	-
	Q3	-	-	-	-	-
	Q4	-	-	-	-	-
1998	Q1	-	-	-	-	-
	Q2	60.2 (2)	-	-	-	60.2 (2)
	Q3	-	-	-	-	-
	Q4	-	-	-	-	-
1999	Q1	14.4 (2)	-	-	-	14.4 (2)
	Q2	-	-	-	-	-
	Q3	10.1 (2)	-	-	-	10.1 (2)
	Q4	129.8 (2)	-	-	-	129.8 (2)

Notes

1. Number of schemes in brackets.

2. Highways Agency Regional Offices: South (comprising Government Office Regions South East and South West), Midlands (comprising Government Office Regions Eastern, West Midlands and East Midlands), and Northern (comprising Government Office Regions North West, Yorkshire & Humberside and Northern). Government Office Regions are defined in Appendix 1.

3. Highways Agency Regional Offices have been restructured. There is no Longer a London Office. The London Schemes are dealt with by the Dorking, Bedford and Birmingham offices.

4. MWU = Motorway Widening Unit

5. There is no longer a MWU from 1997 Jan-Mar onwards.

Source of Data: Roads Procurement Policy Operation, Highways Agency

Contact: Adrian McCabe 020 7921 4333

Table 8.8 Home Office PFI Projects

Project Name	Description	OJEC Published	Invitation to Tender	Bids In	Negotiation	Financial Close	Capital Value (£m)
Contracts Signed:							
Parc (Bridgend) Prison	DCMF Prison	1994	1995	1995	1995	1996	74.0
Altcourse (Fazakerley) Prison	DCMF Prison	1994	1995	1995	1995	1995	88.0
Ashfield (Pucklechurch) Prison	DCMF Prison	1997	1997	1997	1997	1998	30.7
Forest Bank (Agecroft) Prison	DCMF Prison	1997	1997	1997	1997	1998	38.6
IND - Casework	Information Technology	1995	1995	1995	1996	1996	41.0
PABX Prisons	Information Technology	1995	1995	1995	1995	1996	6.0
IND Gatwick	Detention Centre				1995	1996	12.0
Lowdham Grange	DCMF Prison	1995	1995	1996	1996	1996	32.0
Medway STC	Secure Training Centres	1994	1995	1995	1996	1996	10.0
Passport	Data Capture	1996	1997	1997	1996/97	1997	15.0
Passport	Supply and Personalisation	1996	1997	1997	1997	1997	15.0
Rainsbrook (Onley) STC[1]	Secure Training Centres	1995	1997	1997	1998	1998	10.0
Hassockfield STC (Medomsley)	Secure Training Centres	1997	1998	1998	1998	1998	10.0
Prisons/Energy	Boilerhouses	1996	1997			1997	9.0
Onley	DCMF Prison	1997		1998	1998/99	1999	34.7
Marchington	DCMF Prison	1997		1998	1998/99	1999	48.0
Total (contracts agreed):							**474.0**
Tenders Being Sought:							
HOCLAS[1]	Property	1997	1997		1997	2000	200.0
IT 2000	IT and Business Change	1997	2000	2000	2000	2000	100.0
PS HQ (QUANTUM)	Quantum IT	1996	1998/99	1998	1998	2000	60.0
PSRCP	Communications	1996	1996	1996	1999/00	1999/00	500.0
Replacement Harmonsworth	Detention Centre	1999	2000	2000	2000	2001	23.0
Prisons/Energy - Tranche 2		1999	1999				16.0
Ashford		2000	2000	2000	2000	2001	35.0
Peterborough		2000	2000	2000	2000	2001	75.0
Criminal Records Bureau	Informations Systems	2000	2000	2000	2000	2000	
Sharpness	STC						
Aldington Detention Centre		2000	2000	2000	2000	2001	28.0
Total (schemes yet to be agreed):							**1,037.0**
Overall Total:							**1,511.0**

Notes

1. Original contract has expired and a subsequent contract has been assumed.

Source of Data: Home Office Procurement Unit, Home Office

Contact: Donna Sanders 020 7273 2353

Table 8.9 Local Authority PFI Projects

	Description	Pathfinder	OJEC Data	Preferred Bidder Selected	Signed	Expected Contract Value	Expected PFI Credits	Expected Capital Contract Value
Police - England:								
Wiltshire	Air Support				Dec-97	6.0	3.1	3.0
Derbyshire	Ilkeston Police Station	*			Feb-98	5.3	3.8	4.5
Northumbria	Facilities	*			Mar-98	3.0	1.9	3.0
Thames Valley	Force HQ	*	Mar-97	Jan-98	Dec-98	15.0	10.4	7.0
Derbyshire	Divisional HQ	*	May-96	Apr-97	Feb-99	34.5	23.9	16.0
Norfolk	Force HQ	*	Jan-97	Mar-98	Feb-00	45.0	36.3	20.0
Cumbria	Area Facilities	*	Jan-98	Dec-98	Feb-00	8.6	6.6	5.0
Dorset	Dorchester Police Station	*	Sep-97	Jun-98	Feb-00	34.0	24.2	15.0
Cleveland/Durham	Weapons	*	Jan-97	Nov-98	Feb-00	15.0	8.9	9.0
Nottinghamshire	Traffic HQ	*	Aug-97	Jun-98	Jan-00	44.0	25.0	6.0
MPS	Comms & Control		Feb-00			1,264.7	16.5	62.6
Cheshire	Force HQ	*	Oct-98	Jan-00	Apr-00	60.0	41.2	30.0
Hertfordshire	Communications	*	Aug-98	Nov-99	Apr-00	16.0	16.0	8.0
Greater Manchester	Property	*	Jan-99	Oct-99	Apr-00	70.0	47.0	25.0
West Yorkshire	Training Facilities				Apr-00	45.0	30.0	20.0
Sussex	Centralised Custody				Sep-01	80.0	34.0	20.0
Cheshire	Central Custody					30.0	17.0	10.0
Wiltshire	Joint Emergency Control					40.0	35.0	20.0
Kent	Area HQ					35.0	25.0	16.0
Sub-Total:						**1,851.1**	**405.7**	**300.1**
Police - Wales:								
Gwent	Ystrad Mynach Police Station	*	May-97	Mar-98	Nov-99	11.0	6.4	6.0
Dyfed-Powys	Ammanford Police Station		Apr-97	Oct-97	Dec-99	11.0	6.0	6.0
Dyfed-Powys	Lampeter Police Station		May-97	Aug-98	Mar-00	11.0	6.0	6.0
North Wales	Rhyl Police Station, Training Centre & SOCO		Oct-99	Jan-00	Dec-00	17.0	12.0	10.0
Sub-Total:						**50.0**	**30.4**	**28.0**
Sub-Total: Police (England and Wales):						**1,901.1**	**436.1**	**328.1**
Fire - England:								
Greater Manchester	Fire Station	*	1997	1998	Dec-98	6.0	4.8	3.5
North Yorks	Training Centre and Fire Station	*	Mar-98	May-99	Mar-00	8.5	5.3	5.0
Avon Somerset and Gloucester	Training Centre	*	May-99	1999	Feb-00	19.0	9.5	7.0
LFCDA	Vehicles	*		1999	Jan-01	125.0	45.0	60.0
Cornwall	Fire Stations	*	Jan-99		Sep-00		7.0	
Lancashire	Fire Station	*		1999	Oct-00	5.0	3.5	2.5
Sub-Total:						**163.5**	**75.1**	**78.0**
Fire - Wales:								
South Wales	Training Centre	*	1999	1999	Feb-01	16.0	12.4	9.0
Sub-Total:						**16.0**	**12.4**	**9.0**
Sub-Total: Fire (England and Wales):						**179.5**	**87.4**	**87.0**

Table 8.9 Local Authority PFI Projects (continued)

	Description	Pathfinder	OJEC Data	Preferred Bidder Selected	Signed	Expected Contract Value	Expected PFI Credits	Expected Capital Contract Value
Probation - England:								
North Yorks/Humberside					Oct-99	2.0	16.0	1.5
Avon & Somerset						5.8	5.9	4.2
West Midlands						4.5	4.5	1.5
Northumbria						5.6	5.6	1.5
Nottinghamshire						31.0	31.0	1.5
Sub-Total:						**48.9**	**63.0**	**10.2**
Probation - Wales:								
Mid-Glamorgan						13.0	10.0	3.7
Sub-Total:						**13.0**	**10.0**	**3.7**
Sub-Total: Probation (England and Wales):						61.9	73.0	13.9
Sub-Total: (England):						2,063.5	543.8	388.3
Sub-Total: (Wales):						79.0	52.8	40.7
Total: (England and Wales):						2,142.5	596.6	429.0

Source of Data: Home Office Procurement Unit, Home Office
Contact: Donna Sanders 020 7273 235

Table 8.10 Inland Revenue PFI Projects

Project Name	Description	Project Identified	Feasibility Start	Feasibility End	OJEC	ITT	Bids Expected	Bids Received	Deal Closed	Capital Value (£m)	Procurement Team	Winners
Manchester New Office Structure (NOS)	Serviced office accommodation for approximately 1,900 staff	Aug-95	Aug-95	Oct-95	Nov-95	May-96	Jun-96	Jun-96	Contract Signed Mar-97	32	Inland Revenue North West	London & Regional Properties Ltd
St John's House Bootle FICO	Office accommodation on St John's House site for approx. 700 staff	Jan-95	-	-	Aug-95	May-96	Aug-96	Aug-96	Contract signed Dec-98	12	Inland Revenue North West	Mowlem Northern
Glasgow New Office Structure (NOS)	Provision of serviced office accommodation for 290 staff in Glasgow	Jul-95	Oct-95	-	Oct-95	Jul-96	Oct-96	Oct-96	Contract signed Oct-97	10	Inland Revenue Scotland	Gilbert Ash Ltd renamed HBG GA
Edinburgh New Office Structure (NOS)	Office accommodation for 320 staff in a single building in central Edinburgh	Oct-95	Dec-95	-	May-96	Aug-96	Oct-96	Oct-96	Contract signed Nov-97	10	Inland Revenue Scotland	Gilbert Ash Ltd renamed HBG GA
Stockport New Office Structure (NOS)	Serviced office accommodation for approximately 400 staff	Aug-95	Aug-95	Oct-95	Nov-95	May-96	Jun-96	Jun-96	Contract signed Dec-98	6	Inland Revenue North IWest	London Regional Properties Ltd
Newcastle Estate Development (NED)	To provide the DSS with suitable accommodation for it's staff on the Newcastle Estate	Jun-94	Jun-94	Sep-94	Jan-95	Jan-96	Feb-96	Feb-96	Jan-98	164	Contact: NED Contract Team Inland Revenue	Newcastle Estate Partnership (NEP)

Source of Data: Private Finance Unit, Inland Revenue
Contact: Alexandria Barling 020 7438 6859

Table 8.11 Lord Chancellor's Department PFI Projects 1

The Court Service:
The following lists four PFI courthouse projects.

Project Name	Description	Status
Probate Record Centre	Storage of Probate Records (grants of probate) and production of copies on request	Contract signed Aug-99
Sheffield - Family Hearing Centre	A civil justice hearing centre for family related matters	Contract due to be signed Jun-00
Crown Courts in East Anglia	New Crown Court Centres at both Cambridge and Ipswich	Short list selected Feb-00. Contract signing planned for Dec-00
Exeter Combined Court	Service accommodation for civil and criminal cases and the Group Manager's Office	ITN stage. Contract signing planned for Feb-01

Lord Chancellor's Department Headquarters:
The Department has a long term PFI building programme to support local authority delivery of some 50 schemes for new or refurbished magistrates' courts being progressed in phases. The following lists six current projects and their status.

Project Name	Status
Hereford & Worcester	Contract signed Feb-00. Law Courts Partnership Consortia selected as contractor
Manchester	BAFO's received. Preferred bidder being selected. Contract due to be signed Aug-00
Derbyshire	BAFO's due May-00. Contract due to be signed Sep-00
Humberside	Contract signed Mar-00. Mowlems selected as contractor
Merseyside	Issue of OJEC delayed
Avon & Somerset / Bedford	OJEC Issued Feb-00. Contract planned Mar-01

Notes
1. A list of all current and planned LCD PFI projects can be seen on the LCD web pages.
These can be accessed via www.open.gov.uk, find the LCD pages, and look under Courts in the directory.

Source of Data: PFI Unit, Lord Chancellor's Department
Contact: Mark Armstrong 020 7210 8578

Table 8.12 Ministry of Defence PFI Projects

Projects in Procurement:

Project Name & Description	Estimated Capital Cost (£m)	Date of OJEC advertisement (or equivalent)	Status
"C" Vehicles: Provision of "C" vehicles	350		PQQ responses Dec-99. 3 consortia selected for Convergence Phase
Accommodation services in Cyprus: Married quartering, hotel and office accommodation issued for return by 01/02/00. Project approval anticipated Nov-00		Not OJEC. EOI sought prior to approval	Industry day held 02/11/99 and PQQs
Archer Class Support: Partnering/PFI deal for the support of the Archer Class training vessels	Not yet known	26/08/98 MOD Contracts Bulletin	ISOP documents currently being prepared
Astute Trainers: Provision of a training environment for astute class SSNs	50		Preferred bid selection date in Jun 00 Contract signature due Sep-00
AWE Aldermaston: Site Heating Services	18	Bidders' conference held by Hunting Brae on 20/8/97	Project on hold pending the outcome of the management of the GOCO
AWE Management and Operation: GOCO operation and management of the Atomic Weapons Establishments	Not yet known	12/8/98	Tenders received form 3 consortia 31/05/99
Bristol, Bath, Portsmouth and Shrivenham Married Quarters: Provision of serviced accommodation	50-60	5/7/99	ITN return by 17/01/00
Central Britain White Fleet: Provision of vehicles, repair & maintenance services and management information systems for tri-service white fleet vehicles	60	1/7/97	2 companies responded to ITN. Bids evaluated and clarification meetings held
Chelsea Barracks Rationalisation: Rationalisation and redevelopment of Army barracks	35	1/12/96	Final SOR issued 3/12/98. Competition underway to reduce from 4 to 2 bidders
Defence Housing Executive Information Systems: Provision of a range of information systems	80 (whole life)		ITN issued Nov-98. Responses due
DSDA Marketing Partner: The Defence Storage & Distribution Agency is seeking a marketing partner to market surplus and other facilities		28/4/99	
E3D Sentry Aircrew Training Service: Simulators, instructors and maintainers at RAF Waddington	10	6/5/98	Quadrant selected - contract signature due Spring 2000
Electronic Commerce Service (DECS): Secure EM communication between MOD Defence Contractors			Responses to ISOP evaluated and and short-listing done ISDP currently being worked up

Table 8.12 Ministry of Defence PFI Projects (continued)

Project Name & Description	Estimated Capital Cost (£m)	Date of OJEC advertisement (or equivalent)	Status
Expeditionary Camp Infrastructure: Provision of deployable infrastructure	45		Requirement being defined
Field Electrical Power Supplies: Provision of generator sets to support operational electrical requirements in the field	65		2 bids being assessed
Future Cargo Vehicles: Provision of cargo tri-service carrying service including 4 tonne and 8 tonne trunks	350	22/7/98	5 ISOPS returns. Preparing for initial gate
Future Command and Liaison Vehicle: Provision on about 500 lightly armoured wheeled vehicles	370	03/02/99 MOD Contracts Bulletin	Convergence Phase ends Mar-00
Future Fuel Vehicles: Provision of tri-service fuel carrying capability	96	1/1/98	3 ISOP returns - preparing for Initial Gate
Future Strategic Tanker Aircraft: To replace the fleet of air to air refuelling VC10 and Tristar aircraft	2,500	30/12/98 Contracts Bulletin, wider press	ITN due to be issued Aug-00
Future Wheeled Recovery Vehicle: Recovery of battlefield equipment	190		PQQ responses Jan-00
Heavy Equipment Transporters: Provision of heavy equipment transporters to replace existing fleet, and to meet future requirements. Sponsored reserves may also be provided. Treasury significant project	93	1/2/97	BAFO issued to 2 bidders Feb-00
Inmarsat: Provision of replacement Inmarsat terminals and airtime contract for the RN ships. Project scope may be extended to include DCSA requirements for other services	6	8/10/98	ISOP issued to 7 companies on 12/05/99
Joint Rapid Deployment Force (JRDF) Roll-On, Roll-Off Ferries: Provision of 6 RoRo ferries for the MoD	180	01/12/1996 Amended Jan-99	Preferred bidder selection in Spring 2000. Contract signature due Summer 2000
London District White Fleet: Provision of vehicles, repair & maintenance services and Management Information Systems for army white fleet vehicles in London	12	1/8/97	2 responded to ITN. Bids evaluated and clarification meetings held
Marchwood Military Port: Management of the port facilities and possible commercial exploitation	Not yet known	2/12/98	ISOPs issued to 5 groups
Marketing TGDA Ground Training: Potential partnering arrangement to market spare capacity on RAF Training Group Defence Agency ground-based training courses	20		PQQ evaluation

Table 8.12 Ministry of Defence PFI Projects (continued)

Project Name & Description	Estimated Capital Cost (£m)	Date of OJEC advertisement (or equivalent)	Status
MoD-wide Water & Sewerage: Provision of clear water and removal of sewage at all MoD establishments where MoD owns water treatment facilities (140+ sites), or at very large or adjacent sites	100	31/3/99 PIN Jul-98	Scope of project being reviewed. OBC being produced
Northern Calibration Facility: PPP for the use of the Northern Calibration Facility in Faslane	Not yet known	1/6/98	Industry Day held Oct-98
Northwood EDP: Provision of accommodation and support services for JSU Northwood		1/05/98 Planned	Shortlist approved Aug-99. Impact of NCC currently being considered
Porton Down Power Station: Provision of power to Porton Down	4		BAFO negotiation
Records Storage and Management: Reprovision of MoD and other government departments' records archives	15	1/6/98	Shortlisted bidders selected, ITN being prepared
Royal Logistics Corps Training Group: Provision of fully serviced accommodation including DBFO of residential and training facilities	48		Draft ISOPs issued Nov-99
Royal School of Military Engineers: Provision of facilities for Military Engineer Training	49	24/3/99	ISOP responses being evaluated
Royal School of Signals: Partnering for training, particularly for BOWMAN surge and increase in steady state, managed separately from general BOWMAN support	5	16/6/98	ISOP completed, ITN being prepared
Scotland White Fleet: Provision of vehicles, repair & maintenance services and Management Information Systems for tri-service white fleet vehicles in Scotland	35	1/7/97	2 responded to ITN. Bids evaluated and clarification meetings held
Skynet 5: Range of satellite services	750+		Projection Definition study contracts
South West White Fleet: Provision of RN SW white fleet	Not yet known	1/3/98	2 responded to ITN. Bids evaluated and clarification meetings held
Southern Britain White Fleet: Provision of tri-service white fleet in Army 3 & 4 Div areas and RN SE	50	1/11/97	2 responded to ITN. Bids evaluated and clarification meetings held
Surgeon General & Defence Secondary Healthcare: Provision of information systems to support tri-service healthcare provision	Not yet known	19/12/98	ISOP issued, closing date 30/04/99

Table 8.12 Ministry of Defence PFI Projects (continued)

Project Name & Description	Estimated Capital Cost (£m)	Date of OJEC advertisement (or equivalent)	Status
Tri-Service Airfield Support Services: Provision of an airfield support service for all 3 services	350	1/4/98	PQQ evaluation completed. ISOP to be issued
Tri-Service Materials Handling Service: Provision of remaining MHE fleet. Tri-service project	39-50	1/8/97	ITN bids evaluated. Clarification meetings held
Wattisham Married Quarters: Provision of serviced accommodation	50-60	20/9/99	Advert anticipated Sep-99
Projects at Preferred Bidder Stage:			
Colchester Garrison: Redevelopment, rebuilding and refurbishment of Colchester Garrison to provide accommodation and associated services (messing, education, storage workshops etc.)	180		Preferred bidder announced 08/01/99
Defence Animal Centre: Redevelopment of new office and residential accommodation, provision of FM and animal husbandry services and training support	9	1/11/96	Preferred bidder
Main Building Refurbishment: Project to redevelop MOD main building, including temporary decant to other London buildings	175	1/12/96	Preferred bidder announced 08/01/99
Other Possibilities:			
Adjutant General's Corps Training Group: Provision of training facilities and training services at the Schools of Employment Training, Finance & Management, School of Education, Languages, Military Police, Training Support	25		OBC finalised, but awaiting outcome of Defence Training Review and Police Training Study
AFV Training (RAC Bovington): Provision of gunnery and specialised driving synthetic training service	122		OBC being finalised
Aldershot Barracks: Comprehensive redevelopment & refurbishment of Aldershot Garrison to provide modern and efficient accommodation and support services	Not yet known		Scoping Study currently in process, OJEC unlikely this year
Army Training Estate: Opportunities for private sector involvement in provision of land command training ranges services	Not yet known		Feasibility stage
Battle Group Thermal Imaging: Provision of thermal imaging capability to AFVs	140		Examining PPP potential
BOWMAN Training: Provision of conversion training for new generation radio services for the Army	Not yet known		Feasibility study being evaluated

Table 8.12 Ministry of Defence PFI Projects (continued)

Project Name & Description	Estimated Capital Cost (£m)	Date of OJEC advertisement (or equivalent)	Status
Corsham Estate: Accommodation services	Not yet known		Very early consideration underway of project viability
Defence Postal and Courrier Services: Co-location			Scoping and OBC being prepared
Defence Vetting Agency: Provision of accommodation and IT services for Defence vatting Agency	Not yet known		Co-location now to happen at Imphal Barracks, York. Not deemed suitable for partnering
Devonport Support Services: Provision of support services, IT services and fleet accommodation facilities at Devonport Naval Base	Not yet known		Consultation on the feasibility study is being undertaken
DLO Stores Management System: DLO Inventory Management Services. Needs to accommodate different replacement timetables for single service systems			Defining scope of potential requirement
Eurofighter Mission Support Centre: Software modifying and mission data generation facility	55	n/a	PPP agreement made with British
HMS Nelson: Accommodation and support services	Not yet known		Project scope under review
Initial Training Group (Excluding Army Foundation Centre): Rationalisation of the Army Training Regiments and provision of services	Not yet known		OBC being developed
Joint service Parachute Centre: Provision of parachute training centre	Not yet known		Scoping study underway - decision due last quarter 1999
Marine Services Vessels: Long term replacement of marine services vessels for HM Naval Bases	Not yet known		Scoping study underway
Met Office HQ Relocation: Project to relocate and reprovide HQ for Met Office			Pre-feasibility
Multi-engine pilot and rear crew training: Replacement of RAF training capability for pilots and rear crews at present supplied by the Jetstream and Dominie aircraft	Not yet known	By Feb-99 to test market interest for viability	Requirement being drawn up for RAF Board Sep-99
Pay As You Dine: Implementation of Betts Review recommendation to introduce "A La Carte" dining for Armed Forces	25		Development of SOR and procurement strategy

Table 8.12 Ministry of Defence PFI Projects (continued)

Project Name & Description	Estimated Capital Cost (£m)	Date of OJEC advertisement (or equivalent)	Status
Petroleum Supply / Bulk Fuel Installation: Supply depot RAF Akrotiri	17		Local issues and requirement being addressed
RAF Fire Training: RAF Manston fire fighting training facility	Not yet known		Feasibility study
RAF Halton: Redevelopment / refurbishment / provision of new facilities on RAF station	Not yet known		OBC being produced
RAF Infrastructure: Miscellaneous RAF specialist infrastructure	Not yet known		Scoping
RAF Wyton Water Sewerage: Refurbishment of existing facilities to meet regulatory standards	2		Further scoping being undertaken
RMA Sandhurst: Provision of facilities for officer training	Not yet known		Potential ATRA phase 2 project (1999)
Royal School of Artillery, Larkhill: Provision of improved infrastructure and possibly synthetic training and other training facilities	Not yet known		Study underway to be completed by Jun-99
Satellite TV: Provision of satellite TV for recreational purposes in all RN warships	6	Not yet known	SDR commitment to examine feasibility
Submarine Rescue: Provision of submarine rescue service	Not yet known		Requirement being defined
Tidworth Garrison: Redevelopment, rebuilding and refurbishment of Tidworth Garrison to provide accommodation and associated services (messing, storage, workshops, etc.)	150		Scoping project underway in context of SDR movements
Wattisham / Woodbridge Redevelopment: Redevelopment / refurbishment / provision of new facilities required for introduction of AH into service	Not yet known		Feasibility study in progress
Completed Projects:			
Army Foundation College: Provision of a foundation college for the Army		4/6/97	Contract signed 04/02/00
FY 1995/96			
Germany White Fleet: Provision of support vehicles in Germany	52		Completed
Other	30		Completed

Table 8.12 Ministry of Defence PFI Projects (continued)

Project Name & Description	Estimated Capital Cost (£m)	Date of OJEC advertisement (or equivalent)	Status
FY 1996/97			
DHFS: Provision of helicopter training services	118		Completed
HSIS Safety Datasheets: Provision of a safety datasheet system across whole MoD	1		Completed
LISA: Partnering arrangement to provide IS systems within QMG	30		Completed
MHE Vehicles: Provision of material handling equipment	8		Completed
Nelson: Partnering agreement at HMS Nelson, Portsmouth	20		Completed
NRTA: Partnering arrangement for Naval Recruiting and Training Agency	0		Completed
RAF White Fleet: Provision of support vehicles for RAF	35		Completed
TAFMIS: Training management support system	14		Completed
FY 1997/98			
Armed Forces Personnel Administration Agency: Project to implement tri-service pay, personnel & pensions following Betts Review	150		Completed
Armymail: Project to link all Army IS systems	11		Completed
DFTS: Telecommunications services	70		Completed
Hawk Simulator: Provision of simulators to replace existing facilities at RAF Valley	10		Completed
MSHATF: Helicopter training facilities	100		Completed
Tidworth: Provision of water and sewage services to Tidworth Garrison	6		Completed

Table 8.12 Ministry of Defence PFI Projects (continued)

Project Name & Description	Estimated Capital Cost (£m)	Date of OJEC advertisement (or equivalent)	Status
FY 1998/99			
Attack Helicopter Training: Provision of full training package (including 4 simulators) for Attack Helicopter	165		Completed
Joint Service Command and Staff College for the 3 services	68		Completed
Married Quarters at Yeovilton: Accommodation at Yeovilton for aircrew of 2 Lynx Squadrons moved under Project Movit from HMS Osprey at Portland	8		Completed
RAF Basic Flying Training (Bulldog Replacement): Provision of flying training and support services for UAS and AEF tasks	30	Bidders conference held 20/5/96	Contract signed 30/01/99
RAF Cosford/Shawbury - Married quarters: Provision of accommodation	13	1/6/97	Signed 30/03/99
RAF Flyingdales (Power Station): Provision of guaranteed power supply to the missile early warning system	7	Bidders' conference held 09/01/96	Completed - contract signed Dec
RAF Lossiemouth: Redevelopment and reprovision of 279 married quarters	24		Completed
RAF Lyneham Sewage Treatment: Refurbishment of existing facilities to meet regulatory standards. Population served 7000	5		Completed
RAF Mail: Informal messaging services for RAF	12		Completed
FY 1999/00			
Central Scotland Married Quarters: Provision of accommodation	13		Signed 18/08/99
Fire Fighting Training Unit: Provision of fighting training facilities for Naval Recruiting & Training Agency	35		Contract signed 01/04/99
Tornado GR4 Simulator	65		Contract awarded on 30/06/99

Source of Data: Private Finance Unit, Ministry of Defence
Contact: Angela Barratt 0171 218 5951

Table 8.13 Ministry of Defence Non-PPI Construction Expenditure [1,2]

	1995/96	Outturn 1996/97	1997/98	£ Million Estimates 1998/99
Works, Buildings and Land	1,792	1,586	998	1,527
UK Construction	1,288	1,063	819 (p)	1,254 (p)
Overseas Construction	189	93	69 (p)	105 (p)

Notes

1. The table heading claims that this data excludes all PPI construction expenditure, this is only partiallytrue. PPI contracts are relatively new to the MoD and, as such, represent only a small proportion of MoDcontracts. Current MoD accounting practices do not give visibility of PPI contracts. In effect this data will contain some PPI construction expenditure, although relatively insignificant, but may increase significantlyin future years.

2. Outturn figures for 1998/99 will be published in the issue covering 1990-2000.

Source of Data: Defence Analytical Services Agency, Ministry of Defence
Contact: Eric Crane 020 7218 0781

Table 8.14 Scottish Executive PFI Projects

PFI Projects within the responsibility of the Secretary of State for Scotland[1].
The following table categorises projects according to their stage of development and include those which have been identified as potential PFI projects.

Status	Sector/ Procuring Agency/Project Name	Estimated Capital Value (£m)	OBC Submittesd	OJEC Actual (Expected)	ITT Actual (Expected)	Financial Close Actual (Expected)
Completed:						
	Education					
	Falkirk College: Stirling Further Education Centre	3.6		Oct-95	Feb-96	Dec-96
	Health					
	Grampian Health Board: Kincardineshire Healthcare	3.8		May-95	Jan-96	Jan-97
	Lothian & Forth Health Boards: Clinical Waste Disposal	4.5			Mar-94	Oct-94
	Yorkhill Hospitals NHS Trust Glasgow: Health Information System	2.5	Mar-95	Oct-95	Jun-96	Feb-97
	Edinburgh Healthcare NHS Trust: Ferryfield House	2.5	n/k	Jun-95	Jul-95	Feb-96
	Various: Analytical facilities, care for elderly and mentally ill, energy management	31.3			Various	Various
	Perth & Kinross Healthcare NHS Trust: Health Information System	2.3	Apr-95	Apr-95	Mar-96	Jan-97
	Law Hospital NHS Trust: Health Information System	2.5	n/k	n/k	Oct-95	Mar-96
	Northern NHS Trusts / Health Boards: Clinical Waste Disposal	6.0			Jun-94	Oct-94
	Victoria Infirmary NHS Trust: Mearnskirk Hospital Geriatric Beds	2.4	n/k	n/k	Sep-96	Jul-97
	Various: Equipment, IT, Energy Management	11.0		Various	Various	Various
	Edinburgh Healthcare NHS Trust: 60 care of elderly beds	2.0				Jan-99
	Transport					
	DD: Skye Bridge	23.6			Mar-90	Jul-92
	DD: M6 DBFO	103.0		Jun-95	Dec-95	Dec-96
	Highlands and Islands Airports Limited: Inverness Airport Terminal	9.5		Sep-95	Nov-95	Feb-98
	Prisons					
	Scottish Prison Service: Kilmarnock Prison	32.0		Sep-96	Nov-96	Nov-97
Total Completed		**242.5**				
Signed:						
	Education					
	James Watt College: North Ayrshire College, Kilwinning	8.6			Feb-98	Mar-99
	West Lothian College: Livingston Further Education Centre	15.0		Oct-95	Jun-96	Dec-99
	Health					
	Hairmyres and Stonehouse NHS Trust: New Hairmyres Hospital	67.5	Mar-94	Dec-94	Aug-95	Apr-98
	Ayrshire and Arran Community Healthcare NHS Trust: Cumnock Community Hospital	8.6	Sep-95	Oct-95	Dec-95	Mar-99
	RIE NHS Trust: New Royal Infirmary of Edinburgh	180.0	Nov-94	Jan-95	Jan-96	Aug-98
	Law Hospital NHS Trust: Wisham General Hospital	100.0	Mar-94	Jul-95	Nov-95	Jun-98
	Dundee Teaching Hospitals NHS Trust: Energy Management System	2.8		Mar-95		Sep-97

Table 8.14 Scottish Executive PFI Projects (continued)

PFI Projects within the responsibility of the Secretary of State for Scotland[1].
The following table categorises projects according to their stage of development and include those which have been identified as potential PFI projects.

Status	Sector/ Procuring Agency/Project Name	Estimated Capital Value (£m)	OBC Submittesd	OJEC Actual (Expected)	ITT Actual (Expected)	Financial Close Actual (Expected)
	Dundee Teaching Hospitals NHS Trust: Multi-storey Car Park / Car Parking Facilities	3.3		Sep-96		Dec-97
	Tayside Primary Care NHS Trust: Dundee Ninewells Psychiatric Services	10.0	Jun-97	May-97	Feb-98	Jul-99
	Highland Communities NHS Trust: Inverness Psychiatric Unit	16.5	Jun-97	Aug-97	Dec-97	Mar-99
	Southern General Hospital NHS Trust: Southern General HISS	2.4				Jun-99
	Argyll & Clyde Acute Hospitals NHS Trust: Larkfield Geriatric Assessment Facility	10.0	Nov-95	Feb-96	May-96	May-99
	Royal Infirmary of Edinburgh NHS Trust: Hospital Information Support System	12.0	Jun-97	Oct-97	Feb-98	Nov-99
	South Glasgow University Hospitals NHS Trust: SGH Geriatric Medicine and Assessment Facility	11.0	Jul-95	Aug-95	Feb-96	Jul-99
	West Lothian Healthcare NHS Trust: Care of elderly services	2.3				Mar-99
	Local Authorities (Non-Housing)					
	Dundee City Council: Baldovie Waste to Energy Plant	43.0		Dec-93	Feb-94	Oct-97
	Highland Council: IS/IT Services	13.0			Sep-97	Jul-98
	Moray Council: Integrated Education Management Service	5.6	Sep-97	Oct-97	Feb-98	Nov-98
	Falkirk Council: Falkirk Schools	65.0		Apr-97	Aug-97	Aug-98
	Perth and Kinross Council: Office Accommodation	15.0		Sep-97	Dec-97	Mar-99
	Water & Sewerage					
	North of Scotland Water Authority: Inverness Main Drainage / Fort William Sewage Treatment	45.0		Mar-95	Jul-95	Dec-96
	North of Scotland Water Authority: Tay Projects - Dundee, Carnoustie and Arbroath Waste Water Treatment	100.0		Mar-97	Jul-97	Nov-99
	North of Scotland Water Authority: Aberdeen, Stonehaven, Fraserburgh and Peterhead Sewage & Sludge Treatment	80.0		Nov-97	May-98	Dec-99
	West of Scotland Water Authority: Dalmuir Sewage Treatment - Provision of Secondary Treatment	50.0		May-97	Dec-97	May-99
	East of Scotland Water Authority: Almond Valley and Seafield Sewage Scheme	140.0		Jun-96	Nov-96	Mar-99
	East of Scotland Water Authority: Esk Valley Sewage Scheme	20.0		Mar-97	Oct-97	Nov-99
	IT					
	Scottish Children's Reporter Administration: Integrated Information System	3.0	Mar-97	Jul-97	Sep-97	Jul-98
	Law and Order (LA (Non-HRA))					
	Strathclyde Police: Police Force Training centre, East Kilbride	17.0		Jan-97	Apr-97	Dec-99
Total Signed		**1,046.6**				

Table 8.14 Scottish Executive PFI Projects (continued)

PFI Projects within the responsibility of the Secretary of State for Scotland[1].
The following table categorises projects according to their stage of development and include those which have been identified as potential PFI projects.

Status	Sector/ Procuring Agency/Project Name	Estimated Capital Value (£m)	OBC Submittesd	OJEC Actual (Expected)	ITT Actual (Expected)	Financial Close Actual (Expected)
Tenders Invited/Under Negotiation[2]:						
	Water & Sewerage					
	West of Scotland Water Authority: Daldowie / Shieldhall Treatment Centres	65.0		May-96/ Jan-97	Nov-96/ May-97	Mar-00
	West of Scotland Water Authority: Meadowhead (Irvine) Ayr, Stevenston and Inverclyde Sewage Treatment	57.0		Jan-98	Apr-98	Dec-99
	East of Scotland Water Authority: Levenmouth Purification Scheme	47.0		Mar-97	Oct-97	Apr-00
	North of Scotland Water Authority: Moray Coast and Montrose Wastewater Project	50.0		Dec-98	Oct-99	Mar-00
	Transport					
	City of Edinburgh Council: Rapid Transport System (CERT)	49.0		Jun-97	Jan-98	Spring-00
	Health					
	Dumfries and Galloway Acute and Maternity Hospitals NHS Trust: Day Surgery and Maternity Unit	11.0	Sep-95	Dec-95	Mar-96	Jan-00
	East & Midlothian NHS Trust: Midlothian Community Hospital	8.1	Jan-98	Aug-98	Apr-99	May-00
	Local Authorities (Non-Housing)					
	Glasgow Council: Glasgow Schools Project	192.0	Jun-98	Jul-98	Nov-98	Jan-00
	Stirling Council: Balfron School	16.5	Sep-97	Apr-98	Dec-98	Jan-00
	East Renfrewshire Council: Mearns Primary and St. Ninian's High School	12.5	Jun-98	Jan-98	Apr-98	Jan-00
	Aberdeenshire Council: Aberdeenshire Schools Project	14.3	Jun-98	Mar-98	Nov-99	Sep-00
	Argyll & Bute Council: Argyll & Bute Waste Management Project	21.7	Jun-98	Aug-99	Nov-99	Aug-00
	Fife Council: Fife Schools Projects	32.0	Jun-98	Sep-99	Oct-99	Dec-00
	City of Edinburgh Council: Information & Communications Technology Services	30.0		Mar-99	Sep-99	Apr-00
Total Invited/Under Negotiation[2]		**606.1**				
Advertised[1]:						
	Health					
	Glasgow Royal Infirmary University NHS Trust: Car Parking Facilities	6.6	Apr-96	May-97		Dec-99
	Lothian University Hospitals NHS Trust: Finance procurement and servicing of equipment			Nov-99	Jan-00	Sep-00
	Local Authorities (Non-Housing)					
	Dumfries & Galloway Council: Dumfries & Galloway Waste Management / Recycling Project	17.5	Jun-98	Jul-99/ Aug-99	Dec-99	Autumn-00
	Edinburgh Council: Edinburgh Schools Project	80.0	Jun-98	May-99/ Oct-99	Apr-00/ May-00	Mar-01/ Apr-01
	Highland Council - Highland Schools Project	13.7	Jun-98	Nov-99		Oct-00
Total Advertised		**117.8**				

Table 8.14 Scottish Executive PFI Projects

PFI Projects within the responsibility of the Secretary of State for Scotland[1].
The following table categorises projects according to their stage of development and include those which have been identified as potential PFI projects.

Status	Sector/ Procuring Agency/Project Name	Estimated Capital Value (£m)	OBC Submittesd	OJEC Actual (Expected)	ITT Actual (Expected)	Financial Close Actual (Expected)
Potential[2,3]:						
	Transport					
	Strathclyde PTE:					
	Larkhall-Milngavie Rail Route	30.0	Nov-98			Mar-00
	Health					
	Greater Glasgow Community and Mental Health Services NHS Trust:					
	Secure Care Centre	12.5	Dec-97			
	Argyll & Bute NHS Trust:					
	Lochgilphead Hospital	5.0	Spring-99			
	Tayside Health Board and Angus NHS Trust: Replacement of Forfar Infirmary and Whitehalls Hospital	11.0	Jan-98			Mar-00
	Fife Primary Care NHS Trust:					
	North East Fife Health Services	15.0				Mar-00
	Borders Community Health Services NHS Trust: Hawick Cottage Hospital	3.0	Jun-98			Mar-00
	Stobhill NHS Trust:					
	Ambulatory Care Centre, Stobhill	40.0	Spring-99			
	East & Midlothian NHS Trust:					
	NE Edinburgh Continuing Care	3.0				
	East & Midlothian NHS Trust: Acute & Continuing Psychiatric Services	6.0				
	Glasgow Royal Infirmary University NHS Trust:					
	Provision of catering system	1.4				
	Local Authorities (Non-Housing)					
	Angus Council: A92 Upgrading	36.0	Jan-98		Dec-99	Jan-01
	East Renfrewshire Council: Glasgow Southern Orbital Road	37.0	Jun-97			Mar-01
	West Lothian Council: West Lothian Schools Project	27.8	Jun-98	Nov-99	Apr-00	May-01
	Midlothian Council: Midlothian Schools Project	33.0				Dec-00
	Executive Agency					
	National Archives of Scotland: Thomas Thompson House Phase 2	12.0	Mar-00/ Apr-00	2000	2000/01	2000/01
Total Potential		**272.7**				
Grand Total Capital Value		**2,285.6**				

Notes
OBC - Outline Business Case
FBC - Full Business Case
OJEC - Official Journal of EC
PIN - Prior Indicative Notice
1. Public sector comparator.
2. Project value within local delegated authority limits.
3. Estimated in a year.
4. Original capital value £14.1 million. £29.5 million represents the estimated pattern of investment to financial year ending 31 March 2003. This figure is exclusive of any costs (as yet unknown) for the Welsh Assembly.
5. Expected cost of basic service over contract period.

Source of Data: Private Finance Unit, The Scottish Office
Contact: Fiona Mclellan 0131 244 7499

Table 8.15 National Assembly for Wales PFI Projects

Project Name & Description	Capital Cost (Public) £m[1]	Contract Value £m	Project App. Date	OJEC Date	Tender Invite Date	Final WO App. Date	Estimated Start/Comp. Date	Current Position
Local Authority Pathfinder Projects:								
Wrexham: Waste Management				Apr-97			2000	Preferred bidder
Ceredigion: New School				Dec-97			Aug-99/Sep-00	Project signed Sep-99
Pembrokeshire: New Nursery School, Offices							Apr-00	Preferred bidder
Denbighshire: County Council Office Accommodation	11.0			Jul-98		May-00		Prefered bidder
Newport: Southern Distributor Road (SDR)				Mar-99			2002	ITN docs issued
Ysgol Gyfun Cwm Rhymni: Secondary Schools				Feb-99			Sep-00	3 bids currently under evaluation
Bleanau Gwent - Llannilleth Primary School							Sep-02	Preparation of OBC
Bridgend - Maesteg Schools							Sep-02	Prep. of OBC/App. of Advisor
Caerphilly - Sirhowy Enterprise Way							Apr-03	Preparation of OBC
Conway - 3 Secondary School Bundle							Sep-01	ITN to shortlist bidders
Newport - Durham Road School							Dec-02	Prep. of OBC/App. of Advisor
Rhondda Cynon Taff - Rhydfelen & Garth Olwg Schools							Dec-02	Prep. of OBC/App. of Advisor
Projects Approved and Awarded:								
UHW Cardiff: Phase 1 - Car Park[2]	1.0						Completed	Completed Jan-95
UHW Cardiff: Phase 2 - Multi-storey car park and access road	6.5		Dec-94	Apr-95	Feb-95	Dec-95	Completed Jul-96/Feb-98	Contract awarded to APCOA & and Impregilo Contract signed Jul-96. Car park opened Feb-98
UHW Cardiff: Main entrance and retail concourse[2]	2.0						Opened Jun-98	Contract awarded to Rock Eagle Developers
Withybush Hospital: Contract Energy Management[2]	0.3						Completed	Contract awarded to Nedalo UK Ltd

Table 8.15 National Assembly for Wales PFI Projects (continued)

Project Name & Description	Capital Cost (Public) £m[1]	Contract Value £m	Project App. Date	OJEC Date	Tender Invite Date	Final WO App. Date	Estimated Start/Comp. Date	Current Position
Llandough Hospital: Contract Energy Management[2]	0.6	0.4 [3]				Oct-95	Completed	Contract awarded to AHS Emstar
OSIRIS: New office automation system for the WO	33.5 [4]	20 [5]	Jun-96	Sep-94		Jun-96	Jun-96/Nov-97	Contract awarded to Siemens Business Services
Llandough Hospital: Operating Theatres	0.9							
Llandough Hospital: Staff Residences	0.9							
Chepstow Hospital: Neighbourhood Health Unit	6.0	10.0	Nov-95	Nov-95	Feb-96	Dec-96	Sep-98 to Jan-00	Construction underway
Prince Philip Hospital: Contract Energy Management	0.3							
Glan Clwyd Hospital: Contract Energy Management	0.3							
Singleton Hospital: Contract Energy Management	0.7							
Nevill Hall Hospital: Endoscopy and Day Surgery Unit	2.6	2.8	May-95	Jun-95	Sep-95	Mar-97	Aug-98 to Dec-99	
Royal Glamorgan Hospital: Staff Residences	3.7	3.7	Mar-96	Jun-96		Apr-98	Dec-98 to Aug-99	
UHW Cardiff: Sterile Services	1.5	2.2		Sep-96	Mar-97	May-98	Jun-98 to Mar-99	
Royal Gwent Hospital: Contract Energy Management	1.8	1.8				Oct-98		
Projects Under Consideration:								
Baglan Hospital: 250-bed general hospital to serve Neath and Port Talbot	53.0		Dec-96	Feb-97			Nov-98/Dec-02	Preferred SSI chosen Jan -99
Western Cardiff Hospital: Neighbourhood Hospital	16.5		Oct-97	Dec-97		Nov-98		Preferred bidder IMC chosen Jan-99
Bute Avenue: 1.5km transport corridor from the city centre to the inner harbour, plus housing and office development	45.0			Jan-96	Aug-96	Jul-99	Jul-99/Feb-01	Full works commenced Jul-99
A55: Dualling of up to 31kms across Anglesey in North Wales	132.0			Apr-97	Oct-97		Dec-98/Dec-28	Awarded 16 Dec-98
Further and Higher Education Projects								Various projects under early consideration

Notes
OBC - Outline Business Case
FBC - Full Business Case
OJEC - Official Journal of EC
PIN - Prior Indicative Notice
1. Public sector comparator.
2. Project value within local delegated authority limits.
3. Estimated in a year.
4. Original capital value £14.1 million. £29.5 million represents the estimated pattern of investment to financial year ending 31 March 2003. This figure is exclusive of any costs (as yet unknown) for the Welsh Assembly.
5. Expected cost of basic service over contract period.

Source of Data: The National Assembly for Wales
Contact: Lisa Thomas 029 2082 5213

NORTHERN IRELAND EXECUTIVE PFI PLANS

In Northern Ireland some seven projects with a capital value of £31 million were awarded during 1999, bringing the total to date of 17 projects with a total capital value of £53 million. Major works included a sewage treatment works, new school premises, rationalisation of further education teaching accommodation and computerisation of the Land Registry. The latter has been shortlisted for the prestigious PFI 2000 awards.

As well as the £53 million signed off, around £560 million of the other projects are at various stages, of which circa £70 million is construction related. Departments are considering other projects yet to be announced, which would offer substantial opportunities for the construction industry. These are mostly in education including school maintenance, but other areas including the health sector are now also beginning to report some more possible projects.

An Internet wed site providing information on PFI in Northern Ireland is planned for early summer 2000 covering project data and departmental contracts. This will hopefully promote further contact with the private sector, inviting interested parties to identify future opportunities for partnerships with Northern Ireland Departments.

Source: Central Expenditure Division, Department of Finance and Personnel, Northern Ireland Executive
Contact: James McAleer 01247 279279

Table 8.16 Northern Ireland Executive PFI Projects			
Project Name & Description	Capital Cost (public) £m	OJEC Date	Contract Awarded
Completed:			
Holywell Hospital - Contract Energy Scheme	0.2	Feb-95	May-9
Craigavon Area Hospital - Contract Energy Scheme	0.5	Dec-95	Feb-9
Royal Group of Hospitals - Car Parking Facilities	2.0	Jan-96	Oct-9
Belfast City Hospital - Renal Unit	3.3	Apr-95	Dec-9
Antrim Hospital - Renal Unit	2.2	Nov-97	Apr-99
Daisy Hill Hospital - Renal Unit	0.3	Oct-96	May-98
Royal Group of Hospitals - ATICS Equipment	1.8	Feb-96	Jan-98
RGH/BCH Link Laboratory - Laboratory IT System	1.0	n/a	Apr-99
HPSS - Development of Regional Clinical Waste Facility	3.5	May-96	Aug-98
Land Registers IT System - Computerisation	4.0	Nov-97	Jul-99
Planning Service IT System	0.6	Jun-96	Dec-97
Training and Employment Agency IS/IT System	6.5	May-96	May-98
Education and Library Boards Accrual Accounting/Payroll System	3.3	Sep-97	Dec-98
North West Institute of Further & Higher Education - Rationalization of Teaching Accommodation	7.0	May-96	Aug-99
Drumglass High School - New School Premises	5.0	Apr-97	Jun-99
Kinnegar Sewage Treatment Works	11.0	Nov-95	Apr-99
Hydro Electric Scheme	0.9	Jan-97	Jun-99

Table 8.16 Northern Ireland Executive PFI Projects (continued)

Project Name & Description	Capital Cost (public) £m	OJEC Date	Contract Awarded	Current Position
Projects Under Consideration:				
Roads and Transport				
Bus Station and Library Facilities	3.0	Dec-97		BAFO received Jul-98
Education				
Belfast Institute of Further & Higher Education -				
Rationalization of Teaching Accommodation	10.0	Feb-97		BAFO received Sep-98
St. Genevieve's High School - New School Premises	7.0	Jun-97		Preferred bidder identified Feb-99
Wellington College - New School Premises	5.0	Apr-97		Preferred bidder identified Dec-98
Balmoral High School - New School Premises	4.0	Jun-97		Preferred bidder identified Dec-98
Class Room 2000 - Integrated Technical Communications	250.0	Nov-98		Invitation to negotiate issued May-99
Health				
Belfast City Hospital - Car Parking Facilities	2.5	Aug-96		BAFO received Nov-97
Belfast City Hospital - Cancer Treatment	28.0	Jul-99		
HPSS - Unique Patient & Client Identifier System	2.9	Nov-98		Invitation to negotiate issued Jun-99
IT/IS Projects				
DVTA Vehicle Testing Project	11.0	Jun-97		BAFO received Mar-98
Potential:				
Roads and Transport				
NITHC Bus Replacement - Ulsterbus/Citybus Fleet	150.0			Review of potential for PPP
Replacement				completed end of 1999,
Railways Class 80 Rolling Stock Replacement	80.0			still being evaluated
Health				
Altnagelvin Hospital Services Centre - New Pathology				
Pharmacy and Changing Rooms	10.0			Still under consideration
NI Ambulance Service - Ambulances	2.5			Still under consideration

Notes

n/a = not available

BAFO = Best and Final Offer

Source of Data: Central Expenditure Division, Department of Finance and Personnel, Northern Ireland Executive

Contact: James McAleer 01247 279279

CHAPTER 9

Local Authority Expenditure

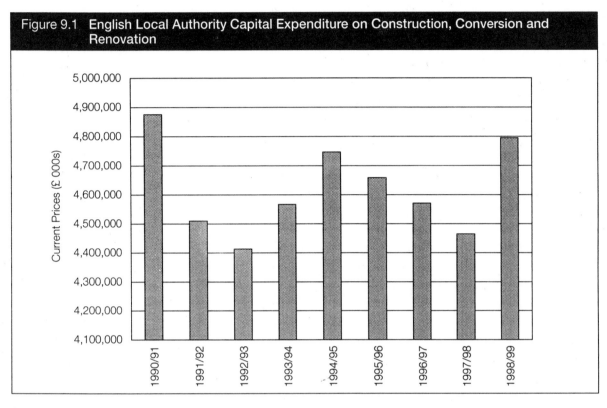

Figure 9.1 English Local Authority Capital Expenditure on Construction, Conversion and Renovation

Source of Data: Table 9.1

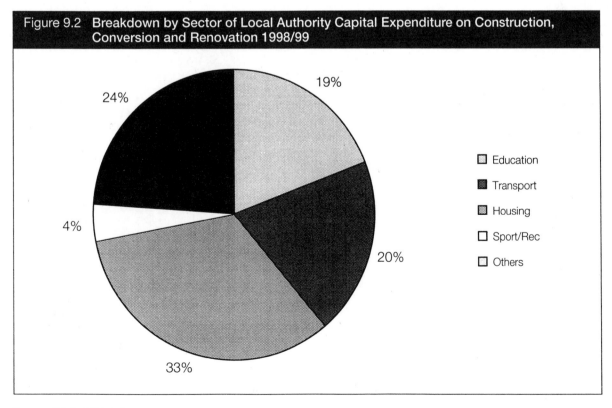

Figure 9.2 Breakdown by Sector of Local Authority Capital Expenditure on Construction, Conversion and Renovation 1998/99

Source of Data: Table 9.1

Table 9.1 English Local Authority Capital Expenditure on Construction Conversion and Renovation

Current Prices £000s

DETR Region[1]	Education	Transport	Housing	Sport/Rec	Others	Total
Financial Year 1990/91						
London	145,502	99,565	438,755	81,363	142,303	907,488
South East	107,123	145,752	282,951	62,677	139,487	737,990
South West	72,308	68,536	168,419	28,665	82,547	420,475
Eastern	71,829	86,947	242,995	57,120	104,982	563,873
West Midlands	67,463	59,047	201,207	24,314	142,684	494,715
East Midlands	42,339	37,171	139,379	26,416	92,942	338,247
North West	86,814	151,342	240,933	21,588	149,885	650,562
Yorkshire & Humberside	46,373	85,761	209,908	29,732	120,317	492,091
Northern	26,484	38,827	120,370	10,560	74,144	270,385
ENGLAND	666,235	772,948	2,044,917	342,435	1,049,291	4,875,826
Financial Year 1991/92						
London	126,721	80,687	464,189	18,649	154,379	844,625
South East	100,276	177,243	188,797	56,396	148,822	671,534
South West	83,193	77,966	126,969	19,841	68,241	376,210
Eastern	79,048	89,688	151,703	31,092	103,007	454,538
West Midlands	61,943	84,122	180,181	14,317	119,343	459,906
East Midlands	54,314	61,089	108,888	21,624	92,414	338,329
North West	94,795	156,537	227,512	19,856	111,338	610,038
Yorkshire & Humberside	61,211	119,232	184,658	14,760	109,005	488,866
Northern	30,047	50,163	123,179	7,137	55,281	265,807
ENGLAND	691,548	896,727	1,756,076	203,672	961,830	4,509,853
Financial Year 1992/93						
London	106,184	93,699	432,904	11,759	128,652	773,198
South East	107,076	197,638	139,275	35,637	128,946	608,572
South West	85,625	91,266	101,440	11,250	70,829	360,410
Eastern	79,924	140,553	122,869	11,139	80,995	435,480
West Midlands	54,790	103,288	184,975	10,534	103,077	456,664
East Midlands	55,024	72,934	113,957	16,436	82,798	341,149
North West	99,010	148,650	254,512	11,877	105,101	619,150
Yorkshire & Humberside	70,873	185,685	180,316	11,804	108,733	557,411
Northern	29,412	49,376	119,777	6,797	56,067	261,429
ENGLAND	687,918	1,083,089	1,650,025	127,233	865,198	4,413,463
Financial Year 1993/94						
London	101,267	128,237	462,988	9,508	131,058	833,058
South East	123,230	206,218	141,286	26,711	103,206	600,651
South West	74,456	112,208	116,347	11,684	72,540	387,235
Eastern	58,612	112,199	119,455	14,883	62,322	367,471
West Midlands	52,676	153,107	206,314	13,738	101,975	527,810
East Midlands	44,771	73,611	136,607	17,040	88,415	360,444
North West	81,288	180,402	268,589	12,797	114,020	657,096
Yorkshire & Humberside	73,791	204,256	195,634	6,723	92,334	572,738
Northern	26,912	58,915	113,665	7,509	53,234	260,235
ENGLAND	637,003	1,229,153	1,760,885	120,593	819,104	4,566,738
Financial Year 1994/95						
London	115,032	132,357	322,801	13,409	250,393	833,992
South East	139,311	131,092	129,251	42,910	116,276	558,840
South West	64,365	100,404	96,410	12,200	111,088	384,467
Eastern	72,707	94,328	119,927	20,299	88,403	395,664
West Midlands	55,019	140,273	127,827	17,542	231,549	572,210
East Midlands	45,018	58,560	98,671	24,584	144,833	371,666
North West	81,856	177,935	185,013	14,915	269,588	729,307
Yorkshire & Humberside	55,019	170,676	150,255	12,813	191,906	580,669
Northern	27,691	51,649	101,350	7,340	132,064	320,094
ENGLAND	656,018	1,057,274	1,331,505	166,012	1,536,100	4,746,909

Table 9.1 English Local Authority Capital Expenditure on Construction Conversion and Renovation (continued)

Current Prices   £000s

DETR Region[1]	Education	Transport	Housing	Sport/Rec	Others	Total
Financial Year 1995/96						
London	123,033	149,623	295,198	11,867	204,177	783,898
South East	168,864	273,893	179,477	65,075	215,504	902,813
South West	74,555	102,019	86,242	16,062	104,574	383,452
Eastern	29,441	45,991	31,737	5,830	41,098	154,097
West Midlands	58,253	152,774	142,137	16,825	139,588	509,577
East Midlands	35,117	67,468	82,338	13,780	137,720	336,423
North West	85,198	173,772	147,542	10,188	266,311	683,011
Yorkshire & Humberside	58,288	159,857	137,162	16,891	195,377	567,575
Northern	35,582	55,964	84,871	7,283	153,560	337,260
ENGLAND	668,331	1,181,361	1,186,704	163,801	1,457,909	4,658,106
Financial Year 1996/97[2]						
London	125,819	134,858	305,892	17,902	284,567	869,038
South East	139,779	238,973	105,666	54,273	158,947	697,638
South West	71,532	84,565	70,411	17,868	87,388	331,764
East	71,908	75,868	88,966	27,098	95,518	359,358
West Midlands	67,410	174,946	123,273	13,506	170,383	549,518
East Midlands	35,930	58,185	77,472	13,543	122,637	307,767
North West	74,063	119,334	119,348	19,783	188,852	521,380
Merseyside	15,796	38,428	26,945	5,323	71,911	158,403
Yorkshire	66,114	121,616	109,561	9,716	198,696	505,703
North East	27,488	44,319	56,305	5,575	136,112	269,799
ENGLAND	695,839	1,091,092	1,083,839	184,587	1,515,011	4,570,368
Financial Year 1997/98[3]						
London	138,240	121,945	419,994	20,560	154,448	855,187
South East	138,534	213,619	118,328	37,152	175,931	683,564
South West	89,435	84,049	85,241	13,456	76,196	348,377
East	77,977	63,765	100,874	26,369	87,881	356,866
West Midlands	67,659	159,174	132,745	19,201	132,887	511,666
East Midlands	48,509	54,045	90,683	13,788	100,644	307,669
North West	78,973	100,347	173,666	25,037	127,214	505,237
Merseyside	22,499	23,470	54,405	3,409	40,645	144,428
Yorkshire	70,812	110,046	165,264	9,364	112,097	467,583
North East	42,390	55,302	89,277	11,624	85,787	284,380
ENGLAND	775,028	985,762	1,430,477	179,960	1,093,730	4,464,957
Financial Year 1998/99						
London	183,462	131,788	497,827	21,074	147,520	981,671
South East	151,671	194,572	125,537	32,688	186,248	690,716
South West	95,177	77,422	103,742	13,127	88,676	378,144
East	83,211	61,304	116,078	34,641	78,150	373,384
West Midlands	88,933	121,182	139,629	19,464	170,801	540,009
East Midlands	68,781	64,447	94,480	19,903	101,112	348,723
North West	85,288	95,864	160,312	34,698	128,570	504,732
Merseyside	27,929	32,588	62,192	4,189	43,435	170,333
Yorkshire	80,240	109,759	172,863	9,684	102,091	474,637
North East	60,080	46,382	112,400	13,769	100,052	332,683
ENGLAND	924,772	935,308	1,585,060	203,237	1,146,655	4,795,032

Notes

1. The DETR Regions are defined in Appendix 1.

2. A change in regional classification from 1996/97 means that data for 1996/97 onwards are not directly comparable with data for previous years. See Appendix 1 for Government Office Regions.

3. Figures for financial years up to and including 1996/97 excluded regenerartion expenditure. For 1997/98 onwards regeneration expenditure is included.

Source of Data: LGF Statistics, Department of the Environment, Transport and the Regions

Contact: Mervion Kirwood 020 7944 4074

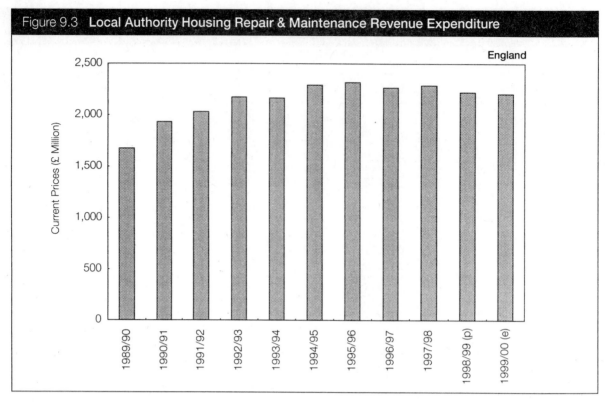

Figure 9.3 Local Authority Housing Repair & Maintenance Revenue Expenditure

Source of Data: Table 9.2

Table 9.2 Local Authority Housing Repair & Maintenance Revenue Expenditure										
Current Prices										**£ Million**
DETR Region[1]	London	South East	South West	Eastern	West Mids	East Mids	North West	Yorks. & Humb.	Northern	England
1989/90	315	213	133	190	210	115	239	150	111	1,676
1990/91	413	205	149	201	245	137	268	189	127	1,933
1991/92	418	222	155	207	261	141	289	201	137	2,031
1992/93	446	217	168	215	291	148	321	220	148	2,175
1993/94	449	207	159	198	293	153	327	222	157	2,165
1994/95	498	210	157	208	309	160	343	238	170	2,293
1995/96	529	212	150	189	296	165	355	243	181	2,320
1996/97	494	188	140	186	309	170	345	246	189	2,267
1997/98	504	192	137	192	294	168	362	252	189	2,290
1998/99 (p)	488	190	125	185	290	164	349	251	181	2,223
1999/00 (e)	508	195	113	189	268	163	347	246	178	2,208

Notes

p = provisional

e = estimated

1. The DETR Regions are defined in Appendix 1.

Source of Data: LAH Division, Department of the Environment, Transport and the Regions
Contact: Dan Varey 020 7944 3598

Figure 9.4 Local Authorities and New Towns: Value of Construction Work Done by Type of Work

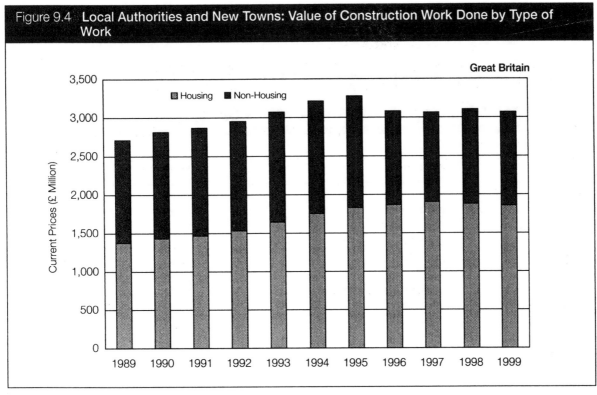

Source of Data: Table 9.3

Table 9.3 Local Authority and New Towns: Value of Construction Work Done by Type of Work

Current Prices (£ Million)					Great Britain
	New Work		Repair & Maintenance		
Year	Housing	Non-Housing	Housing	Non-Housing	All Work
1989	19	203	1,356	1,134	2,711
1990	23	195	1,409	1,186	2,813
1991	18	172	1,449	1,229	2,868
1992	13	187	1,521	1,232	2,952
1993	9	179	1,637	1,245	3,071
1994	9	216	1,745	1,241	3,211
1995	6	208	1,823	1,242	3,279
1996	9	142	1,857	1,071	3,080
1997	12	130	1,891	1,031	3,064
1998	13	147	1,866	1,077	3,102
1999	12	147	1,843	1,065	3,066

Source of Data: Construction Market Intelligence, Department of the Environment, Transport and the Regions
Contact: Neville Price 020 7944 5587

CHAPTER 10

National Lottery Funded Projects

Table 10.1 National Lottery Projects – Fund Application[1]

Town	Development	Value (£m)	June 2000 Stage
Avon			
Bristol	Sports Centre	0.90	Pre-Tender
Bedfordshire			
Bedford	Sports Ground/Pavilion	3.25	Pre-Tender
Bedford	Sports Fields	0.32	Pre-Tender
Berkshire			
Maidenhead	Athletics Track Refurbishment	0.15	Pre-Tender
Maidenhead	Leisure Centre (Extension)	0.50	Contract Awarded
Maidenhead	Sports Pitch	0.15	Pre-Tender
Maidenhead	Infrastructure Works	0.15	Pre-Tender
Reading	Sports Hall	0.80	Pre-Tender
Reading	Sports Pavilion	0.50	Pre-Tender
Buckinghamshire			
Aylesbury	Sports & Leisure Facilities (New/Refurbishment)	9.00	Pre-Tender
Cambridgeshire			
Cambridge	School (Conversion)	0.60	Pre-Tender
Cambridge	Sports Hall	2.00	Contract Awarded
Huntingdon	Swimming Pool	0.95	Contract Awarded
Central Scotland			
Denny	Community Hall	0.50	Bills Called
Central London			
London	Sports Centre Building	3.00	Pre-Tender
London	Sports Hall	3.00	Pre-Tender
London	Sports Clubhouse (Extension) and Synthetic Pitch	0.70	Pre-Tender
London	Academic & Sports Centre	25.00	Pre-Tender
London	Cultural Centre (New/Refurbishment)	22.00	Pre-Tender
London	Sports Centre (Extension)	1.50	Pre-Tender
London	School (Extension)	2.80	Pre-Tender
London	Sports Centre (Extension)	7.00	Pre-Tender
London	Leisure Centre	14.00	Pre-Tender
London	Sports Hall	0.45	Pre-Tender
London	Tennis Courts & Pavilion	0.60	Pre-Tender
London	Watersports Centre (Alterations)	2.20	Start on Site
London	Swimming Pool	3.70	Contract Awarded
London	Flats	1.50	Contract Awarded
London	Performance Space/Restaurant	2.00	Contract Awarded
London	Community Centre	4.50	Pre-Tender
Cheshire			
Congleton	Leisure Park	0.50	Pre-Tender
Cleveland			
Hartlepool	Sports Centre (Extension/Alterations)	0.85	Pre-Tender
Hartlepool	Community Sports Complex	0.80	Pre-Tender
Middlesbrough	Museum (Extension)	0.25	Tender Currently Invited
Co Armagh			
Armagh	University Facility (Conversion/Alterations)	2.00	Contract Awarded
Co Down			
Dromore	Community Hall	0.25	Pre-Tender
Cornwall			
Penzance	School (Extension)	0.60	Tenders Returned
Saltash	Sports Facility	1.20	Contract Awarded
Saltash	Sports Facility	0.90	Contract Awarded

Table 10.1 National Lottery Projects – Fund Application[1] (continued)

June 2000

Town	Development	Vaue (£m)	Stage
Cumbria			
Carlisle	Golf Club House	0.38	Pre-Tender
Cleator Moor	Clubhouse	0.35	Pre-Tender
Cockermouth	Mountain Rescue Team Headquarters	2.00	Pre-Tender
Penrith	Groundworks/Pitches	0.25	Tender Currently Invited
Derbyshire			
Ashbourne	Sports Centre	0.50	Pre-Tender
Derby	Sports Centre (Extension/Alterations)	1.00	Pre-Tender
Derby	Sports Hall (Outline)	1.00	Pre-Tender
Derby	Equestrian Centre	0.30	Start on Site
Derby	Housing (Refurbishment)	4.50	Contract Awarded
Ilkeston	Youth/Community Centre	0.42	Pre-Tender
Devon			
Bideford	Village Hall	0.47	Pre-Tender
Ivybridge	Clubhouse (Alterations/Extension)	0.50	Pre-Tender
Dorset			
Bournemouth	Garden (Refurbishment/Restoration)	0.75	Pre-Tender
Dorchester	Sports Hall (Extension)	1.00	Pre-Tender
Poole	Athletics Track	0.50	Contract Awarded
Sturminster Newton	Sports Centre	0.75	Contract Awarded
East Sussex			
Eastbourne	Sports Pitches	2.00	Pre-Tender
Eastbourne	Sports Park	1.60	Tender Currently Invited
Eastbourne	School (Extension)	1.60	Start on Site
Newhaven	School	25.00	Tenders Returned
Rye	Sports Centre (Extension)	2.20	Pre-Tender
Essex			
Chelmsford	Sports Stadium	2.00	Pre-Tender
Chelmsford	Swimming Pool	3.30	Pre-Tender
Colchester	Park Improvements	2.00	Pre-Tender
Fife			
Dunfermline	Day Care Centre	0.25	Pre-Tender
Gloucestershire			
Cheltenham	School (Extension)	0.82	Pre-Tender
Grampian			
Ballater	Golf Clubhouse	0.40	Start on Site
Ballater	Railway Station (Alterations)	1.50	Pre-Tender
Greater London			
Beckenham	Sports Club (Extension and Alterations)	1.30	Pre-Tender
Brentford	Arts Centre (Re-development)	0.30	Tenders Returned
Carshalton	Sports Pavilion	0.50	Pre-Tender
Croydon	Community/Sports Hall/Synthetic Sports Pitch	1.00	Pre-Tender
Hayes	Clubhouse/Spectator Stand	0.30	Pre-Tender
Ilford	Sports Pavilion	1.20	Pre-Tender
New Malden	Sports Hall	1.00	Pre-Tender
Purley	Bowls Pavilion (Extension/Refurbishment)	0.40	Contract Awarded
Romford	Leisure Centre	7.50	Pre-Tender
Greater Manchester			
Leigh	Xanadu Project	150.00	Pre-Tender
Manchester	Sports Centre/Pitches (Extension/Alterations)	0.60	Pre-Tender
Manchester	Community Centre (Alterations/Extension)	0.30	Pre-Tender
Manchester	Football Pitches/Soccer Pitches/Pavilion/Car Parking	0.50	Pre-Tender
Manchester	Museum	3.50	Pre-Tender

Table 10.1 National Lottery Projects – Fund Application[1] (continued)

June 2000

Town	Development	Value (£m)	Stage
Manchester	Sports Hall & One Stop Shop	0.70	Contract Awarded
Manchester	Clubhouse	0.67	Contract Awarded
Manchester	Museum (Fitting Out)	4.50	Tenders Returned
Manchester	Pitch/Floodlighting/Fencing	0.80	Tenders Returned
Manchester	Holocaust Museum Shoah Centre	7.00	Pre-Tender
Manchester	Sports Centre/Pitches (Extension/Alterations)	0.25	Pre-Tender
Rochdale	Church (Re-modelling)	0.30	Pre-Tender
Salford	Watersports Centre	1.60	Contract Awarded
Stalybridge	Waterway Channel	4.00	Contract Awarded
Gwent			
Tredegar	Community Centre	0.40	Pre-Tender
Hampshire			
Havant	Sports Hall	1.30	Start on Site
Petersfield	Youth Centre/Sports Pavilion	0.25	Contract Awarded
Southampton	Football Stadium & Retail	20.00	Tenders Returned
Waterlooville	Sports Pavilion	0.80	Contract Awarded
Winchester	Village Hall	0.28	Start on Site
Hereford & Worcester			
Malvern	Pavilion	0.36	Contract Awarded
Worcester	Community Centre	0.85	Pre-Tender
Hertfordshire			
Hemel Hempstead	School (Extension)	2.00	Contract Awarded
Welwyn Garden City	Sports Centre (Extension)	1.50	Pre-Tender
Highlands			
Mallaig	Community Study Centre	0.40	Contract Awarded
Isle Of Wight			
Cowes	Youth Centres	0.30	Pre-Tender
Kent			
Canterbury	Offices	4.00	Pre-Tender
Deal	Bowling Hall	0.90	Tenders Returned
Folkestone	Sports Centre	2.00	Pre-Tender
New Romney	Community Sports Centre	2.00	Tenders Returned
Westerham	Museum/Visitors Centre	1.00	Pre-Tender
Lancashire			
Bacup	Swimming Pool/Changing Rooms	2.50	Tenders Returned
Fleetwood	Leisure Centre/Sports Arena	1.00	Pre-Tender
Lancaster	Youth Club (Extension/Alterations)	0.25	Contract Awarded
Lytham St. Anne's	Sports Hall/Changing Area	0.25	Pre-Tender
Preston	Fitness Studio (Improvements/Refurbishment)	0.33	Tender Currently Invited
Thornton Cleveleys	Club House (Extension)	0.25	Contract Awarded
Leicestershire			
Leicester	Community Centre	3.20	Pre-Tender
Loughborough	Pavilion	0.35	Pre-Tender
Lutterworth	Leisure Centre	1.98	Pre-Tender
Lothian			
Edinburgh	Sports Centre	0.80	Pre-Tender
Edinburgh	Opera House/Classical Music Venue (Revitalisation)	15.00	Contract Awarded
Merseyside			
Birkenhead	Marina/Roads	15.00	Pre-Tender
Bootle	Health Centre	0.80	Pre-Tender
Liverpool	Leisure & Health Centre	5.70	Tenders Returned
Liverpool	Sports Hall Complex	4.00	Contract Awarded
Liverpool	Leisure Centre (Extension)	4.00	Pre-Tender
Liverpool	Community/Leisure Centre	5.70	Pre-Tender

Table 10.1 National Lottery Projects – Fund Application[1] (continued)

			June 2000
Town	**Development**	**Value (£m)**	**Stage**
Liverpool	Leisure Centre	5.70	Pre-Tender
Liverpool	Community Centre	2.50	Pre-Tender
Liverpool	Leisure Centre	6.00	Pre-Tender
Norfolk			
Diss	Village Hall	0.75	Pre-Tender
Downham Market	Swimming Pool (Extension)	0.75	Pre-Tender
Norwich	Ski Club House (Extension)	0.40	Pre-Tender
Norwich	Sports Centre	1.00	Pre-Tender
Norwich	Activity Centre	0.60	Tenders Returned
North Yorkshire			
Harrogate	Office Building	0.30	Pre-Tender
Whitby	Sports Centre	1.20	Pre-Tender
York	Cycleway/River Footbridge	2.20	Contract Awarded
York	Community Facility	0.33	Pre-Tender
Northamptonshire			
Northampton	Clubhouse/Sports Hall	2.80	Pre-Tender
Northumberland			
Alnwick	Leisure Centre	10.00	Pre-Tender
Bedlington	Show Piece Sport & Leisure Complex	20.00	Pre-Tender
Blyth	Community Resource Centre	0.40	Pre-Tender
Prudhoe	Leisure Centre (Extension)	1.40	Start on Site
Nottinghamshire			
Mansfield	Sports Hall (Extension)	0.30	Pre-Tender
Nottingham	Castle (Refurbishment)	0.25	Contract Awarded
Oxfordshire			
Abingdon	Clubhouse	0.35	Pre-Tender
Abingdon	Leisure Centre	7.00	Tender Currently Invited
Chipping Norton	Leisure Centre	4.00	Pre-Tender
Witney	Village Hall	0.30	Pre-Tender
Witney	Clubhouse	0.34	Pre-Tender
Powys			
Llanymynech	Village Hall	1.00	Pre-Tender
Shropshire			
Shrewsbury	CCTV Installation	0.30	Pre-Tender
Telford	Golf Clubhouse/Tennis Courts	0.40	Tenders Returned
Telford	Pavilion	0.80	Contract Awarded
Somerset			
Glastonbury	Village Hall	0.40	Pre-Tender
Taunton	Memorial/Village Hall	0.25	Pre-Tender
Wincanton	Leisure Centre	1.24	Start on Site
South Glamorgan			
Cardiff	Arts Centre	13.00	Pre-Tender
South Yorkshire			
Doncaster	Sports Pavilion	1.00	Pre-Tender
Sheffield	Leisure Centre (Extension)	2.00	Contract Awarded
Sheffield	Fine Art/Craft Studio Complex	1.50	Contract Awarded
Staffordshire			
Leek	Leisure Centre (Extension)	3.00	Pre-Tender
Lichfield	Performing Arts Centre Cultural Quarters	0.45	Contract Awarded
Newcastle-Under-Lyme	Sports Hall	0.30	Contract Awarded
Stafford	Sports Hall	1.50	Pre-Tender
Stoke-On-Trent	Sports Hall	1.00	Pre-Tender

Table 10.1 National Lottery Projects – Fund Application[1] (continued)

June 2000

Town	Development	Value (£m)	Stage
Strathclyde			
Clydebank	Golf Clubhouse	0.60	Pre-Tender
Glasgow	Recreation Centre (Extension)	4.00	Pre-Tender
Glasgow	Sports Pavilion/Pitches	2.50	Tender Currently Invited
Glasgow	Sewage Treatment Works	57.00	Start on Site
Isle Of Colonsay	Cafe, Arts & Crafts Centre, Shop & Play Area (Conversion)	0.40	Pre-Tender
Paisley	Sports Pavilion (Alterations/Extension)	0.65	Pre-Tender
West Kilbride	Open Air Amphitheatre	1.00	Pre-Tender
Suffolk			
Ipswich	Offices/Harbour	0.25	Pre-Tender
Surrey			
Addlestone	Leisure Centre (Extension/Refurbishment)	1.00	Pre-Tender
Farnham	School Sports Hall	0.80	Pre-Tender
Godstone	Sports Pavilion	0.50	Pre-Tender
Haslemere	Sports Centre	1.90	Contract Awarded
Hindhead	Sports Pavilion	0.35	Pre-Tender
Tayside			
Dundee	Village Hall	0.25	Tenders Returned
Dundee	Village Hall/Tea Room/Nursery	0.50	Contract Awarded
Tyne & Wear			
Hebburn	Sports Club	0.30	Pre-Tender
Newcastle-Upon-Tyne	Leisure/Health/Community Unit	2.00	Pre-Tender
Warwickshire			
Shipston-On-Stour	Clubhouse	0.38	Tenders Returned
Studley	Village Hall	0.60	Pre-Tender
West Midlands			
Birmingham	Sports Facilities (Refurbishment)	0.35	Start on Site
Coventry	Sports Hall	0.56	Pre-Tender
Walsall	Civic Square	1.75	Contract Awarded
Wednesbury	Community Centre	1.20	Pre-Tender
West Sussex			
Chichester	Village Hall	0.30	Pre-Tender
Chichester	Museum (Extension)	2.00	Pre-Tender
Chichester	Museum	1.50	Tenders Returned
Chichester	Community Sports Hall	0.30	Pre-Tender
West Yorkshire			
Bradford	Community Centre/Cafe	0.25	Tender Currently Invited
Halifax	Community Multi-Sports Centre	1.00	Pre-Tender
Huddersfield	Community Resource Centre	0.30	Contract Awarded
Huddersfield	Sports Hall (Extension)	0.75	Contract Awarded
Leeds	Sports Pitches	1.50	Pre-Tender
Leeds	Equestrian Centre (Extension/Alterations)	1.00	Contract Awarded
Western Isles			
Isle Of Lewis	School Sports Centre (Outline)	1.00	Pre-Tender
Wiltshire			
Calne	Clubhouse/Grandstand/Pitches	0.50	Pre-Tender
Devizes	Leisure Centre (Alterations/Extension)	1.00	Pre-Tender
Salisbury	Trunk Road Upgrading	5.00	Pre-Tender
Trowbridge	Sports Hall	0.75	Pre-Tender

Notes

1. The table covers schemes which are in the process of applying for a lottery grant.

Source of Data: Glenigan Group

Contact: Sarah Wood 01202 432 121

Table 10.2 National Lottery Projects – Funding Received[1]

			June 2000
Town	**Development**	**Value (£m)**	**Stage**
Avon			
Bath	Spa Complex	12.00	Pre-Tender
Bedfordshire			
Bedford	Sports Courts	1.00	Contract Awarded
Berkshire			
Reading	Tennis Centre	0.80	Contract Awarded
Central Scotland			
Stirling	University Sports Facilities (New/Refurbishment)	10.00	Pre-Tender
Stirling	Arts & Cultural Resource Centre	0.25	Pre-Tender
Stirling	Arts Theatre (Extension/Alterations)	3.90	Contract Awarded
Stirling	Arts Centre	5.10	Pre-Tender
Central London			
London	Park Restoration	0.38	Pre-Tender
London	New Piers	3.00	Pre-Tender
London	Dance School	14.50	Tender Currently Invited
London	Park Restoration	1.80	Pre-Tender
London	Church (Envelope Works)	0.55	Bills Called
London	Millennium Dome (Catering Outlets)	6.30	Contract Awarded
London	Sports Centre (Extension/Refurbishment)	2.30	Bills Called
London	Demolition	0.30	Pre-Tender
London	Children's Hospice Phase 1	1.50	Start on Site
London	Sports Centre	3.40	Start on Site
London	Pavilion/Seating Stand/Track	2.00	Start on Site
London	Sports Centre	5.00	Contract Awarded
London	Sports Field	0.35	Contract Awarded
London	Music Training Centre (Conversion)	13.60	Contract Awarded
London	Cathedral (Re-development)	5.00	Contract Awarded
London	Cemetery Restoration	1.00	Contract Awarded
London	National Library of Women (Fawcett Library)	4.00	Contract Awarded
London	Museum (Conversion)	6.70	Contract Awarded
Cheshire			
Warrington	Art Centre (Conversion)	1.50	Contract Awarded
Cleveland			
Stockton-On-Tees	Swimming/Leisure Facility	4.00	Contract Awarded
Co Down			
Holywood	Sports Hall/Leisure Facility	0.80	Contract Awarded
Co Londonderry			
Derry	Millennium Theatre Complex	9.00	Contract Awarded
Cornwall			
Bodmin	Visitor Centre	2.00	Tender Currently Invited
St. Ives	Swimming Pool	3.00	Contract Awarded
County Durham			
Durham	Museum (Refurbishment)	0.90	Contract Awarded
Cumbria			
Barrow-In-Furness	Museum (Extension/Alterations)	0.65	Contract Awarded
Whitehaven	Harbour Improvements	5.00	Contract Awarded
Derbyshire			
Derby	Village Hall	0.25	Tenders Returned
Swadlincote	Community Centre	0.65	Contract Awarded
Swadlincote	The Millennium Discovery Centre	8.50	Start on Site
Devon			
Bideford	Village Hall	0.75	Contract Awarded
Plymouth	Park (Restoration)	1.10	Contract Awarded

Table 10.2 National Lottery Projects – Funding Received[1] (continued)

June 2000

Town	Development	Value (£m)	Stage
Dumfries & Galloway			
Langholm	Sports Complex	1.10	Contract Awarded
Dyfed			
Kidwelly	Multi-Purpose Hall	0.40	Start on Site
East Sussex			
Brighton	Pier (Reconstruction)	40.00	Pre-Tender
Brighton	Brighton Dome (Refurbishment)	28.00	Start on Site
Essex			
Saffron Walden	Sports Hall	1.70	Contract Awarded
Waltham Abbey	Royal Gunpowder Mills Visitor Attraction	5.00	Bills Called
Gloucestershire			
Gloucester	Leisure Centre (Re-development)	12.00	Contract Awarded
Gloucester	Bowling Clubhouse	0.45	Contract Awarded
Gloucester	National Waterways Museum (Extension)	0.50	Contract Awarded
Gloucester	Indoor Bowling Club	1.00	Contract Awarded
Grampian			
Aberdeen	Golf Clubhouse	0.25	Tenders Returned
Aberdeen	Environmental Improvements	1.00	Contract Awarded
Buckie	Church (Restoration)	0.60	Contract Awarded
Greater London			
Southall	College	2.70	Contract Awarded
Wembley	Wembley Stadium (Re-development)	355.00	Pre-Tender
Greater Manchester			
Bolton	Sports/Leisure Complex	1.00	Contract Awarded
Manchester	Sports Institute	7.00	Pre-Tender
Manchester	Imperial War Museum-North	28.50	Contract Awarded
Manchester	Pedestrian Bridge Link	0.35	Contract Awarded
Manchester	Art Gallery (External Cladding Package)	2.00	Contract Awarded
Manchester	City Art Gallery Roof Package	3.00	Tender Currently Invited
Hampshire			
Fordingbridge	2 Houses	0.10	Pre-Tender
Portsmouth	Millennium Tower	27.00	Tender Currently Invited
Portsmouth	Changing Rooms/Facilities	0.30	Start on Site
Portsmouth	Naval Exhibition Centre Conversion	5.80	Contract Awarded
Southampton	Cricket Ground (Re-development) Phase 3	9.00	Contract Awarded
Southampton	Park Improvements Lot 2	0.56	Contract Awarded
Southampton	Park Improvements	0.24	Contract Awarded
Southampton	Park Improvements	1.15	Contract Awarded
Hertfordshire			
Ware	Swimming Pool (Extension)	0.40	Contract Awarded
Highlands			
Alness	Village Hall	0.35	Contract Awarded
Lossiemouth	School (Extension)	1.10	Contract Awarded
Humberside			
Hull	Museum (Alterations)	0.47	Contract Awarded
Scunthorpe	Visual Arts & Crafts Centre (New/Conversion/ Extension)	1.60	Start on Site
Isle Of Wight			
East Cowes	Historic House (Refurbishment)	2.00	Contract Awarded
Sandown	Dinosaur Museum/Visitors Centre	1.50	Contract Awarded
Kent			
Canterbury	Sports Centre & Athletic Track	3.00	Contract Awarded
Canterbury	Pavilion	0.60	Contract Awarded

Table 10.2 National Lottery Projects – Funding Received[1] (continued)

June 2000

Town	Development	Value (£m)	Stage
Lancashire			
Blackpool	Parish Centre	0.70	Contract Awarded
Lancaster	Sports Hall	0.90	Start on Site
Leicestershire			
Leicester	National Space Science Centre	17.00	Start on Site
Loughborough	Sports Hall/Athletics Centre	3.80	Pre-Tender
Lothian			
Edinburgh	The Whale Centre	1.00	Contract Awarded
Edinburgh	Sports Hall (New)	2.00	Start on Site
Merseyside			
Liverpool	Office/Production Space The New Media Factory	9.50	Tender Currently Invited
Liverpool	Amphitheatre	0.25	Pre-Tender
Liverpool	Museum/Art Gallery (Extension/Alterations)	5.00	Start on Site
Liverpool	Civic Squares	12.00	Contract Awarded
Mid Glamorgan			
Merthyr Tydfil	Land Reclamation Scheme	1.00	Start on Site
Porthcawl	Historic House (Restoration)	0.65	Contract Awarded
Norfolk			
Norwich	Library/Business Info Centre Norwich Millennium Project	20.00	Contract Awarded
North Yorkshire			
York	Community Sports Hall	1.20	Contract Awarded
York	Music Centre (Extension/Alterations)	1.00	Start on Site
Northamptonshire			
Northampton	Sports Hall	0.50	Contract Awarded
Northumberland			
Choppington	Community Centre	0.60	Tenders Returned
Nottinghamshire			
Mansfield	Railway Station (Refurbishment)	0.50	Contract Awarded
Nottingham	Museum (Restoration)	4.50	Contract Awarded
Nottingham	Courtyard Re-development	0.45	Contract Awarded
Worksop	Craft Workshops	0.60	Contract Awarded
Oxfordshire			
Banbury	Museum/Heritage Centre	2.50	Contract Awarded
Bicester	Arts Centre (Refurbishment/Alterations)	0.50	Contract Awarded
Kidlington	Village Hall	0.30	Start on Site
Powys			
Montgomery	Canal (Restoration)	0.48	Pre-Tender
South Glamorgan			
Cardiff	Arts Centre (Alterations/Extensions)	0.27	Tender Currently Invited
Cardiff	Millennium Centre For The Arts	70.00	Contract Awarded
South Yorkshire			
Doncaster	Leisure Centre Dearne Valley Leisure Project	4.10	Start on Site
Rotherham	Exhibitions Construction Work	0.70	Tender Currently Invited
Rotherham	Electrical Installation Work	1.10	Pre-Tender
Rotherham	Mechanical Engineering	0.70	Pre-Tender
Rotherham	Architectural Metalwork	0.70	Pre-Tender
Rotherham	Metal/Glass Structure Installation	0.50	Pre-Tender
Rotherham	Miscellaneous Steelwork	0.45	Pre-Tender
Rotherham	Cladding System Installation	0.65	Pre-Tender
Rotherham	Structure Installation Works	0.30	Pre-Tender
Rotherham	Exhibition & Entertainment Complex (Conversion/New)	37.00	Contract Awarded
Sheffield	National Ice Centre	12.00	Pre-Tender

Table 10.2 National Lottery Projects – Funding Received[1] (continued)

June 2000

Town	Development	Value (£m)	Stage
Staffordshire			
Newcastle-Under-Lyme	Sports Hall	0.40	Start on Site
Stoke-On-Trent	Town Hall/Market (Refurbishment/Restoration)	2.40	Start on Site
Stoke-On-Trent	Colliery Regeneration	2.00	Pre-Tender
Strathclyde			
Glasgow	Indoor Bowling Centre	1.40	Bills Called
Glasgow	Community Sports Centre	4.00	Contract Awarded
Glasgow	Country Life Museum	6.00	Contract Awarded
Glasgow	Scottish National Science Centre	78.00	Start on Site
Larkhall	Pavilion	0.60	Contract Awarded
Paisley	Community Centre	2.80	Contract Awarded
Suffolk			
Bury St. Edmunds	Cathedral (Refurbishment)	10.00	Contract Awarded
Tayside			
Dundee	College Theatre	5.00	Start on Site
Perth	Historic Graveyard (Stonework Repairs)	0.40	Tenders Returned
Tyne & Wear			
Gateshead	Regional Music Centre	62.00	Contract Awarded
Gateshead	Contemporary Visual Arts Centre	45.00	Contract Awarded
Newcastle-Upon-Tyne	Synthetic Sports Pitches	0.50	Pre-Tender
Newcastle-Upon-Tyne	Sports Hall (Extension)	4.50	Contract Awarded
Newcastle-Upon-Tyne	New Sports Hall	6.20	Pre-Tender
Sunderland	Museum & Art Gallery (Extension/Refurbishment)	6.00	Start on Site
Sunderland	Sports Centre Complex	5.00	Pre-Tender
Sunderland	Sports Centre	5.00	Start on Site
Warwickshire			
Kenilworth	School (Extension)	1.25	Contract Awarded
West Midlands			
Birmingham	Churchyard (Restoration)	2.20	Pre-Tender
Birmingham	Millennium Campus Millennium Point	50.00	Start on Site
Birmingham	Indoor Cricket School	2.30	Start on Site
Coventry	Phoenix Initiative Phase 1 (Infrastructure)	2.50	Contract Awarded
Wednesbury	Swimming Pool Complex	4.50	Start on Site
West Sussex			
Haywards Heath	Village Hall	0.35	Contract Awarded
West Yorkshire			
Keighley	Golf Clubhouse	0.30	Contract Awarded
Leeds	City Square Re-development	5.40	Contract Awarded
Western Isles			
Stornoway	Arts Centre	4.70	Contract Awarded
Wiltshire			
Devizes	Sports Club (Extension)	0.51	Contract Awarded
Salisbury	School (Extension)	0.25	Contract Awarded
Warminster	Sports Pavilion	0.15	Pre-Tender

Notes

1. The table covers projects which have been granted funding.

Source of Data: Glenigan Group

Contact: Sarah Wood 01202 432 121

CHAPTER 11

Planning Applications and Decisions

Table 11.1 Planning Applications and Decisions by District Planning Authorities[1,2] by Speed of Decision

Year and Quarter		All Applications Received Thousands	All Decisions Thousands	Applications Granted[3]		Number Decided Within 8 Weeks Thousands	Percentage of Total Decisions[4,5]	
				Thousands	Percent[5]		Within 8 weeks	Within 13 Weeks
1991	Q1	126.9	114.9	88.8	80	62.9	55	78
	Q2	137.4	125.7	99.5	82	75.3	60	83
	Q3	128.8	128.2	101.9	83	77.5	60	83
	Q4	118.0	116.7	92.1	82	71.2	61	83
1992	Q1	124.7	111.0	88.1	83	66.7	60	82
	Q2	124.0	119.0	97.2	85	75.2	63	86
	Q3	117.7	115.6	95.2	85	73.7	64	85
	Q4	104.1	105.8	87.0	85	66.8	63	84
1993	Q1	116.1	98.9	81.5	86	62.5	63	84
	Q2	123.6	114.0	96.0	87	76.3	67	88
	Q3	121.3	116.1	97.8	87	76.2	66	87
	Q4	110.9	109.9	92.0	87	70.5	64	85
1994	Q1	123.6	105.8	88.6	87	67.3	64	85
	Q2	125.2	117.1	99.3	88	76.9	66	88
	Q3	123.4	119.8	101.3	88	78.4	65	87
	Q4	109.9	108.7	91.4	87	71.7	66	86
1995	Q1	118.6	105.4	88.3	87	67.0	64	85
	Q2	122.8	115.8	98.4	88	76.2	66	87
	Q3	113.5	113.3	95.5	88	74.0	65	86
	Q4	102.4	101.9	85.1	87	66.1	65	85
1996	Q1	116.9	100.4	83.6	87	63.3	63	84
	Q2	123.5	111.1	94.0	88	73.8	66	87
	Q3	120.4	115.3	97.5	88	74.1	64	86
	Q4	111.9	109.4	91.3	87	69.4	63	84
1997	Q1	117.2	100.9	84.3	88	61.4	61	83
	Q2	136.2	120.1	101.8	88	76.7	64	86
	Q3	130.4	123.7	104.1	88	77.4	63	85
	Q4	110.7	112.4	93.9	88	69.1	62	83
1998	Q1	127.3	105.4	88.0	87	63.1	60	81
	Q2	131.7	120.7	102.4	88	75.6	63	85
	Q3	128.6	123.8	104.6	88	77.3	62	84
	Q4	112.9	115.5	96.5	87	71.6	62	82
1999	Q1	128.0	106.0	88.7	87	65.1	61	81
	Q2	134.3	120.7	102.5	88	77.6	64	85
	Q3	132.4	128.5	109.0	88	80.7	63	84
	Q4	118.2	116.3	97.6	88	73.0	63	83

Notes

1. Includes metropolitan and non-metropolitan districts, unitary authorities, London boroughs, national park authorities and, prior to April 1998, urban development corporations. Figures exclude 'county matters' applications and decisions.

2. Estimates are included for the non-responding authorities.

3. The base for figures in these columns exclude those applications which cannot be granted or refused.

4. The precise definitions of the time bands used throughout this table are 'up to and including 56 days' and 'up to and including 91 days'. The percentages in these two columns are cumulative not additive.

5. Percentages are calculated using unrounded figures.

Source of Data: Planning and Land Use Statistics Division, Department of the Environment, Transport and the Regions
Contact: Ian Rowe 020 7944 5502

| Table 11.2 Planning Decisions by District Planning Authorities[1] by Speed of Decision, Region and Type of Authority – England |

Government Office Region[2]	All Decisions			Applications Granted[3]						Percentage of Total Decisions[4]					
	Thousands			Thousands			Percent			Within 8 Weeks			Within 13 Weeks		
	1997	1998	1999	1997	1998	1999	1997	1998	1999	1997	1998	1999	1997	1998	1999
North East	16.7	16.9	16.9	15.2	15.4	15.3	92	93	93	66	66	70	89	88	89
North West & Merseyside	47.3	47.1	48.8	41.7	41.3	42.8	90	89	90	60	62	64	85	85	86
North West	40.3	39.9	41.2	35.4	34.9	36.1	89	89	89	59	61	63	84	84	85
Merseyside	7.0	7.2	7.5	6.4	6.4	6.8	93	91	92	65	69	70	87	88	88
Yorkshire & Humberside	39.5	39.2	39.0	34.4	33.8	33.7	89	89	89	60	59	62	83	81	82
East Midlands	37.8	36.8	37.6	33.4	32.8	33.5	90	90	90	62	62	63	85	84	84
West Midlands	41.1	42.4	42.1	34.1	35.5	35.7	87	87	88	63	63	63	84	84	85
Eastern	56.4	58.6	59.2	47.4	49.7	50.8	87	88	88	65	65	64	86	85	85
London	64.2	65.8	66.4	47.6	49.5	49.8	84	83	83	58	58	59	80	79	79
South East	91.8	95.8	97.4	77.2	80.3	81.5	88	87	87	63	61	64	84	83	84
South West	62.4	62.7	64.1	53.3	53.3	54.7	89	89	89	65	64	63	85	85	84
ENGLAND[6]	**457.1**	**465.3**	**471.5**	**384.1**	**391.5**	**397.9**	**88**	**88**	**88**	**62**	**62**	**63**	**84**	**83**	**84**
Of which															
London	64.2	65.8	66.4	47.6	49.5	49.8	76	83	83	58	58	59	80	79	79
Other Metropolitan Authorities	63.7	64.1	64.7	55.5	55.5	56.1	89	89	89	61	60	63	83	82	84
Non-Metropolitan Authorities[5]	329.2	335.4	340.3	281.0	286.6	292.0	88	88	89	63	63	64	85	84	84

Notes

1. These statistics are compiled from information received from metropolitan and non-metropolitan districts, London boroughs, unitary authorities, national park authorities and, prior to 31 March 1998, urban development corporations. The figures exclude 'county matters' applications.

2. See Appendix 1 for definition of Government Office regions.

3. Applications which cannot be granted or refused have been excluded from the Applications Granted columns .

4. The precise definitions used are 'up to and including 56 days' and 'up to and including 91 days'. The percentages in these two columns are cumulative not additive.

5. Non-metropolitan authorities includes unitary authorities.

6. Figures may not add to total due to rounding

Source of Data: Planning and Land Use Statistics Division, Department of the Environment, Transport and the Regions

Contact: Ian Rowe 020 7944 5502

Table 11.3 Planning Decisions by District Planning Authorities[1] by Speed of Decision and Type and Size of Development – England

Type of Development	All Decisions Thousands			Applications Granted Thousands			Applications Granted Percent			Percentage of Total Decisions Within 8 Weeks[2]		
	1997	1998	1999	1997	1998	1999	1997	1998	1999	1997	1998	1999
Major Developments												
Dwellings	7.1	6.6	6.2	6.0	5.6	5.2	85	85	84	25	21	20
Offices/R&D/Light Industry	1.1	1.1	1.2	1.0	1.1	1.1	94	94	94	37	34	37
Heavy industry/storage/ warehouses	1.6	1.6	1.5	1.6	1.5	1.4	95	94	95	45	42	40
Retail distribution and services	1.4	1.4	1.3	1.1	1.1	1.1	82	84	86	27	25	25
All other major developments	3.5	3.5	3.2	3.0	3.0	2.9	86	87	89	29	30	28
All major development	**14.7**	**14.1**	**13.4**	**12.7**	**12.3**	**11.7**	**86**	**87**	**87**	**29**	**27**	**26**
Minor Developments												
Dwellings	45.9	45.6	43.8	35.6	34.5	33.3	78	76	76	47	45	44
Offices/R&D/Light Industry	7.7	7.4	7.0	7.1	6.7	6.4	93	92	92	61	58	60
Heavy industry/storage/ warehouses	7.1	6.8	5.8	6.6	6.3	5.4	92	93	93	59	58	58
Retail distribution and services	15.8	15.5	14.0	13.9	13.7	12.5	88	88	89	57	57	58
All other minor developments	63.9	63.5	62.8	57.6	57.1	56.4	90	90	90	59	58	59
All minor development	**140.4**	**138.8**	**133.4**	**120.8**	**118.3**	**114.0**	**86**	**85**	**85**	**55**	**54**	**54**
Change of use	38.8	38.6	38.0	31.6	31.7	31.4	81	82	83	50	50	50
Householder developments	176.0	190.5	203.6	161.0	173.3	185.4	91	91	91	74	73	74
Minerals	0.2	0.2	0.2	0.2	0.1	0.2	89	83	86	39	31	30
All Section 70 Developments[2]	**370.2**	**382.2**	**388.4**	**326.3**	**335.8**	**342.7**	**88**	**88**	**88**	**62**	**62**	**63**
Advertisements	29.5	30.3	29.8	24.1	24.6	24.3	82	81	81	68	68	68
Listed buildings consents	30.8	30.6	30.8	28.2	28.0	28.1	91	92	91	53	52	53
Conservation areas consents	6.1	3.6	3.1	5.5	3.2	2.8	90	89	89	54	47	47
All Development Excluding Other	**436.4**	**446.6**	**452.2**	**384.1**	**391.5**	**397.9**	**88**	**88**	**88**	**62**	**62**	**63**
Other (not included above)[3]	20.5	18.7	19.3	..	..	..	..	..	..	68	64	65
All Developments	**457.1**	**465.3**	**471.5**	..	..	..	..	..	..	**62**	**62**	**63**

Notes

.. = not available

1. Includes metropolitan and non-metropolitan districts, unitary authorities, London boroughs, urban development corporations and national park authorities

2. Decisions under section 70 of the Town and Country Planning Act 1990.

3. Includes applications which cannot be granted or refused.

Source of Data: Planning and Land Use Statistics Division, Department of the Environment, Transport and the Regions
Contact: Ian Rowe 020 7944 5502

Table 11.4 Mineral Planning Decisions by 'County Matters' Authorities[1] – England

Government Office Region[2]	Total Decisions			Percentage Decided Within 13 Weeks[3]			Percentage Granted		
	1997	1998	1999	1997	1998	1999	1997	1998	1999
North East	47	35	38	45	43	71	100	94	89
North West	64	47	55	36	36	29	92	100	75
Yorkshire & the Humber	67	56	51	21	25	41	91	100	98
East Midlands	97	74	84	45	41	42	77	116	98
West Midlands	34	24	29	32	25	24	100	100	79
East of England [2]	88	94	91	48	31	27	85	91	89
London	0	1	1	..	0	0	..	100	100
South East	110	102	77	25	32	32	91	89	92
South West	67	76	53	33	36	36	96	97	96
England	479	509	479	36	34	36	91	98	91

Notes

.. = not available.

1. Includes county councils, unitary and metropolitan councils, London boroughs, urban development corporations and national park authorities.
2. See Appendix 1 for definitions of Government Office Regions. Prior to 1 February 1999, East of England (GOR) was named Eastern (GOR).
3. Decisions involving environmental assesments have been excluded when calculating the percentage decided within 13 weeks.

Source of Data: Planning and Land Use Statistics Division, Department of the Environment, Transport and the Regions
Contact: Ian Rowe 020 7944 5502

Table 11.5 Decisions on Minerals Applications by All 'County Matters' Authorities[1] by Type of Development – England

Type of Development	1996	1997	1998	1999
Chalk	7	12	11	8
China Clay[2]	5	8	8	4
Clay/Shale	43	41	43	34
Coal (Deepmine)	16	15	9	8
Coal (Opencast)	67	56	35	29
Gypsum/Anhydrite	3	2	-	3
Igneous Rock	18	14	9	20
Ironstone	2	-	2	3
Limestone/Dolomite	40	45	42	45
Oil/Gas (Exploration)	8	12	21	11
Oil/Gas (Appraisal)	1	5	1	-
Oil/Gas (Development)	17	20	15	12
Sand/Gravel	204	168	207	159
Sand	44	48	25	32
Sandstone	45	35	28	27
Slate	1	3	2	3
Vein Minerals	11	17	6	12
Recycling plants for secondary aggregates[3]	43	51	32	43
Coal-bed methane extraction[3]	1	-	1	4
Other minerals	43	39	47	52
England	**619**	**591**	**544**	**509**

Notes

- = nil or less than half the final digit shown.

1. Includes decisions made by county councils, the metropolitan councils, London borough councils, unitary and national park authorities and, prior to 1st April 1998, urban development corporations.

2. Prior to 1 April 1995 statistics were collected on 'China clay'.

3. Data on these developments have been collected since 1 April 1995.

Source of Data: Planning and Land Use Statistics Division, Department of the Environment, Transport and the Regions
Contact: Ian Rowe 020 7944 5502

THAMES GATEWAY

Thames Gateway currently extends from Greenford, Stratford and the Royal Docks in London eastwards along both sides of the Thames to Tilbury in Essex and Sheerness in North Kent. Its northern boundary generally follows the line of the A13, and the A207, A2 and M2 broadly defines its southern boundary. The economic health of the area has suffered from a combination of factors including a poor environment, polluting land users and the barrier effect on communications of the River Thames.

Thames Gateway currently has around 4000 hectares ripe for high quality sustainable developments on brownfield sites and its use will be made easier by existing and future infrastructure improvements. These include the Jubilee Line and Docklands Light Railway extension and improvements to the A13 and completion of the M11 link and the Medway towns Northern Relief Road. The construction of Section 2 of the Channel Tunnel Rail Link will start in 2001 and will bring real benefits to the area in terms of jobs and accessibility.

The Thames Gateway Planning Framework published on 14 June 1995 was a major step forward towards reversing the cycle of decline. It assessed the area's potential and provided a vision of the future as well as a framework for how it can be achieved.

The publication in March 2000 of the Draft Regional Planning Guidance for the South East renewed and strengthened the Government's commitment to the Thames Gateway. It identified it as one of the key regeneration areas of the South East and earmarked it for a substantial proportion of the region's new housebuilding and investment. It proposed an extension of the Thames Gateway area further into Essex, bringing more brownfield land for development thereby benefiting more people living in the region. The proposed boundary extension includes more of Thurrock, part of Basildon District, the Boroughs of Castle Point and Southend-on-Sea and Southend Airport in Rochford District. These areas have similar problems and opportunities as the rest of the Gateway area and already have strong links with London.

The Government has proposed a new Strategic Partnership to bring together Government, regional and local bodies to build on existing successes to ensure the areas' long term potential.

The Regional Development Agencies and the new London Development Agency will have an important role in drawing up economic strategies covering the Gateway area. These strategies will provide a framework for further public and private investment and will be vital in identifying land for regeneration and development. The RDAs will continue to work together as part of the Strategic Partnership to ensure a co-ordinated approach to the area is adopted.

Significant investment has already been made in Thames Gateway projects. The outcome of the Government's comprehensive Spending Review 2000 will give RDAs nationally an extra £500 million by 2003/04 to deliver economic regeneration and growth. Decisions on what this will mean for regional and local investment will be made later in the year.

Copies of the Thames Gateway Planning Framework (RPG9A) and Draft Regional Planning Guidance for the South East (RPG9) are available from The Stationery Office.

Source: Regeneration Division, Department of the Environment, Transport and the Regions
Contact: Philip Stables 020 7944 5213

CHAPTER 12

Trends in Employment and the Professions

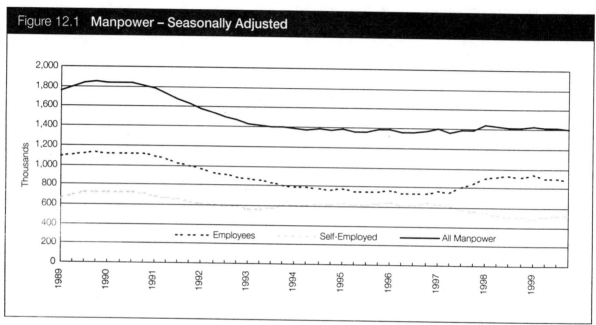

Figure 12.1 Manpower – Seasonally Adjusted

Source of Data: Table 2.1

Table 12.1 Manpower[1] – Seasonally Adjusted

Great Britain				Thousands
Year & Quarter		All Manpower	All Employees	Self-Employed
1989[2]		1,807	1,109	698
1990	Q1	1,838	1,118	720
	Q2	1,841	1,122	719
	Q3	1,835	1,117	718
	Q4	1,812	1,111	701
1991	Q1	1,777	1,094	683
	Q2	1,723	1,057	666
	Q3	1,667	1,019	648
	Q4	1,626	994	632
1992	Q1	1,579	962	617
	Q2	1,533	932	601
	Q3	1,494	909	585
	Q4	1,475	890	585
1993	Q1	1,432	873	559
	Q2	1,409	852	557
	Q3	1,403	831	572
	Q4	1,398	803	595
1994	Q1	1,394	795	599
	Q2	1,379	784	595
	Q3	1,390	778	612
	Q4	1,375	766	609
1995	Q1	1,394	773	621
	Q2	1,362	749	613
	Q3	1,362	743	619
	Q4	1,382	750	632
1996	Q1	1,392	758	634
	Q2	1,354	741	613
	Q3	1,355	741	614
	Q4	1,378	739	639
1997	Q1	1,399	767	632
	Q2	1,361	754	607
	Q3	1,384	806	578
	Q4	1,392	839	553
1998	Q1	1,447	901	546
	Q2	1,426	906	520
	Q3	1,419	921	498
	Q4	1,418	912	506
1999	Q1	1,429	933	496
	Q2	1,409	899	510
	Q3	1,418	897	521
	Q4 (p)	1,403	885	518

Notes

p = provisional

1. There are certain differences in the definition and coverage from the series published by the Office for National Statistics.

2. Average of four quarters.

Source of Data: Construction Market Intelligence, Department of Environment, Transport and the Regions
Contact: Neville Price 020 7944 5587

Table 12.2 Manpower

Great Britain									Monthly Averages in Thousands

Year & Quarter	Employees in Employment							Self Employed[6]	All Manpower
	Contractors on register[2]		Public Authorities		All Employees on Register[2]	Estimated Employees not on Register[4]	All[5]		
	Operatives	APTCs	Operatives	APTCs[3]					
1989[1]	533	254	134	71	1,042	67	1,109	698	1,806
1990 Q1	582	261	126	66	1,035	80	1,115	720	1,835
Q2	587	267	128	64	1,046	72	1,118	719	1,837
Q3	587	266	127	67	1,047	73	1,120	718	1,838
Q4	575	270	127	67	1,039	77	1,115	701	1,816
1991 Q1	550	268	124	65	1,006	85	1,091	683	1,774
Q2	519	252	121	66	958	95	1,053	666	1,719
Q3	508	250	118	60	936	87	1,023	648	1,671
Q4	487	236	115	63	901	94	994	632	1,626
1992 Q1	460	229	111	58	858	102	960	617	1,577
Q2	438	222	113	56	829	99	928	601	1,529
Q3	432	217	112	55	816	96	912	585	1,497
Q4	419	214	112	53	798	95	893	585	1,478
1993 Q1	404	212	106	51	774	97	870	559	1,429
Q2	399	205	103	51	757	93	850	557	1,407
Q3	384	201	99	49	734	95	829	572	1,401
Q4	374	198	97	48	717	93	810	595	1,405
1994 Q1	368	198	92	44	702	91	793	599	1,392
Q2	359	197	91	41	688	93	781	595	1,376
Q3	360	201	89	42	692	85	776	612	1,388
Q4	357	195	89	42	683	89	772	609	1,381
1995 Q1	357	198	89	40	685	86	770	621	1,391
Q2	347	198	86	39	671	75	746	613	1,359
Q3	347	199	85	37	668	74	741	619	1,360
Q4	348	200	85	38	671	85	756	632	1,388
1996 Q1	348	202	83	39	672	84	756	634	1,390
Q2	349	201	78	36	663	75	739	613	1,351
Q3	351	202	71	26	650	90	739	614	1,353
Q4	352	199	70	26	647	98	745	640	1,384
1997 Q1	357	201	71	26	655	110	765	632	1,396
Q2	353	201	71	26	651	101	752	607	1,359
Q3	408	199	71	25	702	102	804	578	1,382
Q4	418	200	66	25	710	136	846	553	1,399
1998 Q1	421	199	69	25	714	185	879	546	1,445
Q2	429	209	69	25	732	170	903	521	1,423
Q3	421	204	69	25	718	201	919	498	1,417
Q4	451	207	64	24	746	174	920	506	1,425
1999 Q1	476	204	62	23	764	166	930	496	1,426
Q2	449	201	58	24	732	164	896	509	1,406
Q3	499	207	60	23	789	106	895	522	1,417
Q4 (p)	513	210	60	22	806	88	894	517	1,411

Notes

p = provisional

1. Average of four quarters.
2. Estimates by the DETR based on returns from contractors and public authorities.
3. APTC are administrative, professional, technical and clerical staff.
4. Estimates of employees not on DETR register of firms or misclassified as working proprietors.
5. This series has a different coverage from the series published by the Office for National Statistics.
6. Estimates are based on the Office for National Statistics Labour Force Survey.

Source of Data: Construction Market Intelligence, Department of Environment, Transport and the Regions
Contact: Neville Price 020 7944 5587

Table 12.3 Earnings and Hours in the Construction Industry

(a) Full time male manual workers, on adult rates, whose pay was not affected by absence

April Each Year	Average Gross Weekly Earnings (£)				Percentage of Employees who received		
	Total	of which Overtime Pay	PBR, etc., Payments[1]	Premium Payments[2]	Overtime Pay	PBR, etc., Payments[1]	Premium Payments[2]
1989	214.2	32.4	26.9	0.9	56.1	51.5	2.6
1990	245.7	35.2	30.0	1.7	54.4	50.0	3.0
1991	257.1	36.1	28.6	1.7	51.5	48.6	3.0
1992	274.7	36.3	26.8	3.8	50.3	47.2	3.6
1993	274.3	35.2	25.5	1.5	49.3	39.1	3.6
1994	277.4	37.8	23.6	1.4	51.3	39.9	3.0
1995	294.7	41.9	23.0	1.9	53.0	38.3	3.9
1996 (r)	308.2	44.5	24.0	1.8	51.9	37.2	4.3
1997 (r)	324.8	53.6	18.0	1.8	55.0	25.6	4.7
1998	342.3	53.2	17.3	1.7	51.3	22.6	3.3
1999	351.3	52.7	15.1	1.6	51.3	18.7	3.4

	Distribution of Weekly Earnings (£)			Average hourly earnings exc. the effect of overtime pay and hours (pence)	Average Weekly Hours	
	10% earned less than	50% earned less than	10% earned more than		Total, inc. Overtime	Overtime
1989	131.6	194.2	318.6	456.1	46.0	6.4
1990	144.9	220.3	374.9	523.9	46.0	6.3
1991	156.6	231.0	385.1	551.3	45.4	5.8
1992	165.6	246.9	418.1	595.0	45.1	5.6
1993	169.4	249.1	414.9	602.0	44.7	5.3
1994	176.0	255.3	407.9	601.0	45.1	5.6
1995	180.6	271.8	437.0	632.0	45.9	6.0
1996 (r)	191.3	280.7	462.7	657.0	45.8	5.8
1997 (r)	196.6	296.6	491.6	676.0	46.9	6.8
1998	206.0	306.3	517.0	711.0	46.9	6.3
1999	214.5	322.8	525.2	735.0	46.4	5.9

(b) Full-time male non-manual workers, on adult rates, whose pay was not affected by absence

April Each Year	Average Gross Weekly Earnings (£)		Percentage of Employees who received Overtime Pay	Distribution of Weekly Earnings (£)			Average hourly earnings exc. the effect of overtime pay and hours (pence)	Average Weekly Hours	
	Total	of which Overtime Pay		10% earned less than	50% earned less than	10% earned more than		Total, inc. Overtime	Overtime
1989	312.6	9.1	15.7	175.6	280.6	464.7	774.1	40.3	1.4
1990	346.8	11.3	16.2	195.5	316.8	529.7	855.0	40.2	1.5
1991	368.2	9.4	14.0	201.6	337.6	557.3	910.9	40.0	1.1
1992	390.0	9.1	13.8	211.0	353.4	593.4	961.0	40.3	1.2
1993	401.0	9.1	11.6	224.7	363.6	608.2	991.0	40.0	1.0
1994	414.5	10.2	12.6	223.2	377.0	636.1	1006.0	40.3	1.2
1995	431.6	13.2	13.7	236.8	392.0	662.2	1053.0	40.9	1.4
1996	445.8	9.4	11.2	238.7	408.6	668.7	1085.0	40.7	0.9
1997 (r)	460.0	12.2	12.0	257.4	422.3	684.4	1112.0	41.3	1.1
1998	474.1	10.4	13.0	269.3	440.1	729.4	1144.0	41.6	1.1
1999	508.5	11.5	13.9	278.3	465.3	767.1	1230.0	41.4	1.0

Notes

r = revised

1. Payment by results (e.g. piecework), bonuses (including profit-sharing), commission and other incentive payments.

2. Premium pay (not total pay) for shiftwork, and for nightwork or weekend work where these are not treated as overtime.

Source of Data: Construction Market Intelligence, Department of Environment, Transport and the Regions
Price 020 7944 5587

Contact: Neville

159

Table 12.4 Earnings in the Construction Industry by Craft[1]

Full time male workers, on adult rates, whose pay was not affected by absence

April Each Year	Foremen, Building & Civil Engineering[2]		Bricklayers & Masons		Carpenters & Joiners		Road Construction & Maintenance Workers		Craftsmen: Mates to Building Trade Workers		Craftsmen: Other Building & Civil Engineering Workers	
	Average Gross Weekly Earnings (pounds)	Average hourly earnings exc. the effect of overtime pay and hours (pence)	Average Gross Weekly Earnings (pounds)	Average hourly earnings exc. the effect of overtime pay and hours (pence)	Average Gross Weekly Earnings (pounds)	Average hourly earnings exc. the effect of overtime pay and hours (pence)	Average Gross Weekly Earnings (pounds)	Average hourly earnings exc. the effect of overtime pay and hours (pence)	Average Gross Weekly Earnings (pounds)	Average hourly earnings exc. the effect of overtime pay and hours (pence)	Average Gross Weekly Earnings (pounds)	Average hourly earnings exc. the effect of overtime pay and hours (pence)
1989	252.2	552.1	195.7	440.6	205.4	449.8	208.2	446.5	176.2	379.7	..	..
1990	..	..	232.5	529.1	233.7	508.5	239.8	512.7	202.7	442.7	205.9	444.0
1991	..	..	227.9	533.2	240.3	535.4	245.3	532.0	200.2	448.4	210.9	469.0
1992	..	..	237.7	569.0	250.8	564.0	273.0	583.0	201.0	467.0	233.8	519.0
1993	..	..	242.0	575.0	252.1	579.0	289.0	624.0	196.8	467.0	234.0	523.0
1994	..	..	252.4	576.0	260.5	583.0	295.3	630.0	213.2	494.0	241.2	527.0
1995	..	..	262.7	607.0	274.0	616.0	320.0	662.0	..	..	250.2	532.0
1996 (r)	..	..	274.6	632.0	287.5	637.0	320.8	677.0	234.9	501.0	255.1	544.0
1997	..	..	303.5	681.0	294.5	653.0	328.7	697.0	..	..	271.7	592.0
1998	..	..	310.4	714.0	319.4	690.0	347.5	732.0	258.9	561.0	291.3	615.0
1999	..	..	318.4	738.0	339.6	737.0	347.1	737.0	261.3	585.0	300.9	646.0

Notes

r = revised

.. = not available; standard error too high or sample number too low for a reliable estimate.

1. Includes some working outside the construction industry.
2. Figures unavailable after 1989. Foreman have been reclassified under the new Standard Occupational Classification (S.O.C.).

Source of Data: Construction Market Intelligence, Department of Environment, Transport and the Regions
Contact: Neville Price 020 7944 5587

Table 12.5 Earnings and Hours in the Construction Industry and in All Industries and Services

Full time male manual workers, on adult rates, whose pay was not affected by absence

April Each Year	Full Time Male Manual Workers						Full Time Male Non-Manual Workers					
	Construction Industry			All Industries and Services			Construction Industry			All Industries and Services		
	Average Gross Weekly Earnings (pounds)	Average hourly earnings exc. the effect of overtime pay and hours (pence)	Average weekly hours inc. overtime (hours)	Average Gross Weekly Earnings (pounds)	Average hourly earnings exc. the effect of overtime pay and hours (pence)	Average weekly hours inc. overtime (hours)	Average Gross Weekly Earnings (pounds)	Average hourly earnings exc. the effect of overtime pay and hours (pence)	Average weekly hours inc. overtime (hours)	Average Gross Weekly Earnings (pounds)	Average hourly earnings exc. the effect of overtime pay and hours (pence)	Average weekly hours inc. overtime (hours)
1989	214.2	456.1	46.0	217.8	466.2	45.3	312.6	774.1	40.3	323.6	824.0	38.8
1990	245.7	523.9	46.0	237.2	508.5	45.2	346.8	855.0	40.2	354.9	902.1	38.7
1991	257.1	551.3	45.4	253.1	554.1	44.4	368.2	910.9	40.0	375.7	956.3	38.7
1992	274.7	595.0	45.1	268.3	589.0	44.5	390.0	961.0	40.3	400.4	1023.0	38.6
1993	274.3	602.0	44.7	274.3	605.0	44.3	401.0	991.0	40.0	418.2	1069.0	38.6
1994	277.4	601.0	45.1	280.7	614.0	44.7	414.5	1006.0	40.3	428.2	1093.0	38.9
1995	294.7	632.0	45.9	291.3	625.0	45.2	431.6	1053.0	40.9	443.3	1136.0	39.0
1996 (r)	305.4	657.0	45.8	301.3	651.0	44.8	444.4	1085.0	40.7	464.5	1187.0	39.1
1997 (r)	318.4	676.0	46.9	309.9	679.0	45.1	459.0	1112.0	41.3	480.6	1239.0	39.1
1998	342.3	711.0	46.9	328.5	710.0	45.0	474.1	1144.0	41.6	506.1	1294.0	39.1
1999	351.3	735.0	46.4	335.0	736.0	44.4	508.5	1230.0	41.4	525.5	1352.0	39.0

Notes

r = revised

Source of Data: Construction Market Intelligence, Department of Environment, Transport and the Regions
Contact: Neville Price 020 7944 5587

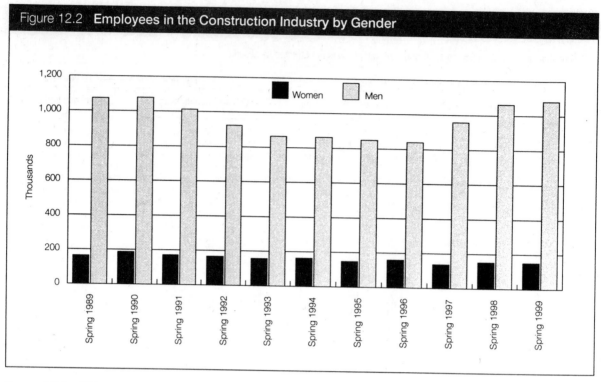

Figure 12.2   Employees in the Construction Industry by Gender

Source of Data: Table 12.6

Table 12.6   Employment in the Construction Industry by Gender[1,2]

United Kingdom   **Not Seasonally Adjusted: Thousands**

	Employees			Self-Employed		
	All	Women	Men	All	Women	Men
Spring 1989	1,239	164	1,074	922	18	904
Spring 1990	1,265	187	1,078	918	23	895
Spring 1991	1,183	171	1,012	837	20	818
Spring 1992	1,089	166	923	757	18	739
Spring 1993	1,021	157	864	746	17	728
Spring 1994	1,025	163	862	797	17	780
Spring 1995	996	148	849	802	14	788
Spring 1996	1,000	160	840	799	19	779
Spring 1997	1,095	137	958	749	15	734
Spring 1998	1,217	152	1,064	659	14	645
Spring 1999	1,235	151	1,083	664	12	652

Notes

1. SIC 92 was introduced in Winter 1993/4; estimates for earlier periods have been adjusted to be broadly comparable at the aggregate level.

2. Figures are for Spring of each year; data for other seasons are obtainable from the Office for NationalStatistics.

Source of Data: Labour Force Survey, Office for National Statistics
Contact: Labour Market Statistics Helpline 020 7533 6094

Table 12.7 Employees and Self-Employed in the Construction Industry by Ethnic Origin

Great Britain — Thousands

	All	White	Non-White	Black	Indian	Pakistani/ Bangladeshi	Chinese	Other/ Mixed
Four quarter average: Spring 1999 to Winter 1999/2000								
All	1,871	1,829	42	19	11	. .	. .	. .
Female	162	157	. .	. .	. .	. .	. .	. .
Male	1,710	1,672	38	17	10	. .	. .	. .

Great Britain — Percent

	All	White	Non-White	Black	Indian	Pakistani/ Bangladeshi	Chinese	Other/ Mixed
Four quarter average: Spring 1999 to Winter 1999/2000								
All	100	97.7	2.3	1.0	0.6	. .	. .	. .
Female	100	97.0	. .	. .	. .	. .	. .	. .
Male	100	97.8	2.2	1.0	0.6	. .	. .	. .

Notes

. . = not available; sample size too small for reliable estimate.

Source of Data: Labour Force Survey, Office for National Statistics
Contact: Andrew Risdon 020 7533 6145

Table 12.8 Royal Institution of Chartered Surveyors – Total Workload Survey[1]

Year & Quarter		Total Workloads			
		More %	Same %	Less %	Net Balance %
1994	Q2	28	51	21	7
	Q3	31	50	19	11
	Q4	26	50	23	3
1995	Q1	21	52	27	–6
	Q2	17	56	27	–11
	Q3	12	56	32	–20
	Q4	24	53	23	1
1996	Q1	19	59	23	–4
	Q2	24	55	21	2
	Q3	28	54	18	9
	Q4	22	65	13	10
1997	Q1	26	54	19	7
	Q2	34	55	11	22
	Q3	26	62	12	14
	Q4	27	57	16	11
1998	Q1	26	62	12	14
	Q2	25	63	12	13
	Q3	25	63	12	12
	Q4	29	52	19	10
1999	Q1	28	49	22	6
	Q2	35	53	12	23
	Q3	32	53	15	17
	Q4	34	55	11	23

Notes

1. The table shows the percentage of chartered surveyors reporting a rise, fall, or unchanged workloads, compared to the previous quarter. By subtracting the number reporting a fall, a net balance figure (either positive or negative) is derived.

Source of Data: The Royal Institution of Chartered Surveyors
Contact: Milan Khatri 020 7334 3774

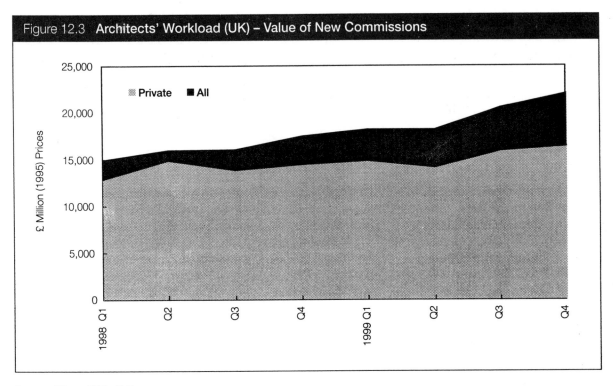

Figure 12.3 Architects' Workload (UK) – Value of New Commissions

Source of Data: Table 12.9

Table 12.9 Architects' Workload: Total Value of Construction Work in which Architects are Employed

Estimated value of new commissions analysed by building type (£ Million) United Kingdom.   1995 Prices

Year & Quarter		HOUSING			NON-HOUSING			TOTAL		
		Private	Public	All	Private	Public	All	Private	Public	All
1995		6,782	3,158	9,940	29,564	3,213	32,777	36,346	6,371	42,717
1996		6,636	2,762	9,399	31,091	4,788	35,879	37,728	7,550	45,278
1997		14,366	2,824	17,191	32,809	8,959	41,768	47,175	11,784	58,959
1998		8,539	2,498	11,037	49,601	4,925	54,526	58,140	7,423	65,563
1999		11,428	2,894	14,322	47,158	15,137	62,296	58,586	18,032	76,618
1998	Q1	4,279	1,583	5,862	8,538	537	9,075	12,817	2,120	14,937
	Q2	3,338	730	4,068	11,522	402	11,924	14,860	1,132	15,992
	Q3	3,282	670	3,952	10,552	1,560	12,112	13,835	2,230	16,064
	Q4	2,208	705	2,912	12,215	2,343	14,558	14,422	3,048	17,470
1999	Q1	1,982	496	2,478	12,832	2,890	15,723	14,815	3,386	18,201
	Q2	2,512	527	3,039	11,610	3,562	15,172	14,122	4,089	18,211
	Q3	2,797	1,031	3,828	13,123	3,581	16,704	15,920	4,612	20,532
	Q4	3,530	1,339	4,869	12,842	4,324	17,166	16,372	5,663	22,035

Source of Data: Mirza & Nacey
Contact: Aziz Mirza 01243 551302

Table 12.10 Architects' Workload: Total Value of Construction Work in which Architects are Employed

Estimated value of <u>production drawings</u> analysed by building type (£ Million) United Kingdom.   1995 Prices

Year & Quarter		HOUSING			NON-HOUSING			TOTAL		
		Private	Public	All	Private	Public	All	Private	Public	All
1995		3,247	2,124	5,371	14,865	3,209	18,074	18,112	5,333	23,445
1996		3,998	1,511	5,509	17,098	3,616	20,714	21,096	5,127	26,223
1997		7,485	1,487	8,972	18,643	4,235	22,878	26,128	5,722	31,850
1998		4,073	1,226	5,298	32,704	3,261	35,966	36,777	4,487	41,264
1999		6,228	1,355	7,583	20,756	3,719	24,475	26,984	5,074	32,057
1998	Q1	1,735	510	2,245	6,238	970	7,208	7,972	1,480	9,453
	Q2	1,274	369	1,643	7,877	571	8,449	9,152	940	10,092
	Q3	1,528	395	1,923	6,920	1,539	8,459	8,448	1,935	10,382
	Q4	1,115	335	1,449	8,049	1,304	9,352	9,163	1,638	10,801
1999	Q1	1,096	273	1,369	5,940	1,421	7,360	7,035	1,694	8,729
	Q2	1,065	301	1,366	6,398	1,169	7,567	7,463	1,470	8,933
	Q3	1,598	397	1,995	5,813	848	6,661	7,411	1,245	8,656
	Q4	2,438	374	2,812	4,082	772	4,854	6,520	1,146	7,666

Source of Data: Mirza & Nacey
Contact: Aziz Mirza 01243 551302

Table 12.11 Consulting Engineers' Fees[1] for Work in the UK

	Gross Fees Billed (£ Millions Current Prices)	RPI Adjusted Amount (from 1982) (£ Millions)
1990	1,241	788
1991	1,208	734
1992	1,298	768
1993	1,248	725
1994	1,162	660
1995	1,091	602
1996	1,083	579
1997	1,279	660
1998	1,677	847
1999	1,834	936

Notes

1. Estimated from returns (about 10% of ACE members), by grossing up within firm size bands, to reflect the fees of the entire ACE membership for their work in the UK.

Source of Data: The Association of Consulting Engineers
Contact: Craig Beaumont 020 7222 6557

Table 12.12 Consulting Engineers' Source of New Commissions

Firm Size – Staff based in the UK	Percentage of New Commissions (by Value)		
	Public	Private	PFI
Year 1999			
>500	65	34	1
100-499	25	74	1
10-99	34	66	0
<10	34	66	0
Total	**37**	**63**	**Negligible**

Source of Data: The Association of Consulting Engineers
Contact: Craig Beaumont 020 7222 6557

Table 12.13 CITB Trainee Numbers Survey: First-Year Intake for 1998/99[1,2]

Great Britain

| | New Entrant Trainees[3] | | Others | TOTAL |
	CITB	Non-CITB		
General Building Operatives	229	217	110	556
Carpenters & Joiners	5,637	2,978	2,605	11,220
Bench Joiners	724	445	416	1,585
Wood Machinists	414	60	359	833
Shopfitters	110	31	18	159
Bricklayers	2,862	1,711	2,110	6,683
Stonemasons	83	10	79	172
Painters & Decorators	1,826	1,185	1,450	4,461
Plasterers (Solid & Fibrous)	515	196	568	1,279
Total Building	**12,400**	**6,833**	**7,715**	**26,948**
Specialist Building	**434**	**74**	**160**	**668**
Plant Mechanics	185	28	65	278
Scaffolders	196	0	13	209
Others	79	5	6	90
Total Civil Engineering	**460**	**33**	**84**	**577**

Table 12.14 CITB Trainee Numbers Survey: First-Year Intake for 1997/98[1,2]

Great Britain

| | New Entrant Trainees[3] | | Others | TOTAL |
	CITB	Non-CITB		
General Building Operatives	7	246	246	499
Carpenters & Joiners	3,408	1,515	4,407	9,330
Bench Joiners	253	137	574	964
Wood Machinists	49	152	504	705
Shopfitters	91	10	95	196
Bricklayers	1,614	800	2,887	5,301
Stonemasons	53	9	48	110
Painters & Decorators	1,280	468	1,926	3,674
Plasterers	313	92	566	971
Total Building	**7,068**	**3,429**	**11,253**	**21,750**
Specialist Building	**378**	**45**	**388**	**811**
Plant Mechanics	258	47	146	451
Scaffolders	277	20	0	297
Others	51	0	0	51
Total Civil Engineering	**586**	**67**	**146**	**799**

Notes

1. The CITB Trainee Numbers Survey formerly known as the College Survey.

2. The CITB Trainee Numbers Survey is a voluntary survey. The figures may, therefore, exclude trainees in a number of colleges and other training centres.

3. New Entrant Trainees include National Traineeships, Modern Apprentices and Construction Apprentices.

4. Others include students under the age of 18 on full time college courses, adult trainees and employed trainees on day or block release.

Source of Data: Construction Industry Training Board
Contact: Martin Turner 01485 577640

Table 12.15 Claimant Unemployed for Skilled Construction Trades[1,2]

	Claimant Unemployed			Claimant Unemployed	
	Last Occupation[3]	Occupation Sought[4]		Last Occupation[3]	Occupation Sought[4]
North			**South West**		
Jan-98	5,181	5,476	Jan-98	5,005	5,105
Apr-98	4,500	4,757	Apr-98	4,138	4,261
Jul-98	3,767	4,016	Jul-98	3,742	3,944
Oct-98	3,374	3,630	Oct-98	3,411	3,526
Jan-99	4,432	4,746	Jan-99	3,999	4,189
Apr-99	3,966	4,229	Apr-99	3,642	3,945
Jul-99	3,483	3,751	Jul-99	3,142	3,399
Oct-99	3,041	3,365	Oct-99	2,798	2,959
Yorkshire & Humberside			**South East**		
Jan-98	5,945	6,417	Jan-98	10,077	10,397
Apr-98	5,111	5,584	Apr-98	8,555	8,925
Jul-98	4,448	4,883	Jul-98	7,572	7,930
Oct-98	3,991	4,446	Oct-98	6,933	7,311
Jan-99	4,940	5,513	Jan-99	8,314	8,811
Apr-99	4,531	5,147	Apr-99	7,495	7,945
Jul-99	3,839	4,452	Jul-99	6,597	7,158
Oct-99	3,530	4,100	Oct-99	5,809	6,397
North West			**Greater London**		
Jan-98	6,633	7,145	Jan-98	11,768	12,794
Apr-98	5,762	6,257	Apr-98	11,293	12,371
Jul-98	5,173	5,721	Jul-98	10,563	11,695
Oct-98	4,848	5,353	Oct-98	9,910	10,879
Jan-99	5,732	6,405	Jan-99	10,396	11,497
Apr-99	5,078	5,743	Apr-99	9,858	10,976
Jul-99	4,753	5,450	Jul-99	9,370	10,543
Oct-99	4,242	4,959	Oct-99	8,465	9,721
East Midlands			**England**		
Jan-98	3,096	3,367	Jan-98	55,765	59,398
Apr-98	2,650	2,925	Apr-98	49,067	52,714
Jul-98	2,359	2,630	Jul-98	44,096	47,835
Oct-98	2,188	2,469	Oct-98	40,759	44,275
Jan-99	2,710	3,040	Jan-99	47,556	51,876
Apr-99	2,511	2,831	Apr-99	43,638	47,985
Jul-99	2,198	2,504	Jul-99	39,303	43,785
Oct-99	1,944	2,217	Oct-99	35,115	39,613
West Midlands			**Wales**		
Jan-98	6,274	6,759	Jan-98	3,785	3,889
Apr-98	5,593	6,043	Apr-98	3,167	3,239
Jul-98	5,192	5,597	Jul-98	2,925	3,023
Oct-98	4,875	5,297	Oct-98	2,715	2,861
Jan-99	5,460	5,917	Jan-99	3,311	3,516
Apr-99	5,120	5,579	Apr-99	2,964	3,112
Jul-99	4,673	5,156	Jul-99	2,550	2,745
Oct-99	4,191	4,699	Oct-99	2,274	2,514
East Anglia			**Scotland**		
Jan-98	1,786	1,938	Jan-98	6,349	6,600
Apr-98	1,465	1,591	Apr-98	5,210	5,491
Jul-98	1,280	1,419	Jul-98	4,555	4,855
Oct-98	1,229	1,364	Oct-98	4,276	4,590
Jan-99	1,573	1,758	Jan-99	5,187	5,537
Apr-99	1,437	1,590	Apr-99	4,457	4,770
Jul-99	1,248	1,372	Jul-99	4,032	4,368
Oct-99	1,095	1,196	Oct-99	3,675	4,003

Table 12.15 Claimant Unemployed for Skilled Construction Trades[1,2] (continued)					
	Claimant Unemployed			Claimant Unemployed	
	Last Occupation[3]	Occupation Sought[4]		Last Occupation[4]	Occupation Sought[4]
Great Britain					
Jan-98	65,899	69,887			
Apr-98	57,444	61,444			
Jul-98	51,576	55,713			
Oct-98	47,750	51,726			
Jan-99	56,054	60,929			
Apr-99	51,059	55,867			
Jul-99	45,885	50,898			
Oct-99	41,064	46,130			

Notes

1. All included in Standard Occupational Classification 50 namely bricklayers, plasterers, painters & decorators, roofers, glaziers, floorers, scaffolders, builders, building contractors and other construction trades not elsewhere specified.

2. The standard regions are defined in Appendix 1.

3. 'Last Occupation' gives the total number of people unemployed whose last paid job was in the given occupation.

4. 'Occupation Sought' gives the total number of unemployed seeking employment in the given occupation.

Source of Data: Taylor Associates
Contact: David Taylor 020 8926 0473

Table 12.16 Job Centre Vacancies in Skilled Construction Trades[1,2]

	Vacancies Notified in Previous 3 Months	Stock of Unfilled Vacancies at end of Period		Vacancies Notified in Previous 3 Months	Stock of Unfilled Vacancies at end of Period
North			**South West**		
Jan-98	391	110	Jan-98	953	343
Apr-98	573	175	Apr-98	1,397	464
Jul-98	796	330	Jul-98	1,316	620
Oct-98	965	334	Oct-98	1,566	583
Jan-99	564	225	Jan-99	1,161	410
Apr-99	659	102	Apr-99	1,389	521
Jul-99	893	282	Jul-99	1,575	674
Oct-99	1,224	381	Oct-99	1,812	685
Yorkshire & Humberside			**South East**		
Jan-98	685	200	Jan-98	1,876	589
Apr-98	923	279	Apr-98	2,070	709
Jul-98	1,031	458	Jul-98	2,294	778
Oct-98	1,166	454	Oct-98	2,737	849
Jan-99	826	317	Jan-99	1,909	464
Apr-99	775	332	Apr-99	1,884	633
Jul-99	991	307	Jul-99	2,327	907
Oct-99	1,403	470	Oct-99	3,132	1,162
North West			**Greater London**		
Jan-98	703	271	Jan-98	1,908	802
Apr-98	912	361	Apr-98	1,516	846
Jul-98	1,032	469	Jul-98	1,675	983
Oct-98	1,125	470	Oct-98	2,020	1,021
Jan-99	799	320	Jan-99	1,398	830
Apr-99	863	364	Apr-99	1,229	566
Jul-99	1,054	421	Jul-99	1,672	650
Oct-99	1,503	477	Oct-99	2,571	987
East Midlands			**England**		
Jan-98	479	159	Jan-98	7,890	2,814
Apr-98	678	241	Apr-98	9,240	3,531
Jul-98	737	326	Jul-98	10,332	4,637
Oct-98	723	337	Oct-98	11,955	4,902
Jan-99	547	214	Jan-99	8,511	3,506
Apr-99	454	205	Apr-99	8,597	3,597
Jul-99	594	204	Jul-99	10,575	4,400
Oct-99	761	268	Oct-99	14,420	5,292
West Midlands			**Wales**		
Jan-98	555	198	Jan-98	679	227
Apr-98	803	319	Apr-98	862	285
Jul-98	920	468	Jul-98	651	334
Oct-98	1,104	590	Oct-98	772	300
Jan-99	988	605	Jan-99	563	210
Apr-99	1,003	688	Apr-99	684	271
Jul-99	1,023	777	Jul-99	876	274
Oct-99	1,419	663	Oct-99	934	293
East Anglia			**Scotland**		
Jan-98	340	142	Jan-98	1,118	375
Apr-98	368	137	Apr-98	1,784	504
Jul-98	531	205	Jul-98	2,243	602
Oct-98	549	264	Oct-98	2,088	685
Jan-99	319	121	Jan-99	1,453	631
Apr-99	341	186	Apr-99	1,844	592
Jul-99	446	178	Jul-99	1,939	689
Oct-99	595	199	Oct-99	1,970	706

Table 12.16 Job Centre Vacancies in Skilled Construction Trades[1,2]					
	Vacancies Notified in Previous 3 Months	Stock of Unfilled Vacancies at end of Period		Vacancies Notified in Previous 3 Months	Stock of Unfilled Vacancies at end of Period
Great Britain					
Jan-98	9,687	3,416			
Apr-98	11,886	4,320			
Jul-98	13,226	5,573			
Oct-98	14,815	5,887			
Jan-99	10,527	4,347			
Apr-99	11,125	4,460			
Jul-99	13,390	5,363			
Oct-99	17,324	6,291			

Notes

1. All included in Standard Occupational Classification 50 namely bricklayers, plasterers, painters & decorators, roofers, glaziers, floorers, scaffolders, builders, building contractors and other construction trades not elsewhere specified.

2. The standard regions are defined in Appendix 1.

Source of Data: Taylor Associates
Contact: David Taylor 020 8926 0473

Table 12.17 Claimant Unemployed for Carpenters and Joiners[1]

	Claimant Unemployed			Claimant Unemployed	
	Last Occupation[2]	Occupation Sought[3]		Last Occupation[2]	Occupation Sought[3]
North			**South West**		
Jan-98	1,404	1,548	Jan-98	1,000	1,055
Apr-98	1,286	1,429	Apr-98	837	879
Jul-98	1,085	1,201	Jul-98	802	868
Oct-98	917	1,006	Oct-98	756	812
Jan-99	1,248	1,344	Jan-99	862	929
Apr-99	1,238	1,356	Apr-99	791	869
Jul-99	992	1,113	Jul-99	675	776
Oct-99	787	888	Oct-99	626	672
Yorkshire & Humberside			**South East**		
Jan-98	1,704	1,900	Jan-98	1,885	1,865
Apr-98	1,579	1,751	Apr-98	1,599	1,619
Jul-98	1,349	1,504	Jul-98	1,509	1,527
Oct-98	1,145	1,295	Oct-98	1,340	1,365
Jan-99	1,608	1,819	Jan-99	1,637	1,698
Apr-99	1,437	1,635	Apr-99	1,464	1,517
Jul-99	1,187	1,379	Jul-99	1,271	1,350
Oct-99	993	1,205	Oct-99	1,021	1,120
North West			**Greater London**		
Jan-98	1,897	2,122	Jan-98	2,116	2,266
Apr-98	1,531	1,737	Apr-98	1,959	2,114
Jul-98	1,392	1,610	Jul-98	1,845	1,997
Oct-98	1,308	1,505	Oct-98	1,674	1,879
Jan-99	1,664	1,921	Jan-99	1,797	2,016
Apr-99	1,429	1,700	Apr-99	1,669	1,917
Jul-99	1,221	1,481	Jul-99	1,554	1,816
Oct-99	1,035	1,257	Oct-99	1,373	1,683
East Midlands			**England**		
Jan-98	765	832	Jan-98	12,456	13,435
Apr-98	686	733	Apr-98	10,903	11,871
Jul-98	613	676	Jul-98	9,970	10,933
Oct-98	555	627	Oct-98	8,926	9,890
Jan-99	745	825	Jan-99	11,046	12,206
Apr-99	658	719	Apr-99	10,076	11,288
Jul-99	585	653	Jul-99	8,790	10,057
Oct-99	498	565	Oct-99	7,443	8,655
West Midlands			**Wales**		
Jan-98	1,284	1,423	Jan-98	1,042	1,092
Apr-98	1,110	1,244	Apr-98	873	938
Jul-98	1,067	1,198	Jul-98	805	861
Oct-98	973	1,085	Oct-98	700	757
Jan-99	1,155	1,267	Jan-99	860	938
Apr-99	1,083	1,192	Apr-99	760	840
Jul-99	1,031	1,145	Jul-99	651	743
Oct-99	859	964	Oct-99	601	690
East Anglia			**Scotland**		
Jan-98	401	424	Jan-98	2,630	2,766
Apr-98	316	365	Apr-98	2,171	2,293
Jul-98	308	352	Jul-98	1,874	2,027
Oct-98	258	316	Oct-98	1,636	1,773
Jan-99	330	387	Jan-99	2,072	2,226
Apr-99	307	383	Apr-99	1,610	1,773
Jul-99	274	344	Jul-99	1,482	1,685
Oct-99	251	301	Oct-99	1,191	1,376

Table 12.17 Claimant Unemployed for Carpenters and Joiners[1] (continued)

| | Claimant Unemployed | | | Claimant Unemployed | |
	Last Occupation[2]	Occupation Sought[3]		Last Occupation[2]	Occupation Sought[3]
Great Britain					
Jan-98	16,128	17,293			
Apr-98	13,947	15,102			
Jul-98	12,649	13,821			
Oct-98	11,262	12,420			
Jan-99	13,978	15,370			
Apr-99	12,446	13,901			
Jul-99	10,923	12,485			
Oct-99	9,235	10,721			

Notes

1. The standard regions are defined in Appendix 1.
2. 'Last Occupation' gives the total number of people unemployed whose last paid job was in the given occupation.
3. 'Occupation Sought' gives the total number of unemployed seeking employment in the given occupation.

Source of Data: Taylor Associates
Contact: David Taylor 020 8926 0473

Table 12.18 Overseas Service Trade by the Type of Service being Supplied Overseas (Exports) or being Purchased from Overseas (Imports) – 1998

			£ Million
Technical Services	**Exports**	**Imports**	**Net**
Architectural	59	12	48
Surveying	41	26	15
Construction	298	111	187

Notes

1. Figures for 1999 will be published in the issue covering 1990-2000.

Source of Data: Office for National Statistics
Contact: Chris Peggie 01633 813458

Table 12.19 Overseas Trade in Services by the Industrial Classification of the Business – 1998

			£ Million
Industry	**Exports**	**Imports**	**Net**
Property Management	32	17	15
Architectural Activities	51	12	39
Quantity Surveying Activities	53	28	26
Consultant Engineering and Design Activities	1,522	609	913

Notes

1. Figures for 1999 will be published in the issue covering 1990-2000.

Source of Data: Office for National Statistics
Contact: Chris Peggie 01633 813458

Table 12.20 Professional and Scientific Services[1]

United Kingdom

| | | | | Persons engaged as: | | | |
	Description	Number of Businesses	Total Turnover[2] (£ Million)	Surveyors (Thousand)	Architects (Thousand)	Draughtsmen (Thousand)	Consultant Engineers (Thousand)
1995: Number of professional persons engaged by selected SIC (92)							
74.20	Architectural and engineering activities and related technical consultancy	45,927	15,123	32.8	20.3	27.7	53.3
1996: Number of professional persons engaged by selected SIC (92)							
74.20	Architectural and engineering activities and related technical consultancy	48,884	18,057	27.3	21.1	26.9	58.6
1997: Number of professional persons engaged by selected SIC (92)[3]							
74.20	Architectural and engineering activities and related technical consultancy	51,986	22,662	..	..	..	..
1998: Number of professional persons engaged by selected SIC (92)[3]							
74.20	Architectural and engineering activities and related technical consultancy	57,289	24,615	..	..	..	..

Notes

.. = not available

1. Figures for 1999 will be published in the issue covering 1990-2000.

2. Exclusive of VAT.

3. Changes have been made to the system running the results and the methodology used to calculate the aggregate data. Consequently data for the split between different professions are no longer available.

Source of Data: Business Statistics Group, ONS

Contact: Keri J Hodson 01633 812264

Table 12.21 Stoppages in the Construction Industry in the UK[1]

Year	Month	Number of Stoppages[2]	Workers Involved[2,3]	Working Days Lost[3]
1989		40	20,100	128,000
1990		12	4,500	14,000
1991		18	6,200	14,000
1992		12	3,900	10,000
1993		4	1,200	1,000
1994		4	800	5,000
1995		9	1,700	10,000
1996		11	3,100	8,000
1997		11	12,600	16,900
1998		13	2,400	13,000
1999		19	17,800	48,800
1990	Jan	0	0	0
	Feb	0	0	0
	Mar	3	500	4,000
	Apr	1	1,000	1,000
	May	1	..	..
	Jun	2	300	1,000
	Jul	1	100	..
	Aug	1	100	1,000
	Sep	2	1,000	1,000
	Oct	1	100	..
	Nov	1	1,500	5,000
	Dec	1	100	..
1991	Jan	1	1,800	4,000
	Feb	1	200	..
	Mar	3	600	3,000
	Apr	4	2,000	2,000
	May	0	0	0
	Jun	2	200	1,000
	Jul	4	600	1,000
	Aug	1	100	..
	Sep	2	800	4,000
	Oct	0	0	0
	Nov	1	100	..
	Dec	0	0	0
1992	Jan	0	0	0
	Feb	1	..	..
	Mar	3	1,100	4,000
	Apr	1	100	..
	May	1	100	1,000
	Jun	2	500	3,000
	Jul	1	..	..
	Aug	2	1,100	1,000
	Sep	1	..	..
	Oct	2	1,100	1,000
	Nov	0	0	0
	Dec	0	0	0
1993	Jan	0	0	0
	Feb	0	0	0
	Mar	1	500	1,000
	Apr	0	0	0
	May	0	0	0
	Jun	1	100	..
	Jul	0	0	0
	Aug	0	0	0
	Sep	0	0	0
	Oct	1	..	..
	Nov	0	0	0
	Dec	1	100	..

Table 12.21 Stoppages in the Construction Industry in the UK[1] (continued)

Year	Month	Number of Stoppages[2]	Workers Involved[2,3]	Working Days Lost[3]
1994	Jan	0	0	0
	Feb	0	0	0
	Mar	0	0	0
	Apr	1	..	..
	May	1	100	..
	Jun	1	600	4,000
	Jul	0	0	0
	Aug	0	0	0
	Sep	0	0	0
	Oct	1	100	..
	Nov	0	0	0
	Dec	0	0	0
1995	Jan	0	0	0
	Feb	0	0	0
	Mar	2	500	5,000
	Apr	3	200	1,000
	May	1	..	..
	Jun	2	..	1,000
	Jul	1	200	..
	Aug	0	0	0
	Sep	1	500	..
	Oct	0	0	0
	Nov	1	300	2,000
	Dec	1	300	1,000
1996	Jan	..	..	..
	Feb	5	2,300	5,000
	Mar	1	..	..
	Apr	1	300	3,000
	May	1	100	..
	Jun	2	300	..
	Jul	..	..	..
	Aug	..	..	..
	Sep	..	..	..
	Oct	1	100	..
	Nov	..	..	..
	Dec	..	..	..
1997	Jan	..	..	..
	Feb	..	..	..
	Mar	..	..	..
	Apr	1	..	1,100
	May	3	1,800	1,600
	Jun	..	..	..
	Jul	..	..	..
	Aug	..	..	..
	Sep	..	..	..
	Oct	4	3,000	5,300
	Nov	2	6,300	6,300
	Dec	3	3,500	2,700
1998	Jan	4	900	1,500
	Feb	4	2,500	9,400
	Mar	1	200	1,00
	Apr	2	200	300
	May	1	100	100
	Jun	..	..	..
	Jul	..	..	..
	Aug	..	..	..
	Sep	..	..	..
	Oct	1	100	100
	Nov	1	400	400
	Dec	1	100	300

Table 12.21 Stoppages in the Construction Industry in the UK[1] (continued)

Year	Month	Number of Stoppages[2]	Workers Involved[2,3]	Working Days Lost[3]
1999	Jan	1	..	100
	Feb	3	500	600
	Mar	2	200	200
	Apr	..	..	..
	May	2	4,500	25,400
	Jun	..	..	..
	Jul	3	1,400	3,200
	Aug	2	100	400
	Sep	1	8,000	16,00
	Oct	1	300	300
	Nov	3	1,000	1,000
	Dec	2	2,100	1,700

Notes

.. = not available; less than 50 workers involved or less than 50 working days lost.

1. Statistics cover stoppages of work in progress in the UK during a year caused by labour disputes between employers and workers, or between workers and other workers, connected with terms and conditions of employment. Stoppages invloving fewer than ten workers or lasting less than one day are also excluded, unless the total number of working days lost in the dispute is 100 or more.

2. The monthly figures are snap-shots and will be shown for each month of the strike, but only once in the yearly figure.

3. Figures for workers involved and working days lost are rounded to the nearest 100.

Source of Data: Office for National Statistics
Contact: Jackie Davies 01928 792825

Table 12.22 Injuries in the Construction Industry Reported to All Enforcement Authorities[1] Analysed by Severity of Injury

Based on the findings of the Labour Force Survey   Great Britain

	1991/92	1992/93	1993/94	1994/95	1995/96	1996/97[5]	1997/98	1998/99(p)
Employees								
Fatal injuries[2]	83	70	75	58	62	66	58	48
Non-fatal major injuries[3]	2,583	2,061	1,806	1,872	1,806	3,227	3,860	4,263
Over 3 day injuries[4]	14,998	11,428	9,497	9,642	8,305	8,637	9,756	9,236
All reported injuries	17,664	13,559	11,378	11,572	10,173	11,930	13,674	13,547
Self-employed								
Fatal injuries[2]	17	26	16	25	17	24	22	18
Non-fatal major injuries[3]	729	684	768	755	671	827	466	356
Over 3 day injuries[4]	1,231	1,291	1,576	1,532	1,390	1,029	509	373
All reported injuries	1,977	2,001	2,360	2,312	2,078	1,880	997	747
Members of the public								
Fatal injuries[2]	6	5	6	5	3	3	6	4
Non-fatal major injuries[3]	148	104	116	121	117	405	339	382
All reported injuries[4]	154	109	122	126	120	408	345	386
Total								
Fatal injuries[2]	106	101	97	88	82	93	86	70
Non-fatal major injuries[3]	3,460	2,849	2,690	2,748	2,594	4,459	4,665	5,001
Over 3 day injuries[4]	16,229	12,719	11,073	11,174	9,695	9,666	10,265	9,609
All reported injuries	19,795	15,669	13,860	14,010	12,371	14,218	15,016	14,680

Notes

p = provisional

1. Reported to Enforcement Authorities under the Reporting of Injuries, Diseases and Dangerous Occurrences Regulations 1985 and from 1996/97 1995 (RIDDOR).

2. The definition of a fatal injury includes a death occurring up to a year after the accident.

3. As defined in Regulation 3(2) of RIDDOR 1985 and Regulation 2(1) of RIDDOR 1995.

4. These are defined as injuries not falling in the previous two categories, necessitating absence from work on more than three consecutive working days.

5. Injury figures from 1996/97 cannot be directly compared to previous years due to the introduction of RIDDOR 1995.

Source of Data: Health and Safety Executive
Contact: Operations Unit 0151 951 4341

Table 12.23 Injuries to Employees in the Construction Sector Reported to HSE's Field Operations Division Inspectorates and Local Authorities during 1998/99 (p)

Great Britain

Kind of Accident	Fatal	Major	Severity of Injury Over 3 Days	Total
Contact with moving machinery or material being machined	2	124	266	392
Struck by moving including flying/falling object	7	773	1,681	2,461
Struck by moving vehicle	9	124	134	267
Struck against something fixed or stationary	–	144	454	598
Injured whilst handling, lifting or carrying	–	342	3,257	3,599
Slip, trip or fall on same level	–	874	1,599	2,473
Fall from a height:	–	1,574	1,248	2,844
Up to and including two metres	–	764	797	1,561
Over two metres	21	706	315	1,042
Height not stated	1	104	136	241
Trapped by something collapsing or overturning	3	56	48	107
Drowning or asphyxiation	1	13	5	19
Exposure to or contact with harmful substance	1	85	195	281
Exposure to fire	–	7	16	23
Exposure to an explosion	1	13	17	31
Contact with electricity or an electrical discharge	2	70	75	147
Injured by an animal	–	–	26	26
Injuries caused by assault or violence	–	15	40	55
Other kind of accident	–	43	160	203
Injuries not classified by kind	–	6	15	21
Total	**48**	**4,263**	**9,236**	**13,547**

Notes

p = provisional

Source of Data: Health and Safety Executive
Contact: Operations Unit 0151 951 4341

Table 12.24 Construction v All Industries: Injuries to Employees Reported to All Enforcement Authorities

Great Britain

Rates per 100,000 employees	1991/92	1992/93	1993/94	1994/95	1995/96	1996/97	1997/98	1998/99(p)
Fatal and major:								
Construction	269	238	223	228	232	411	388	399
All Industries	83	82	80	81	78	128	129	119
All reported injuries:								
Construction	1,785	1,516	1,351	1,368	1,262	1,490	1,354	1,254
All Industries	791	751	721	738	685	709	718	666

Notes

p = provisional

Source of Data: Health and Safety Executive
Contact: Operations Unit 0151 951 4341

Table 12.25 HSE Enforcement Actions: Construction Industry

Enforcement notices issued by HSE's Field Operations Division Inspectorates 1991/92 – 1998/99 (p)
The most common type of notice served in the construction industry was an immediate prohibition notice which stops a work activity until an imminent risk of personal injury is eliminated. Nearly all other notices issued in this industrial sector were improvement notices which require employers to put right a contravention of health and safety legislation within a specified time limit.

Great Britain

	1991/92	1992/93	1993/94	1994/95	1995/96	1996/97	1997/98	1998/99 (p)
Notices issued	2,169	2,410	2,156	2,599	1,864	1,976	2,028	2,647

Proceedings taken by HSE's Field Operations Inspectorates by result 1991/92 – 1998/99 (p)
These figures are based on the informations laid by inspectors before Magistrates in England and Wales, and charges preferred in Scottish Courts. Each information laid or charge preferred relates to a breach of an individual legal requirement, and one case may involve one or more of these breaches.

	1991/92	1992/93	1993/94	1994/95	1995/96	1996/97	1997/98	1998/99 (p)
Informations laid	747	702	531	630	598	508	719	699
Convictions	605	573	415	494	476	385	544	556
Average penalty per conviction (£)	1,035	1,300	3,384 [1]	2,697 [2]	2,232	3,934 [3]	3,123	5,741 [4]

Notes
p = provisional
1. Includes the fine of £150,000 against J Murphy and Sons Ltd. The average fine without this conviction would be £3,030.
2. Includes the fine of £200,000 against BP Chemicals. The average fine without this conviction would be £2,677.
3. Includes the fine of £100,000 against Cheetham Hill Construction Ltd and the fine of £125,000 against TE Scudder Ltd. The average fine without these convictions would be £3,367.
4. Includes the fine of £1.2m against Balfour Beatty Civil Engineering Ltd and the fine of £500,000 against Geoconsult ZT GES MBH. The average fine without these convictions would be £2,693.

Source of Data: Health and Safety Executive
Contact: Operations Unit 0151 951 4341

CHAPTER 13

Northern Ireland, Wales and Scotland

Figure 13.1 Northern Ireland – Index of Construction Output[1,2]

Source of Data: Table 13.1

Table 13.1 Northern Ireland – Index of Construction Output[1,2]

Year	Quarter	Unadjusted	Index 1995=100 Seasonally Adjusted
1995 (r)	Q1	108	102
	Q2	96	100
	Q3	100	100
	Q4	96	98
1996 (r)	Q1	108	102
	Q2	98	103
	Q3	106	105
	Q4	103	106
1997 (r)	Q1	108	102
	Q2	98	103
	Q3	100	100
	Q4	105	107
1998 (r)	Q1	114	108
	Q2	105	110
	Q3	110	109
	Q4	106	108
1999 (p)	Q1	103	98
	Q2	102	106
	Q3	104	103
	Q4	99	101

Notes

p = provisional

r = revised

1. The Index has been rebased from 1990 to 1995 prices.

2. The Index was revised in May 2000.

Source of Data: Central Statistics and Research Branch, Department for Regional Development (NI)

Contact: Rodney Redmond 028 9054 0799

Table 13.2 Northern Ireland – Housing Executive Starts by District Council

District Council Area	1993 Total	1994 Total	1995 Total	1996 Total	1997 Total	1998 Total	1999 Total	1999 Q1	Q2	Q3	Q4
Antrim	26	0	2	4	0	0	0	0	0	0	0
Ards	53	85	64	88	38	0	2	0	0	2	0
Armagh	0	18	30	0	19	15	0	0	0	0	0
Ballymena	57	10	0	0	56	0	0	0	0	0	0
Ballymoney	0	12	0	0	31	0	0	0	0	0	0
Banbridge	22	20	0	3	10	14	0	0	0	0	0
Belfast	148	301	251	343	237	112	43	4	4	23	12
Carrickfergus	0	7	31	0	34	0	0	0	0	0	0
Castlereagh	30	0	0	5	0	0	0	0	0	0	0
Coleraine	32	12	0	0	0	0	0	0	0	0	0
Cookstown	0	0	2	0	0	6	0	0	0	0	0
Craigavon	37	37	42	42	22	14	0	0	0	0	0
Derry	327	136	121	53	45	22	0	0	0	0	0
Down	35	53	40	52	69	0	32	12	7	11	2
Dungannon	4	10	62	13	21	0	0	0	0	0	0
Fermanagh	71	6	28	22	35	7	4	3	0	1	0
Larne	53	0	16	0	5	0	0	0	0	0	0
Limavady	13	19	2	0	0	0	0	0	0	0	0
Lisburn	21	138	126	82	25	23	2	0	0	2	0
Magherafelt	62	11	44	12	0	0	0	0	0	0	0
Moyle	27	8	27	0	7	6	6	6	0	0	0
Newry & Mourne	36	52	34	123	74	0	0	0	0	0	0
Newtownabbey	0	48	1	6	14	34	0	0	0	0	0
North Down	24	0	14	101	6	2	0	0	0	0	0
Omagh	8	10	37	31	6	3	0	0	0	0	0
Strabane	46	78	20	35	63	3	6	0	4	2	0
TOTAL	**1,132**	**1,071**	**994**	**1,015**	**817**	**261**	**95**	**25**	**15**	**41**	**14**

Source of Data: Central Statistics and Research Branch, Department for Regional Development (NI)
Contact: Rodney Redmond 028 9054 0799

Table 13.3 Northern Ireland – Private Sector Housing Starts by District Council

District Council Area	1993 Total	1994 Total	1995 Total	1996 Total	1997 Total	1998 Total	1999 Total	1999 Q1	Q2	Q3	Q4
Antrim	274	271	238	246	341	196	69	16	24	13	16
Ards	436	509	429	499	443	575	631	132	127	200	172
Armagh	235	260	314	265	390	385	319	107	70	55	87
Ballymena	326	276	297	303	337	266	441	134	158	68	81
Ballymoney	135	117	178	196	316	194	210	51	75	41	43
Banbridge	274	297	304	372	359	478	258	33	60	28	137
Belfast	172	80	158	68	500	350	826	213	114	247	252
Carrickfergus	302	286	237	232	300	199	300	56	116	55	73
Castlereagh	390	430	447	486	210	148	218	27	95	69	27
Coleraine	458	496	479	468	531	512	621	113	175	128	205
Cookstown	130	196	161	131	110	145	244	85	69	53	37
Craigavon	427	435	380	467	588	511	235	24	71	41	99
Derry	492	536	620	520	343	374	390	128	80	137	45
Down	259	307	349	375	451	412	384	153	134	46	51
Dungannon	50	85	182	332	385	350	264	72	61	96	35
Fermanagh	316	328	343	396	314	535	307	96	98	68	45
Larne	106	200	251	170	194	153	281	61	55	91	74
Limavady	181	214	162	196	190	246	294	22	123	84	65
Lisburn	494	559	601	498	466	384	307	61	60	101	85
Magherafelt	159	224	237	295	335	271	251	60	72	65	54
Moyle	67	120	102	132	243	165	195	46	57	66	26
Newry & Mourne	574	510	503	563	530	681	473	121	171	61	120
Newtownabbey	376	377	372	417	433	453	615	98	215	157	145
North Down	209	235	214	274	258	259	375	110	87	55	123
Omagh	142	136	209	142	325	417	349	45	129	85	90
Strabane	86	205	208	207	223	210	118	36	38	26	18
TOTAL	**7,070**	**7,689**	**7,975**	**8,250**	**9,115**	**8,869**	**8,975**	**2,100**	**2,534**	**2,136**	**2,205**

Source of Data: Central Statistics and Research Branch, Department for Regional Development (NI)
Contact: Rodney Redmond 028 9054 0799

Table 13.4 Northern Ireland – Housing Association Starts by District Council

District Council Area	1993 Total	1994 Total	1995 Total	1996 Total	1997 Total	1998 Total	1999 Total	1999 Q1	Q2	Q3	Q4
Antrim	0	0	0	0	0	13	0	0	0	0	0
Ards	53	51	114	0	0	16	51	30	0	0	21
Armagh	0	54	0	0	6	110	4	0	0	0	4
Ballymena	56	95	27	38	14	20	65	63	2	0	0
Ballymoney	0	0	0	40	22	0	0	0	0	0	0
Banbridge	0	0	0	19	21	0	32	32	0	0	0
Belfast	130	174	139	259	298	357	496	315	1	108	72
Carrickfergus	0	48	21	17	19	46	98	71	0	0	27
Castlereagh	40	0	0	2	22	37	59	59	0	0	0
Coleraine	11	1	13	13	62	16	1	1	0	0	0
Cookstown	0	0	10	1	0	12	0	0	0	0	0
Craigavon	15	25	53	29	7	29	57	57	0	0	0
Derry	0	84	274	167	176	209	159	54	0	0	105
Down	33	0	0	0	44	75	59	47	0	0	12
Dungannon	25	31	0	74	35	8	10	10	0	0	0
Fermanagh	0	43	18	13	6	44	50	40	4	0	6
Larne	0	26	22	32	17	0	0	0	0	0	0
Limavady	0	0	16	28	32	41	30	30	0	0	0
Lisburn	20	100	0	4	44	87	170	113	29	15	13
Magherafelt	0	0	0	3	13	16	0	0	0	0	0
Moyle	0	30	0	0	24	20	8	8	0	0	0
Newry & Mourne	28	6	39	51	72	42	67	23	34	0	10
Newtownabbey	77	38	12	10	75	17	14	0	0	0	14
North Down	0	111	0	0	63	12	18	7	11	0	0
Omagh	1	10	46	11	6	91	37	16	0	0	21
Strabane	0	0	6	10	16	11	59	59	0	0	0
TOTAL	489	927	810	821	1,094	1,329	1,544	1,035	81	123	305

Source of Data: Central Statistics and Research Branch, Department for Regional Development (NI)
Contact: Rodney Redmond 028 9054 0799

Table 13.5 Wales – New Orders obtained by Contractors

£ Million

Year	Quarter	New Housing			Other New Work					All New Work
		Public	Private	All New Housing	Infra-structure	Public	Private Industrial	Private Commercial	All Other New Work	
1989		42	310	352	189	154	161	219	722	1,073
1990	Q1	22	100	122	50	27	40	74	192	313
	Q2	22	119	141	29	61	38	63	192	332
	Q3	10	58	68	36	60	44	52	192	260
	Q4	8	59	66	38	41	27	43	150	216
1991	Q1	7	40	47	110	46	43	45	244	291
	Q2	9	49	59	35	19	44	72	170	229
	Q3	15	61	76	129	47	61	26	263	339
	Q4	14	43	57	99	51	22	43	216	273
1992	Q1	24	68	92	51	68	24	31	173	265
	Q2	12	63	75	38	34	11	29	113	187
	Q3	19	34	53	47	33	12	25	117	169
	Q4	13	38	51	43	28	29	24	124	175
1993	Q1	18	52	70	69	33	13	22	137	207
	Q2	6	56	62	14	20	17	28	79	141
	Q3	12	51	63	34	41	23	52	150	213
	Q4	19	50	69	117	64	23	46	249	318
1994	Q1	30	64	94	126	65	13	42	246	340
	Q2	37	51	88	136	58	42	43	280	367
	Q3	10	62	71	47	76	45	62	229	300
	Q4	9	32	41	35	98	31	52	215	256
1995	Q1	17	60	77	77	28	17	42	164	241
	Q2	9	60	69	57	42	30	54	183	253
	Q3	7	45	52	59	40	28	45	171	223
	Q4	16	47	63	105	61	18	32	215	279
1996	Q1	17	54	71	63	38	49	111	262	332
	Q2	13	53	66	30	36	45	36	147	213
	Q3	10	57	68	48	34	37	37	157	224
	Q4	19	50	69	53	32	47	85	217	286
1997	Q1	19	76	95	49	39	44	149	281	376
	Q2	7	76	83	82	31	43	84	240	324
	Q3	4	50	54	24	45	64	71	205	259
	Q4	7	50	57	61	48	38	49	196	253
1998	Q1	19	55	74	52	42	50	74	218	292
	Q2	12	62	74	87	24	29	57	197	271
	Q3	10	50	60	41	47	43	66	198	258
	Q4	10	51	60	60	39	19	101	219	279
1999	Q1	30	45	75	34	22	35	51	143	217
	Q2	5	60	64	57	18	24	50	148	213
	Q3	2	58	60	74	28	56	97	255	315
	Q4	6	52	58	91	35	24	41	191	249

Source of Data: Construction Market Intelligence, Department of the Environment, Transport and the Regions
Contact: Neville Price 020 7944 5587

Table 13.6 Wales – Output obtained by Contractors

£ Million

Year & Quarter		New Housing		Other New Work Non-Housing				All New Work	Repair & Maintenance				All Work
					Other New Work Exc. Infrastructure								
		Public	Private	Infra-structure	Public	Private Indus-trial	Private Com-mercial		Housing	Other Work		Repair & Main-tenance	
										Public	Private		
1989		42	313	187	189	208	236	1,175	517	102	172	791	1,966
1990	Q1	13	67	44	41	57	64	286	138	27	49	214	500
	Q2	11	73	61	51	45	70	311	138	29	48	214	526
	Q3	17	94	67	49	43	65	335	135	27	47	210	544
	Q4	16	69	50	46	65	62	308	136	26	56	219	527
1991	Q1	15	42	60	56	42	57	273	130	29	50	209	482
	Q2	15	41	65	49	45	53	269	120	28	53	201	470
	Q3	13	48	82	42	54	49	289	133	29	45	206	495
	Q4	10	58	97	42	53	48	307	130	29	47	205	512
1992	Q1	18	61	91	52	50	57	329	122	31	48	201	530
	Q2	19	64	101	57	42	49	332	117	29	49	195	527
	Q3	20	66	105	59	32	37	318	129	29	45	203	521
	Q4	19	59	93	52	25	44	291	133	28	52	214	505
1993	Q1	19	48	93	57	15	29	261	119	29	46	194	455
	Q2	17	50	76	50	29	31	252	121	25	49	196	448
	Q3	16	44	67	41	27	36	231	124	30	49	203	434
	Q4	14	56	91	40	39	38	278	132	29	51	211	489
1994	Q1	15	52	87	49	32	47	282	135	32	49	215	497
	Q2	21	58	98	50	28	47	302	132	30	49	211	513
	Q3	23	59	98	61	52	58	351	137	32	51	220	572
	Q4	19	57	104	56	69	59	366	138	30	56	225	591
1995	Q1	19	55	110	73	44	55	357	141	32	51	224	581
	Q2	16	61	144	75	39	56	391	141	28	54	223	615
	Q3	15	54	124	74	42	66	374	148	31	58	237	612
	Q4	14	51	111	68	32	56	333	146	30	62	237	570
1996	Q1	12	59	107	70	27	51	325	145	39	63	246	572
	Q2	16	41	99	67	41	47	312	143	31	60	233	545
	Q3	19	61	102	76	48	57	364	153	33	63	249	612
	Q4	17	65	93	57	47	64	344	154	36	67	258	601
1997	Q1	16	49	78	53	53	68	316	149	33	66	248	564
	Q2	18	65	81	54	67	80	365	149	29	69	248	613
	Q3	17	69	68	75	64	98	391	161	32	71	263	654
	Q4	13	75	128	52	62	89	418	170	43	77	290	708
1998	Q1	12	71	76	55	60	97	371	156	36	80	271	642
	Q2	12	83	68	54	50	103	371	151	36	69	255	626
	Q3	12	66	71	62	48	109	368	160	31	71	262	630
	Q4(p)	12	72	126	62	67	110	450	171	46	77	294	743
1999	Q1(p)	10	71	67	68	62	120	398	159	34	53	245	644
	Q2(p)	10	83	69	63	53	131	409	148	35	64	247	656
	Q3(p)	12	68	74	75	48	121	400	145	42	73	259	659
	Q4(p)	9	64	82	56	60	120	391	157	36	56	249	640

Notes

p = provisional

Source of Data: Construction Market Intelligence, Department of the Environment,Transport and the Regions
Contact: Neville Price 020 7944 5587

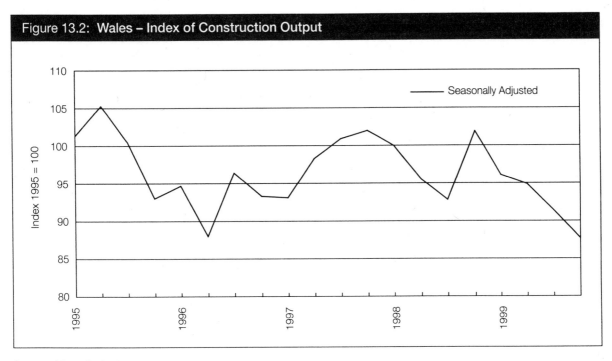

Figure 13.2: Wales – Index of Construction Output

Source of Data: Table 13.7

Table 13.7 Wales – Index of Construction Output

Seasonally Adjusted   Index 1995 = 100

Year	Quarter	Construction Index
1991	Q1	86
	Q2	84
	Q3	86
	Q4	91
1992	Q1	97
	Q2	97
	Q3	94
	Q4	91
1993	Q1	84
	Q2	85
	Q3	81
	Q4	90
1994	Q1	95
	Q2	96
	Q3	101
	Q4	101
1995	Q1	101
	Q2	105
	Q3	100
	Q4	93
1996	Q1	95
	Q2	88
	Q3	96
	Q4	93
1997	Q1	93
	Q2	98
	Q3	101
	Q4	102
1998	Q1	100
	Q2	96
	Q3	93
	Q4	102
1999	Q1	96
	Q2	95
	Q3	91
	Q4	88

Source of Data: National Assembly for Wales
Contact: Kathryn Hughes 029 2082 5017

Table 13.8 Wales – Planning Applications

Year	Quarter	Number Granted	Number Decided	% Granted	% Decided Within 13 Weeks
For Welsh Planning Authorities					
1991	Q1	5,650	7,030	80	75
	Q2	6,338	8,148	78	79
	Q3	6,898	9,182	75	81
	Q4	6,053	7,666	79	82
1992	Q1	6,065	7,488	81	81
	Q2	6,388	7,601	84	86
	Q3	6,394	7,595	84	85
	Q4	6,227	7,359	85	85
1993	Q1	5,972	6,941	86	85
	Q2	6,520	7,484	87	89
	Q3	6,679	7,696	87	88
	Q4	6,166	7,158	86	87
1994	Q1	6,130	7,138	86	86
	Q2	6,456	7,432	87	89
	Q3	6,661	7,635	87	89
	Q4	6,100	7,024	87	88
1995	Q1	5,835	6,719	87	86
	Q2	6,184	7,057	88	90
	Q3	6,420	7,356	87	89
	Q4	5,982	6,811	88	87
1996	Q1 [1]	5,599	6,269	89	85
For Welsh Unitary Authorities					
	Q2 [2,3]	4,416	4,900	90	86
	Q3 [4]	5,432	5,998	91	85
	Q4	5,494	6,177	89	83
1997	Q1	5,201	5,890	88	81
	Q2	6,066	6,760	90	87
	Q3	6,272	6,893	91	87
	Q4	5,970	6,647	90	85
1998	Q1	5,562	6,175	90	85
	Q2	6,276	6,993	90	88
	Q3	6,413	7,115	90	88
	Q4	5,749	6,467	89	87
1999	Q1	5,236	5,870	89	84
	Q2 [5]	5,837	6,403	91	88

Notes

1. No data available for Lliw Valley or Port Talbot.
2. Data only supplied for the geographical areas of the former Boroughs of Dinefwr and Llanelli (data unavailable for the former District of Carmarthen).
3. No data available for Denbighshire, Flintshire or Neath Port Talbot.
4. No data available for Flintshire.
5. No data available for Brecon Beacons National Park or Conwy.

Source of Data: National Assembly for Wales
Contact: Kathryn Hughes 029 2082 5017

Table 13.9 Wales – Mineral Planning Applications

Year	Quarter	Number Granted	Number Decided	% Granted	% Decided Within 13 Weeks
Major mineral planning applications for Welsh County Councils					
1991	Q1	17	18	94	22
	Q2	17	18	94	6
	Q3	21	22	95	32
	Q4	11	13	85	54
1992	Q1	32	33	97	36
	Q2	47	49	96	35
	Q3	25	29	86	45
	Q4	26	28	93	46
1993	Q1	13	13	100	31
	Q2	25	26	96	38
	Q3	34	34	100	38
	Q4	21	21	100	52
1994	Q1	20	23	87	35
	Q2	24	28	86	50
	Q3	23	23	100	48
	Q4	29	32	91	47
1995	Q1	15	20	75	30
	Q2	18	19	95	47
	Q3	12	13	92	54
	Q4	17	18	94	17
1996	Q1 [1]	..	..	..	..
All mineral planning applications for Welsh Unitary Authorities					
1996	Q2 [2,3]	5	6	83	50
	Q3 [4]	15	22	68	36
	Q4	16	18	89	50
1997	Q1	39	44	89	45
	Q2	22	26	85	62
	Q3	48	49	98	82
	Q4	53	57	93	72
1998	Q1	22	23	96	70
	Q2	22	25	88	56
	Q3	32	33	97	76
	Q4	15	17	88	59
1999	Q1	26	31	84	39
	Q2 [5]	33	35	94	63

Notes

.. = not available

1. No complete data available for eight former county councils.
2. Data only supplied for the geographical areas of the former Boroughs of Dinefwr and Llanelli (data unavailable for the former District of Carmarthen).
3. No data available for Denbighshire, Flintshire or Neath Port Talbot.
4. No data available for Flintshire.
5. No data available for Brecon Beacons National Park or Conwy.

Source of Data: National Assembly for Wales
Contact: Kathryn Hughes 029 2082 5017

Table 13.10 Wales – Detailed Analysis of Capital Account

£000s

New Construction and Conversion	1991/2	1992/3	1993/4	1994/5	1995/6	1996/97	1997/98	1998/99
Education and Libraries etc:								
Education	47,309	48,442	35,819	46,548	61,177	54,295	57,887	55,361
Library Service	1,881	1,646	1,798	1,705	813	1,507	1,005	1,561
Museums and Art Galleries	1,285	485	820	3,228	2,133	771	3,797	4,051
Personal Social Services	11,299	10,943	13,907	13,897	14,313	10,505	7,209	8,385
Port Health	0	0	0	0	0	0	0	0
Law, Order and Protective Services:								
Fire	2,439	2,443	1,315	1,476	3,038	1,171	1,534	2,144
Civil Service	74	57	389	61	33	0	0	0
Police (including crossing patrols)	4,510	4,878	5,082	4,575	4,151	5,531	4,386	4,443
Administration of Justice	1,852	3,243	3,578	1,671	2,135	1,188	318	694
Local Transport:								
Public Passenger Transport	1,557	2,019	3,161	1,351	874	529	2,127	10,780
Highways and other Transport	93,935	104,832	109,141	124,031	140,406	98,681	77,630	66,850
Employment:								
Consumer Protection	61	56	119	262	4	0	1	2
Careers Service Administration	0	0	0	0	0	0	0	0
Sheltered Employment and Workshops	220	92	717	509	1,256	69	129	129
Local Environment Services:								
Waste Collection/Disposal	4,010	2,711	3,539	2,929	1,431	2,072	5,277	4,173
Parks, Sport and Recreation	29,514	31,769	32,261	41,582	51,064	30,457	15,809	21,522
Environmental Health	2,110	2,258	4,086	2,447	4,051	1,920	1,838	2,140
Planning and Economic Development	40,325	46,601	47,104	56,263	48,642	25,515	46,757	47,024
Agriculture and Fisheries	11,504	6,252	6,647	7,130	7,099	8,374	8,054	5,211
Miscellaneous Services	7,785	5,079	2,007	2,611	2,027	1,848	3,285	3,095
Council Tax Preparation	21	517	89	0	107	0	0	0
General Administration	11,280	15,323	7,675	5,380	10,456	25,750	20,816	16,301
Trading Services Accounts	10,040	18,002	20,545	12,705	10,143	5,383	15,221	17,295
Housing:								
General Fund Housing	21,763	28,819	18,961	25,276	13,367	9,770	..	..
Housing to which the HRA relates	94,284	90,880	94,059	87,612	91,867	70,449	..	..
Housing Total¹	116,047	119,699	113,020	112,888	105,234	80,219	77,402	100,895
All Services	399,058	427,347	412,819	443,249	470,587	355,785	350,482	372,056

Notes

. . = not available

1. No split of housing is available for 1997/98 onwards.

Source of Data: National Assembly for Wales
Contact: Kathryn Hughes 029 2082 5017

Table 13.11 Wales – 1996 Based Household Projections

Type of Household	1996	2001	2006	2011	2016	2021
						£000s
Married Couple	629	606	587	576	569	561
Cohabiting Couple	65	83	99	110	115	118
Lone Parent	69	73	75	76	75	74
One Person	318	345	371	397	427	457
Other Multi-Person	89	96	104	114	122	127
All Households	1,170	1,203	1,236	1,273	1,308	1,337
Private Household Population	2,881	2,907	2,925	2,948	2,976	2,999
Average Household Size (Number)	2.46	2.42	2.37	2.32	2.28	2.24
						Percentage
Married Couple	53.8	50.4	47.5	45.2	43.5	42.0
Cohabiting Couple	5.6	6.9	8.0	8.6	8.8	8.8
Lone Parent	5.9	6.1	6.1	6.0	5.7	5.5
One Person	27.2	28.7	30.0	31.2	32.6	34.2
Other Multi-Person	7.6	8.0	8.4	9.0	9.3	9.5
All Households	100.0	100.0	100.0	100.0	100.0	100.0

Source of Data: National Assembly for Wales
Contact: Kathryn Hughes 029 2082 5017

Table 13.12 Scotland – New Orders obtained by Contractors

£ Million

Year & Quarter		New Housing			Other New Work				All Other New Work	All New Work
		Public	Private	All New Housing	Infra-structure	Public	Private Industrial	Private Commercial		
1989		133	526	659	300	390	206	729	1,624	2,283
1990	Q1	19	144	163	71	63	52	100	285	448
	Q2	16	190	206	60	75	82	169	386	592
	Q3	16	132	147	50	98	56	127	331	479
	Q4	37	156	193	75	57	67	154	353	546
1991	Q1	27	168	195	59	87	36	114	296	491
	Q2	37	153	191	72	82	73	111	338	529
	Q3	18	106	124	48	49	32	178	307	431
	Q4	31	120	151	60	138	33	91	322	473
1992	Q1	48	106	154	119	82	41	129	372	527
	Q2	32	121	153	84	78	41	114	317	470
	Q3	31	111	142	69	109	39	125	342	485
	Q4	33	89	122	109	83	35	107	333	455
1993	Q1	57	107	164	110	122	57	69	358	522
	Q2	35	157	192	98	43	59	96	295	487
	Q3	43	137	180	99	61	42	169	370	550
	Q4	69	131	200	59	96	35	95	286	486
1994	Q1	43	149	192	81	94	41	89	305	497
	Q2	44	212	256	55	73	43	100	271	526
	Q3	29	151	180	98	95	69	225	488	668
	Q4	41	126	167	78	86	46	102	313	480
1995	Q1	51	161	212	100	119	215	109	543	755
	Q2	71	147	218	71	88	53	163	374	592
	Q3	38	145	183	105	98	83	133	419	602
	Q4	26	96	121	71	159	38	100	369	490
1996	Q1	72	127	199	117	136	44	210	507	706
	Q2	52	125	177	75	74	50	346	544	721
	Q3	34	147	181	61	59	126	160	406	587
	Q4	28	169	197	152	85	48	112	396	593
1997	Q1	58	164	222	96	120	78	171	464	686
	Q2	34	190	225	114	49	125	190	477	702
	Q3	26	162	188	122	53	72	152	400	588
	Q4	45	170	214	90	121	69	167	447	662
1998	Q1	47	191	238	101	64	85	210	461	698
	Q2	22	171	193	121	155	114	225	616	809
	Q3	19	182	200	96	81	53	306	535	735
	Q4	34	110	144	132	101	46	248	527	671
1999	Q1	33	150	183	101	99	54	263	518	701
	Q2	36	172	208	157	124	63	344	689	897
	Q3	31	178	209	122	58	62	232	474	683
	Q4	50	174	225	107	105	71	177	459	684

Source of Data: Construction Market Intelligence, Department of the Environment, Transport and the Regions
Contact: Neville Price 020 7944 5587

Table 13.13 Scotland – Output obtained by Contractors

£ Million

Year & Quarter		New Housing		Other New Work Non-Housing				All New Work	Repair & Maintenance				
				Infra-struc-ture	Other New Work Exc. Infrastructure					Other Work		All Repair & Mainte-nance	All Work
		Public	Private		Public	Private Indus-trial	Private Comm-ercial		Housing	Public	Private		
1989		115	466	388	416	199	557	2,141	997	234	361	1,592	3,733
1990	Q1	32	140	95	118	70	158	612	270	64	93	427	1,038
	Q2	32	155	94	126	72	165	644	259	62	98	419	1,063
	Q3	27	152	106	128	70	181	663	259	65	96	420	1,083
	Q4	27	185	109	112	56	153	643	253	66	103	422	1,066
1991	Q1	26	147	101	120	60	153	606	240	64	100	404	1,009
	Q2	29	180	112	135	69	167	692	225	62	100	387	1,079
	Q3	30	147	105	126	64	150	621	248	64	98	409	1,031
	Q4	34	120	119	124	58	143	598	229	63	103	394	992
1992	Q1	35	123	108	154	53	144	617	228	66	99	393	1,010
	Q2	39	129	128	149	58	155	659	220	62	100	382	1,041
	Q3	41	133	141	136	60	158	668	249	71	100	420	1,088
	Q4	42	111	143	123	42	150	612	241	63	107	411	1,022
1993	Q1	39	113	162	119	39	134	607	239	74	101	414	1,021
	Q2	46	133	165	131	53	159	687	237	65	106	408	1,095
	Q3	44	148	148	122	60	144	667	238	71	110	419	1,086
	Q4	45	156	105	99	67	140	611	260	68	110	438	1,049
1994	Q1	50	162	101	108	65	130	617	278	82	115	475	1,092
	Q2	54	181	106	103	65	129	638	260	76	109	445	1,083
	Q3	55	163	110	99	54	133	613	268	73	115	456	1,069
	Q4	46	155	124	96	56	117	594	272	76	124	472	1,066
1995	Q1	50	145	92	127	85	118	616	278	91	125	494	1,110
	Q2	58	160	124	132	98	141	713	274	79	136	489	1,201
	Q3	62	142	112	123	145	146	730	277	72	146	495	1,226
	Q4	58	144	127	107	130	167	731	275	79	149	503	1,234
1996	Q1	60	141	124	149	97	145	716	271	82	139	492	1,208
	Q2	64	161	109	145	91	186	757	272	76	156	504	1,260
	Q3	57	153	102	152	86	193	744	278	76	159	512	1,256
	Q4	55	176	119	125	90	196	761	281	68	157	505	1,266
1997	Q1	49	159	128	133	80	209	758	292	82	142	516	1,274
	Q2	46	193	139	108	106	223	815	279	76	153	508	1,323
	Q3	42	176	134	100	108	216	776	277	73	150	500	1,276
	Q4	41	204	124	104	102	213	787	291	79	160	530	1,318
1998	Q1	45	176	134	90	101	169	716	285	79	169	533	1,249
	Q2	43	193	134	120	109	198	799	275	72	153	501	1,299
	Q3	39	190	152	122	102	233	838	287	73	180	540	1,378
	Q4 (p)	36	197	122	124	110	266	856	296	85	161	542	1,397
1999	Q1 (p)	38	176	117	113	104	211	759	292	70	150	513	1,272
	Q2 (p)	37	193	135	139	115	253	873	277	62	165	504	1,377
	Q3 (p)	37	196	160	148	103	260	904	278	75	177	530	1,434
	Q4 (p)	48	222	176	141	111	267	965	285	84	198	566	1,532

Notes

p = provisional

Source of Data: Construction Market Intelligence, Department of the Environment, Transport and the Regions

Contact: Neville Price 020 7944 5587

CHAPTER 14

Building Materials

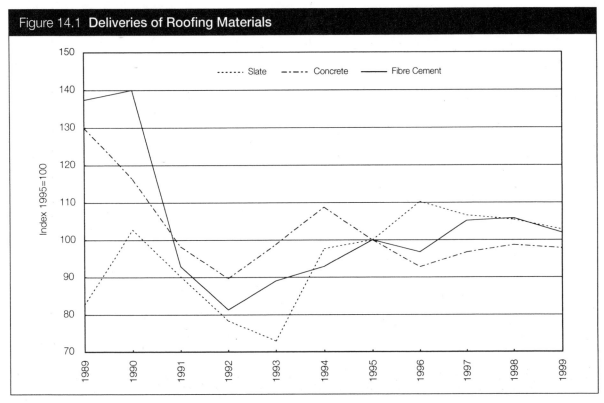

Figure 14.1 **Deliveries of Roofing Materials**

Source of Data: Construction Market Intelligence, Department of Environment, Transport and the Regions
Contact: David Williams 020 7944 5593

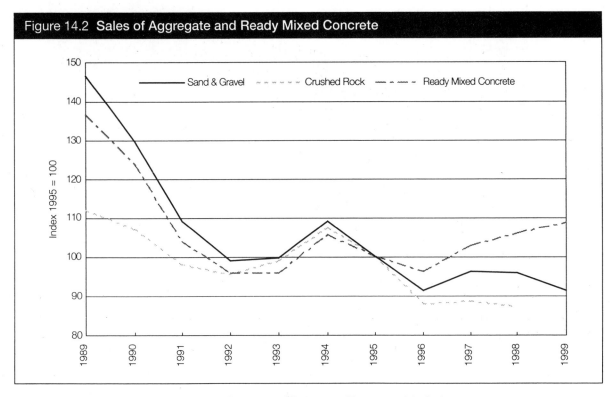

Figure 14.2 **Sales of Aggregate and Ready Mixed Concrete**

Source of Data: Construction Market Intelligence, Department of Environment, Transport and the Regions
Contact: David Williams 020 7944 5593

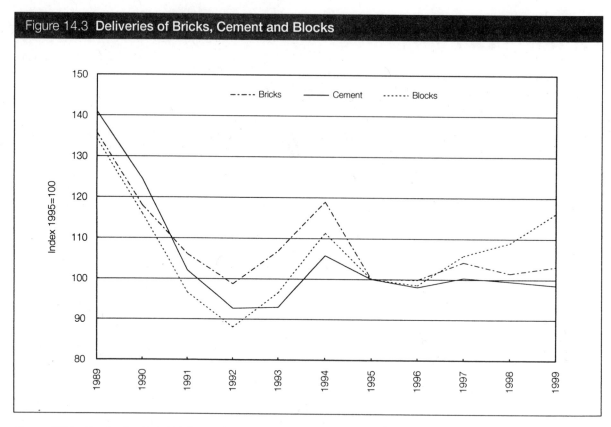

Figure 14.3 Deliveries of Bricks, Cement and Blocks

Source of Data: Construction Market Intelligence, Department of Environment, Transport and the Regions
Contact: David Williams 020 7944 5593

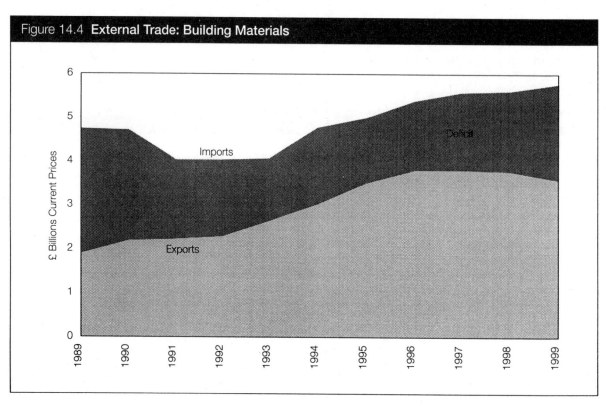

Figure 14.4 External Trade: Building Materials

Source of Data: Construction Market Intelligence, Department of Environment, Transport and the Regions
Contact: David Williams 020 7944 5593

Table 14.1 Price Indices¹ of Construction Materials – Annual Averages

United Kingdom   **1995=100**

	1991	1992	1993	1994	1995	1996	1997	1998	1999
Aggregates									
Coated Roadstone	91.3	87.0	90.3	93.7	100.0	104.7	108.4	111.5	112.1
Crushed Rock	87.6	86.1	90.2	91.1	100.0	105.6	114.7	121.5	127.9
of which									
Limestone	88.4	86.0	88.8	92.2	100.0	105.5	111.8	116.9	120.3
Sand & Gravel	109.0	99.9	91.1	89.2	100.0	106.2	113.5	117.5	118.0
Cement & Concrete									
Cement	86.3	90.1	92.6	97.2	100.0	101.6	105.6	109.5	113.1
Plasterboard etc	80.4	87.4	92.8	95.6	100.0	102.9	103.2	105.0	106.8
Pre-cast Concrete Products	93.4	93.0	93.6	96.6	100.0	99.7	102.6	109.1	114.6
of which									
Blocks, Bricks, Tiles & Flagstones	91.6	91.6	92.3	96.0	100.0	99.4	102.0	108.6	114.5
of which									
Concrete Blocks & Bricks	84.6	85.6	86.4	93.8	100.0	102.3	107.2	112.2	114.0
Concrete Roofing Tiles	96.7	97.7	97.8	99.4	100.0	99.8	103.3	114.0	125.3
Clay Products									
Ceramic Tiles	91.6	92.8	95.1	96.1	100.0	100.7	100.8	103.2	106.4
Ceramic Sanitaryware	90.3	90.8	93.6	95.6	100.0	101.8	103.1	107.8	106.5
Timber & Joinery									
Imported Sawn Wood	74.8	73.8	81.7	93.3	100.0	98.2	95.9	88.8	
of which									
Hardwood	73.4	76.1	88.0	97.1	100.0	100.2	94.2	89.5	91.3
Sawn Wood	84.7	81.0	84.1	94.1	100.0	93.3	96.1	89.5	88.1
Builders Woodwork	88.0	88.4	91.3	93.5	100.0	100.5	103.4	105.6	104.2
of which									
Doors & Windows	80.1	82.6	87.6	93.3	100.0	100.5	106.5	110.3	111.2
Metal Products									
Concrete Reinforcing Bars	86.6	76.2	90.4	99.1	100.0	89.2	86.0	75.3	70.5
Fabricated Structural Steel	95.4	99.8	98.8	98.0	100.0	99.3	99.7	101.4	101.8
Doors & Windows	93.3	93.7	94.5	97.0	100.0	104.0	107.6	109.1	113.0
Door Locks etc	83.8	88.6	92.5	95.8	100.0	104.6	108.3	111.5	117.6
Central Heating Boilers	91.0	95.1	97.5	98.8	100.0	104.3	109.1	112.9	114.9
Central Heating Pumps	85.0	88.9	93.2	96.2	100.0	104.1	108.5	110.2	109.8
Taps & Valves (Domestic)	83.1	86.5	90.0	93.3	100.0	104.4	108.1	112.6	115.4
Metal Sanitaryware	82.3	86.1	90.5	92.7	100.0	106.3	106.1	108.2	110.5
Copper Pipes	61.9	65.5	70.7	82.4	100.0	99.7	116.4	111.4	72.3
Plastic Products									
Pipes and Fittings (Rigid)	66.7	69.9	74.0	83.8	100.0	105.3	105.8	110.1	112.9
Pipes and Fittings (Flexible)	92.6	94.5	93.1	92.1	100.0	93.9	95.2	95.8	95.3
Sanitaryware	85.2	86.8	89.5	92.2	100.0	107.2	111.4	112.9	113.0
Doors & Windows	92.8	94.5	92.9	94.5	100.0	102.2	103.3	100.7	101.4

Table 14.1 Price Indices' of Construction Materials – Annual Averages (continued)

United Kingdom   1995=100

	1991	1992	1993	1994	1995	1996	1997	1998	1999
Other Building Materials									
Slate Products	109.2	111.7	110.5	99.2	100.0	106.8	108.9	106.8	103.7
Asphalt Products (Not in Rolls)	92.1	87.7	94.0	91.7	100.0	103.2	108.1	112.5	112.6
Insulating Materials									
(Thermal or Acoustic)	80.5	84.1	90.7	94.2	100.0	105.5	108.2	106.3	106.2
Paint (Aqueous)	103.7	105.6	109.1	105.7	100.0	105.3	107.2	101.5	102.2
Paint (Non-aqueous)	91.9	96.5	100.4	98.9	100.0	104.8	106.7	105.7	106.7
Lighting Equipment for Buildings	84.4	89.4	94.3	96.9	100.0	102.9	105.4	108.8	112.2
Electric Heating Apparatus	89.4	92.9	96.1	97.6	100.0	105.0	107.7	110.6	111.9
Electric Water Heaters	84.4	89.7	93.5	95.6	100.0	103.4	105.3	109.1	111.8
Kitchen Furniture	94.2	97.6	97.4	97.9	100.0	99.7	98.7	99.1	97.6
Wallpaper	97.9	100.9	101.1	97.5	100.0	104.4	103.8	102.3	102.7

Notes

1. These indices, which are compiled by the Office for National Statistics, are reproduced here as information for the construction industry. They are a selection of Producer Price Indices which cover all products in the manufacturing sector, and the complete list is published monthly by The Stationery Office in Business Monitor MM22. When the indices were re-based to 1995=100, the EU harmonised classification of products known as PRODCOM was used which, for some products, is not consistent with the classification used for 1990=100.

Source of Data: Construction Market Intelligence, Department of Environment, Transport and the Regions
Contact: David Williams 020 7944 5593

Table 14.2 Bricks[1]

Millions

	1989	1990	1991	1992	1993	1994	1995	1996	1997	1998	1999
(a) Great Britain											
Use Type											
Commons											
Production	1,166	878	642	497	436	464	480	401	422	385	367
Deliveries	970	845	627	522	508	539	439	402	408	375	375
Stock at end of period	280	306	293	264	186	107	152	151	172	184	168
Facings											
Production	3,215	2,690	2,340	2,266	1,978	2,421	2,546	2,430	2,386	2,411	2,369
Deliveries	2,754	2,402	2,251	2,128	2,381	2,699	2,288	2,296	2,457	2,393	2,420
Stock at end of period	649	988	1,121	1,241	827	555	812	913	837	828	766
Engineering											
Production	273	233	230	237	225	229	230	216	190	204	204
Deliveries	253	210	229	244	242	247	204	227	187	205	227
Stock at end of period	36	56	59	51	33	17	44	37	43	48	27
Material Type											
Clay											
Production	4,129	3,441	2,940	2,776	2,396	2,845	3,025	2,849	2,828	2,830	2,759
Deliveries	3,470	3,093	2,827	2,654	2,889	3,219	2,704	2,720	2,881	2,800	2,845
Stock at end of period	902	1,296	1,426	1,520	1,013	643	966	1,068	1,029	1,042	943
Sandlime											
Production	127	80	58	44	50	55	40	31	–	–	–
Deliveries	122	81	59	45	50	53	39	34	–	–	–
Stock at end of period	13	10	9	7	7	9	11	8	–	–	–
Concrete											
Production	398	281	213	181	192	214	191	166	169	171	180
Deliveries	385	283	221	195	193	213	188	172	171	173	176
Stock at end of period	50	45	38	29	26	27	32	26	23	18	18
All Bricks											
Production	4,654	3,802	3,212	3,000	2,639	3,114	3,256	3,046	2,997	3,000	2,939
Deliveries	3,977	3,457	3,107	2,893	3,132	3,485	2,930	2,926	3,052	2,973	3,021
Stock at end of period	965	1,350	1,473	1,556	1,046	679	1,008	1,101	1,052	1,060	961
(b) Deliveries from Standard Regions, Wales and Scotland[2]											
North	271	222	201	182	193	212	175	149	163	155	165
Yorkshire & Humberside	320	312	312	300	306	317	234	228	235	206	198
East Midlands	431	400	370	337	350	449	362	446	475	526	509
East Anglia	407	362	317	285	359	394	350	323	296	267	262
South East	943	718	615	548	554	574	500	494	517	514	564
South West	209	160	144	144	174	189	154	167	158	156	157
West Midlands	528	514	498	506	595	671	578	551	593	580	599
North West	366	327	292	280	295	317	280	285	307	294	273
England	3,475	3,014	2,748	2,580	2,826	3,122	2,633	2,643	2,744	2,698	2,727
Wales	196	147	118	100	101	127	104	102	115	106	117
Scotland	305	296	241	213	205	236	194	181	193	169	177
Great Britain	**3,977**	**3,457**	**3,107**	**2,893**	**3,132**	**3,485**	**2,930**	**2,926**	**3,052**	**2,973**	**3,021**

Notes

– = nil or less than half the final digit shown.

1. From 1997 sand-lime bricks are included with clay bricks.

2. Deliveries by brickworks situated in each region.

Source of Data: Construction Market Intelligence, Department of Environment, Transport and the Regions

Contact: David Williams 020 7944 5593

Table 14.3: Concrete Building Blocks

Thousand Square Metres

	1989	1990	1991	1992	1993	1994	1995	1996	1997	1998	1999
(a) Great Britain											
Dense Aggregate											
Production	45,564	39,297	32,456	29,732	30,116	36,997	36,933	34,996	37,250	39,439	41,951
Deliveries	44,055	38,365	32,483	29,653	30,879	36,223	36,029	35,403	37,142	39,024	41,835
Stocks at end of period	4,387	4,771	4,905	4,475	3,543	4,248	4,884	4,473	4,507	4,568	4,713
Lightweight Aggregate											
Production	31,041	23,768	18,581	17,479	19,235	22,048	18,147	16,316	17,783	19,110	21,486
Deliveries	29,945	23,354	18,697	17,509	19,386	21,761	18,130	16,539	17,768	18,847	21,499
Stocks at end of period	3,371	3,362	3,016	2,526	2,070	2,349	2,361	1,959	1,957	2,226	1,970
Aerated Concrete											
Production	31,394	28,089	23,594	20,984	24,936	28,503	23,207	24,554	27,505	26,113	27,607
Deliveries	30,119	28,180	23,733	21,097	24,674	28,324	23,382	24,436	27,131	26,615	26,768
Stocks at end of period	2,544	2,401	2,105	1,949	2,212	2,444	2,232	2,232	2,606	2,106	2,820
All Concrete Blocks											
Production	107,999	91,154	74,631	68,194	74,287	87,548	78,287	75,866	82,537	84,662	91,045
Deliveries	104,119	89,899	74,913	68,259	74,939	86,308	77,541	76,378	82,042	84,486	90,101
Stocks at end of period	10,301	10,534	10,027	8,950	7,824	9,042	9,477	8,665	9,069	8,900	9,503
(b) Deliveries from Standard Regions, Wales and Scotland[1]											
North	5,197	4,416	3,625	3,214	3,544	4,063	3,297	3,103	2,840	4,720	7,416
Yorkshire & Humberside	13,447	11,060	9,935	9,580	11,720	13,303	11,942	13,187	13,992	13,338	13,679
East Midlands	8,334	7,231	6,525	5,978	6,361	7,414	7,176	7,941	8,802	8,842	9,086
East Anglia	3,858	3,066	2,791	2,734	3,031	3,798	2,462	2,384	2,586	2,610	2,724
South East	25,735	23,287	19,366	16,423	19,027	22,302	19,696	18,513	20,691	20,956	22,939
South West	15,309	11,729	8,396	7,635	8,186	10,019	10,327	10,853	12,132	11,378	12,050
West Midlands	11,940	12,447	10,525	9,504	9,456	9,573	9,388	8,199	7,968	8,174	7,815
North West	6,232	5,844	4,847	4,295	4,229	4,977	3,250	2,974	3,665	3,445	5,199
England	90,052	79,080	66,010	59,363	65,554	75,449	67,538	67,154	72,676	73,463	80,908
Wales	9,065	6,602	4,984	5,160	5,517	6,608	5,993	5,531	5,234	5,480	4,832
Scotland	5,001	4,217	3,919	3,736	3,867	4,251	4,011	3,693	4,132	5,543	4,361
Great Britain	104,119	89,899	74,913	68,259	74,939	86,308	77,541	76,378	82,042	84,486	90,101

Table 14.4 Sales[1] of Aggregate by use from Standard Regions, Wales and Scotland: 1999

(a) Sand and Gravel (including marine dredging)
Thousand Tonnes

	Sand			Gravel			Sand Gravel and Hoggin for Fill	All	Of Which Marine Dredged Material
	Building Sand		Concreting Sand	For Coating	Concrete Aggregate	Other Uses (Exc. Fill)			
	For Coating	Other Uses (Exc. Fill)							
North	..	619	1,045	–	732	..	507	3,056	476
Yorkshire & Humberside	..	429	1,931	–	1,518	..	417	4,381	177
East Midlands	..	1,042	4,205	..	3,881	42	992	10,416	–
East Anglia	212	318	2,059	69	2,323	72	926	5,979	..
South East	727	3,253	9,117	..	13,679	..	2,706	29,637	9,894
South West	..	986	2,266	..	1,650	122	841	6,166	673
West Midlands	..	741	3,664	..	3,954	–	831	9,721	–
North West	..	1,530	1,467	–	134	–	339	3,579	..
England	1,991	8,919	25,753	408	27,870	436	7,559	72,935	11,694
Wales	45	768	1,282	–	532	-	333	2,959	1,258
Scotland	447	1,153	3,210	79	1,968	198	3,020	10,074	–
Great Britain	2,483	10,840	30,244	487	30,369	634	10,911	85,968	12,952

(b) Crushed Rock
Thousand Tonnes

	Roadstone			Fill & Ballast	Concrete Aggregate	All
	Sold Coated	For Coating at Remote Plants	For Use Uncoated			
North	765	738	3,732	3,741	1,270	10,246
Yorkshire & Humberside	512	1,733	4,691	4,064	2,745	13,745
East Midlands	2,150	3,828	6,592	10,059	4,304	26,933
East Anglia	–	–	75	532	–	607
South East	–	–	277	1,068	13	1,358
South West	2,186	1,868	7,607	7,186	4,565	23,411
West Midlands	1,345	..	..	2,184	185	6,028
North West	360	..	..	4,248	921	6,348
England	7,318	8,758	25,516	33,080	14,003	88,675
Wales	1,971	1,347	4,222	8,445	3,919	19,903
Scotland	2,198	1,540	7,077	10,098	2,224	23,138
Great Britain	11,486	11,645	36,816	51,623	20,146	131,716

Notes

.. = not available; for reasons of disclosure these figures are not quoted.

– = nil or less than half the final digit shown.

1. Sales of materials from production site: marine dredged by region of landing.

Source of Data: Construction Market Intelligence, Department of Environment, Transport and the Regions

Contact: David Williams 020 7944 5593

Table 14.5 Other Building Materials

	1989	1990	1991	1992	1993	1994	1995	1996	1997	1998	1999
										Thousand Tonnes	
Cement[1]											
Production	16,849	14,740	12,297	11,006	11,039	12,307	11,805	12,214	12,638	12,409	12,697
Home deliveries	16,791	14,826	12,160	11,046	11,081	12,600	11,914	12,808	12,965	13,113	12,885
Stocks at end of period	411	351	422	349	305	288	346	361	423	417	403
Clinker[1]											
Production	15,234	13,199	10,845	9,872	9,996	11,521	11,371	11,609	12,141	12,372	11,816
Stocks at end of period	1,024	1,144	918	743	586	658	828	897	828	952	718
Crushed Rock											
Production											
Coated Roadstone	23,733	26,430	26,387	26,647	27,238	28,512	28,972	26,270	23,906	23,131	. .
Uncoated Roadstone	66,015	61,742	60,748	53,471	54,412	51,121	49,307	40,893	40,186	36,816	. .
Fill & Ballast	59,689	54,640	45,669	48,919	52,141	65,779	56,140	50,982	51,396	51,623	. .
Concrete Aggregate	19,356	18,804	15,203	14,929	15,786	16,345	16,419	14,748	18,300	20,146	. .
All	168,794	161,615	148,007	143,967	149,576	161,757	150,838	132,894	133,787	131,716	. .
Sand & Gravel[2] (Sales)											
Building Sand	23,290	20,948	18,079	16,769	17,406	18,534	17,389	14,655	15,337	13,810[r]	12,848
Concreting Sand	41,223	37,213	31,239	28,573	28,021	30,977	29,390	28,659	30,130	30,244[r]	29,276
Gravel (inc. Hoggin)	66,718	58,010	48,598	43,557	44,043	48,161	42,877	38,683	40,899	41,914[r]	39,703
All	131,232	116,172	97,918	88,898	89,470	97,672	89,656	81,997	86,366	85,968[r]	81,827
										Thousand Cubic Metres	
Ready Mixed Concrete[1]											
Deliveries	29,596	26,782	22,527	20,776	20,771	22,931	21,676	20,892	22,327	22,983	23,550
										Tonnes	
Slate[3]											
Production	83,976	97,494	94,951	78,067	71,731	90,192	99,821	109,764	106,307	104,899	98,870
Deliveries	80,468	99,961	87,718	76,163	70,898	94,989	97,245	107,164	103,637	102,549	99,957
Stocks at end of period	6,594	8,648	15,007	16,577	15,274	11,383	14,312	16,211	20,046	21,270	20,013
										Thousand Square Metres	
Concrete Roofing Tiles											
Production	35,787	31,510	26,359	21,490	24,574	28,149	26,118	24,651	24,958	24,981	25,972
Deliveries	33,692	30,165	25,468	23,245	25,606	28,191	25,926	24,046	25,075	25,587	25,365
Stocks at end of period	5,861	7,183	7,716	5,823	4,822	4,490	4,518	4,961	4,267	3,442	3,543
										Thousand Tonnes	
Fibre Cement Products											
Production	221	235	134	121	129	154	161	146	164	161	156
Deliveries	213	217	144	126	138	144	155	150	163	164	158
Stocks at end of period	45	48	31	26	16	28	34	35	39	36	34

Notes

. . = not available

r = revised

1. Information is for the United Kingdom.
2. The 1997 figures are based on the quarterly sample.
3. Figures comprise tiles and powder and granules but exclude slate waste used for fill.

Source of Data: Construction Market Intelligence, Department of Environment, Transport and the Regions
Contact: David Williams 020 7944 5593

Table 14.6 Value of Overseas Trade in Selected Materials and Components for Constructional Use: Imports (cif) & Exports (fob)

United Kingdom — Current Prices (£000s)

		1989	1990	1991	1992	1993	1994	1995	1996	1997	1998	1999
All Building Materials & Components	Imports	4,734,830	4,695,771	4,030,915	4,036,711	4,070,565	4,761,937	4,988,495	5,386,440	5,576,799	5,628,078	5,766,736
	Exports	1,891,818	2,205,257	2,240,312	2,305,275	2,650,999	3,024,988	3,523,753	3,802,944	3,804,927	3,761,025	3,583,836
	Balance	-2,843,012	-2,490,514	-1,790,603	-1,731,436	-1,419,566	-1,736,949	-1,464,742	-1,583,496	-1,771,872	-1,867,053	-2,182,900
All Raw Materials	Imports	42,574	47,797	43,944	39,952	47,315	71,545	86,170	81,308	86,889	85,079	78,379
	Exports	13,779	15,684	20,362	34,219	41,666	47,617	48,109	52,573	49,687	45,772	51,409
	Balance	-28,795	-32,113	-23,582	-5,733	-5,649	-23,928	-38,061	-28,735	-37,202	-39,307	-26,970
of which Aggregates	Imports	15,298	16,158	15,590	12,040	10,735	18,361	18,221	12,495	12,260	13,326	12,687
	Exports	9,337	11,378	15,249	28,346	27,334	35,147	43,283	46,953	44,648	40,932	44,254
Unprocessed Stone	Imports	8,807	13,523	10,691	10,018	9,846	17,628	18,821	20,929	34,569	34,648	31,671
	Exports	322	495	690	951	11,550	9,180	579	722	812	869	3,065
Unprocessed Slate	Imports	475	582	1,618	1,377	1,460	1,498	2,282	2,876	2,985	4,006	4,961
	Exports	783	1,046	1,015	1,256	577	237	828	840	626	853	550
Timber	Imports	13,509	12,556	11,950	11,823	20,791	25,332	34,772	37,668	28,883	26,552	19,758
	Exports	2,560	1,994	2,256	2,067	1,560	2,263	2,776	2,588	2,235	2,053	2,268
All Semi-Manufactures	Imports	2,222,982	2,094,874	1,667,995	1,714,972	1,812,324	2,071,179	2,033,278	2,068,479	2,110,853	1,944,379	1,878,021
	Exports	268,428	303,967	302,375	323,724	339,963	412,656	528,634	507,143	520,598	473,934	440,719
	Balance	-1,954,554	-1,790,907	-1,365,620	-1,391,248	-1,472,361	-1,658,523	-1,504,644	-1,561,336	-1,590,255	-1,470,445	-1,437,302
of which Clinker	Imports	47,812	12,081	6,695	7,412	6,325	7,472	4,783	6,999	19,907	15,123	22,768
	Exports	250	1,314	54	603	2,973	1,747	1,797	4,473	12,061	13,481	12,720
Sawn & Laminated Wood	Imports	1,459,493	1,407,624	1,056,489	1,049,924	1,097,248	1,346,655	1,155,465	1,152,578	1,215,066	1,104,210	1,107,069
	Exports	18,246	24,575	23,100	32,829	26,909	43,114	50,970	53,812	55,579	49,128	53,886
Hardboard & Fibreboard	Imports	56,152	54,792	52,240	64,449	68,159	80,929	88,451	97,942	107,858	117,364	109,780
	Exports	17,994	23,306	22,459	19,393	30,589	29,828	40,867	40,563	38,219	28,171	25,642
Metal Sheets etc	Imports	139,186	136,232	116,517	116,000	108,005	132,096	163,083	165,424	159,285	159,052	135,968
	Exports	67,401	74,349	78,946	86,355	85,074	99,905	122,898	115,894	113,918	111,745	93,110
Plastic Sheets & Profiles	Imports	198,227	215,698	205,857	208,506	185,487	217,396	295,383	280,624	260,902	260,005	256,150
	Exports	79,478	93,460	100,788	107,172	120,833	135,009	178,570	187,679	170,570	166,522	157,849

Table 14.6 Value of Overseas Trade in Selected Materials and Components for Constructional Use: Imports (cif) & Exports (fob) (continued)

United Kingdom		1989	1990	1991	1992	1993	1994	1995	1996	1997	1998	1999
All Products & Components	Imports	2,469,274	2,553,100	2,318,976	2,281,787	2,210,926	2,619,213	2,869,047	3,236,653	3,379,057	3,598,620	3,810,336
	Exports	1,609,611	1,885,606	1,917,575	1,947,332	2,269,368	2,564,715	2,947,010	3,243,228	3,234,642	3,241,319	3,091,708
	Balance	-859,663	-667,494	-401,401	-334,455	58,442	-54,498	77,963	6,575	-144,415	-357,301	-718,628
of which												
Building Stone	Imports	62,624	79,554	64,426	49,092	31,033	40,567	43,679	51,945	57,518	61,429	58,868
	Exports	12,783	16,031	16,458	15,311	20,689	23,895	19,301	24,804	27,091	22,086	21,462
Slate Products	Imports	14,997	15,105	10,863	9,634	10,968	14,629	17,508	18,895	21,377	21,614	23,848
	Exports	4,101	4,081	3,802	2,850	4,132	5,538	7,800	8,228	9,694	8,541	9,698
Portland Cement	Imports	73,899	73,049	56,460	46,363	33,912	42,481	42,253	49,310	38,951	47,137	46,978
	Exports	6,682	9,877	15,617	14,289	15,185	21,836	25,882	28,819	27,385	30,734	29,102
Clay Bricks	Imports	18,390	9,213	7,685	5,503	5,108	9,948	8,891	6,756	7,600	6,444	9,806
	Exports	4,125	7,408	7,921	4,558	4,138	8,398	11,213	25,925	7,099	7,005	7,470
Clay Roofing Tiles	Imports	2,729	2,887	4,326	5,047	898	1,037	3,418	4,967	4,262	7,857	6,001
	Exports	989	1,361	1,755	1,513	1,577	1,400	1,442	3,756	2,962	3,630	3,490
Concrete Blocks & Bricks	Imports	6,111	4,237	5,398	2,519	1,281	2,324	2,302	2,259	3,976	5,173	4,372
	Exports	4,483	4,504	4,446	4,826	5,251	7,034	8,780	13,308	13,416	12,633	12,067
Concrete Roofing Tiles etc	Imports	811	165	231	645	1,161	1,498	1,753	719	415	711	1,393
	Exports	849	1,109	995	1,324	1,002	3,365	2,192	2,548	3,031	2,887	4,445
Concrete Paving etc	Imports	4,427	3,146	2,336	5,070	7,396	4,587	3,143	4,219	3,703	6,298	7,690
	Exports	1,633	2,902	3,052	3,619	5,806	12,066	16,638	19,511	19,217	29,563	26,861
Fibre Cement Products	Imports	31,552	39,670	25,762	21,090	17,558	21,497	18,473	14,696	14,572	11,655	12,162
	Exports	5,748	6,115	7,398	6,482	3,110	5,140	5,543	8,556	15,810	12,908	14,456
Plasterboard	Imports	19,327	14,006	10,481	9,486	4,923	5,959	7,007	5,589	6,256	6,231	6,276
	Exports	7,241	10,534	11,013	11,479	15,394	13,578	12,958	20,713	22,565	21,195	15,996
Unglazed Ceramic Tiles	Imports	8,035	8,521	6,966	7,775	7,895	10,021	10,626	12,643	15,235	18,790	19,229
	Exports	6,245	5,902	5,775	5,859	4,183	5,308	5,617	5,464	5,175	5,636	4,631
Glazed Ceramic Tiles	Imports	121,849	116,540	109,969	115,544	97,502	120,973	130,441	177,871	205,218	231,477	204,059
	Exports	9,225	13,674	12,789	13,048	17,902	16,491	16,662	18,288	19,548	15,770	13,999
Cork Tiles	Imports	9,865	9,600	7,078	6,354	7,289	9,180	7,849	9,436	8,680	9,391	6,603
	Exports	2,015	2,344	2,310	2,112	1,524	1,602	1,873	1,940	2,381	1,592	1,656
Plastic Floor Covering	Imports	43,679	40,192	39,616	42,677	46,263	51,339	52,910	54,663	58,858	61,311	70,346
	Exports	81,741	82,298	87,284	86,336	94,779	109,049	121,005	126,169	120,393	122,285	112,625
Vinyl Floor Covering	Imports	15,546	19,849	20,014	23,094	24,027	34,118	33,068	34,525	46,667	57,631	68,957
	Exports	13,876	12,421	14,602	16,676	15,976	15,308	17,992	25,265	35,594	44,614	61,610

Current Prices (£000s)

Table 14.6 Value of Overseas Trade in Selected Materials and Components for Constructional Use: Imports (cif) & Exports (fob) (continued)

United Kingdom

Current Prices (£000s)

		1989	1990	1991	1992	1993	1994	1995	1996	1997	1998	1999
Ceramic Sanitaryware	Imports	17,541	16,910	13,722	16,250	14,101	20,646	22,957	25,492	32,545	33,911	41,277
	Exports	22,655	31,582	30,603	30,982	27,961	34,011	35,772	38,806	39,639	40,325	41,936
Plastic Sanitaryware	Imports	22,371	18,425	16,673	14,756	13,545	19,177	18,137	20,292	22,701	29,587	32,245
	Exports	29,033	35,328	38,504	45,701	47,071	59,638	60,977	72,194	66,852	64,293	62,264
Iron & Steel Sanitaryware	Imports	32,957	31,982	24,740	22,996	20,833	25,278	24,485	26,010	29,753	34,780	45,340
	Exports	12,129	16,324	16,780	16,607	15,759	18,212	18,782	20,082	19,402	21,236	22,225
Copper Pipes	Imports	12,761	12,058	11,162	10,598	11,830	12,185	16,781	17,343	16,727	14,984	12,741
	Exports	9,347	9,340	10,111	10,224	12,682	15,729	18,826	17,621	15,750	12,894	11,671
Plastic Pipes	Imports	25,224	25,604	26,729	30,576	31,778	39,941	43,516	47,383	54,088	54,280	52,907
	Exports	38,946	41,465	41,892	43,869	62,230	72,465	78,885	90,831	95,593	104,111	108,187
Other Plastic Building Products	Imports	44,489	45,777	41,266	41,426	46,673	59,108	64,521	65,548	70,743	68,880	71,840
	Exports	21,711	27,809	29,453	35,039	39,617	48,642	66,326	74,355	86,389	96,930	126,435
Builders Ironmongery	Imports	294,402	307,418	297,499	321,358	348,194	382,855	425,252	467,671	478,876	484,307	531,133
	Exports	112,023	124,132	121,638	123,491	152,896	192,943	230,024	256,133	274,726	281,269	276,205
Nails, Screws etc	Imports	60,671	59,846	53,491	56,086	62,153	74,994	90,952	97,933	96,977	97,869	96,274
	Exports	27,801	30,015	28,401	30,769	36,039	45,432	54,244	56,910	60,168	63,554	61,636
Mastics, Putty etc	Imports	40,245	48,678	53,506	61,819	65,570	66,857	82,426	81,611	75,415	73,343	75,035
	Exports	37,418	41,264	46,014	49,045	54,739	62,369	69,134	67,385	67,313	65,925	74,934
Paint & Varnish	Imports	69,934	72,571	70,154	72,149	82,325	87,546	102,668	115,746	120,116	130,813	129,496
	Exports	120,763	136,288	139,389	150,232	156,909	183,078	197,642	210,287	196,425	196,704	190,632
Windows (Wood)	Imports	12,917	19,671	16,709	14,432	16,467	22,502	22,332	20,650	25,088	26,437	33,548
	Exports	2,751	3,336	3,905	3,826	4,863	8,746	12,070	20,101	16,443	18,909	19,343
Doors (Wood)	Imports	94,930	84,213	91,042	81,672	90,914	106,570	95,621	102,035	106,314	106,420	117,178
	Exports	3,362	4,898	5,346	6,653	7,136	12,770	12,907	9,719	20,207	21,049	22,360
Doors & Windows (Steel)	Imports	23,383	24,769	29,460	23,634	16,350	20,342	17,535	22,381	23,188	23,274	28,523
	Exports	13,889	14,536	13,437	20,084	14,314	15,176	16,419	19,571	19,518	17,557	13,527
Doors & Windows (Aluminium)	Imports	35,567	38,910	27,126	22,677	25,921	26,973	31,677	30,703	29,233	39,968	38,753
	Exports	22,683	23,618	19,913	22,037	17,782	22,157	21,716	20,047	22,866	24,006	22,501
Flat Glass	Imports	25,550	27,341	21,940	19,454	18,302	20,049	21,659	18,914	17,009	13,331	12,278
	Exports	7,182	7,928	8,391	8,752	10,282	11,756	13,153	9,211	9,210	11,252	7,913
Glass Blocks & Glass Fibre	Imports	23,201	23,275	21,992	24,385	18,216	21,432	24,065	25,464	24,297	22,830	23,707
	Exports	17,676	21,758	26,481	26,416	21,197	25,275	27,476	30,979	30,029	27,538	28,114

Table 14.6 Value of Overseas Trade in Selected Materials and Components for Constructional Use: Imports (cif) & Exports (fob) (continued)

United Kingdom — Current Prices (£000s)

		1989	1990	1991	1992	1993	1994	1995	1996	1997	1998	1999
Wood Veneers	Imports	42,376	45,934	32,055	32,053	38,019	45,398	43,062	48,198	44,933	40,954	39,048
	Exports	6,248	7,053	6,657	7,466	7,927	10,632	11,660	12,806	9,970	11,696	11,144
Wood Mouldings	Imports	26,547	29,081	23,069	25,778	32,725	37,278	38,618	44,948	57,688	52,103	57,405
	Exports	699	645	559	734	1,072	2,141	1,614	2,376	2,485	2,547	3,369
Other Builder's Woodwork	Imports	35,693	41,297	32,906	38,541	33,348	47,660	45,218	55,813	60,200	74,951	91,307
	Exports	6,217	9,144	10,471	8,314	7,199	10,026	7,858	10,713	9,609	9,870	10,145
Prefabricated Buildings (Wood)	Imports	12,062	11,758	6,777	6,120	3,307	4,197	5,614	6,573	3,421	5,805	7,064
	Exports	6,534	11,502	17,549	7,513	7,355	8,742	9,937	16,881	20,576	19,511	18,487
Prefabricated Buildings (Steel)	Imports	11,217	11,270	8,479	7,089	5,539	9,122	7,768	18,832	13,752	14,979	23,507
	Exports	31,308	34,252	34,979	32,699	34,486	35,940	56,594	61,906	59,023	61,471	57,140
Prefabricated Buildings (Other)	Imports	15,621	23,375	14,717	8,142	6,984	7,651	7,638	8,096	9,171	11,023	12,810
	Exports	8,629	10,420	12,830	15,744	5,397	12,148	21,031	28,771	29,514	28,830	24,202
Structural Units (Steel)	Imports	187,485	182,408	177,469	134,267	83,970	89,795	104,903	94,478	121,000	161,754	157,779
	Exports	160,420	188,044	203,727	203,122	266,003	279,194	359,957	329,370	339,442	397,201	289,733
Structural Units (Aluminium)	Imports	72,517	66,380	54,023	44,985	28,459	34,699	56,163	62,073	71,718	71,609	82,387
	Exports	32,748	62,108	61,296	40,240	39,684	43,988	44,823	64,503	54,980	51,221	109,630
Concrete Reinforcing Bars	Imports	79,017	74,163	53,957	48,733	54,086	54,464	37,290	37,559	40,858	31,619	59,551
	Exports	62,792	94,012	103,118	88,008	135,069	120,909	52,659	79,034	61,210	53,624	46,508
Tubes & Pipes (Iron & Steel)	Imports	55,911	72,413	63,010	65,651	52,516	71,070	83,663	80,378	61,394	78,919	62,425
	Exports	107,742	106,982	98,838	110,613	118,298	133,367	176,549	156,147	158,989	151,232	119,592
Angles & Sections (Iron & Steel)	Imports	31,917	26,236	22,645	19,510	22,289	29,264	35,045	26,610	32,216	31,991	26,844
	Exports	52,099	69,584	73,694	70,775	88,481	95,240	95,227	86,271	63,636	63,868	45,834
Zinc Building Products	Imports	4,137	3,140	4,868	5,543	4,409	3,561	5,353	5,519	4,866	4,039	4,951
	Exports	6,506	9,202	7,384	4,969	6,304	9,847	9,720	9,559	9,341	10,858	11,242
Bituminous Roofing Felt	Imports	17,588	21,980	18,366	20,695	16,670	17,396	16,578	17,586	18,587	20,150	19,905
	Exports	15,685	17,371	17,360	20,169	25,090	23,587	24,772	27,043	28,079	31,825	22,585
Mineral Insulating Materials	Imports	37,456	41,194	35,388	35,418	25,331	28,472	28,640	25,957	31,478	32,640	32,506
	Exports	22,540	27,116	33,186	37,025	39,360	39,870	53,352	52,302	49,505	43,765	42,191
Central Heating Boilers	Imports	51,162	59,000	70,711	88,429	75,006	92,422	106,646	136,071	163,212	160,562	189,116
	Exports	13,697	13,743	14,855	17,374	18,497	28,823	27,516	31,904	43,871	32,444	33,070

Table 14.6 Value of Overseas Trade in Selected Materials and Components for Constructional Use: Imports (cif) & Exports (fob) (continued)

United Kingdom											Current Prices (£000s)	
		1989	1990	1991	1992	1993	1994	1995	1996	1997	1998	1999
Radiators	Imports	50,546	48,875	50,944	46,703	43,378	50,835	47,608	51,305	49,653	54,484	70,631
	Exports	6,314	6,612	8,639	10,181	14,365	14,796	13,744	20,994	19,000	12,214	15,648
Air Conditioning	Imports	142,048	162,858	134,837	122,124	150,179	202,056	227,214	315,863	318,213	357,345	348,057
Units	Exports	70,679	85,483	78,484	87,647	135,460	139,907	193,713	239,034	222,687	216,871	240,856
Electric Wires	Imports	128,355	130,464	131,566	143,578	151,931	191,791	241,968	277,056	265,056	263,403	281,874
	Exports	104,929	129,081	107,603	108,562	121,341	150,767	184,636	188,978	191,003	186,855	177,006
Lamps & Fittings	Imports	99,517	108,446	98,554	113,780	122,932	140,526	142,213	170,071	178,474	202,634	230,500
	Exports	81,363	91,678	84,925	87,626	96,176	100,118	121,196	159,906	183,401	190,472	167,191
Wallpaper	Imports	18,220	20,414	20,603	20,949	16,201	20,920	20,995	26,925	25,954	23,447	23,406
	Exports	108,032	106,272	111,604	119,796	158,097	171,691	190,334	216,358	203,423	159,157	124,443

Source of Data: Construction Market Intelligence, Department of Environment, Transport and the Regions

Contact: David Williams 020 7944 5593

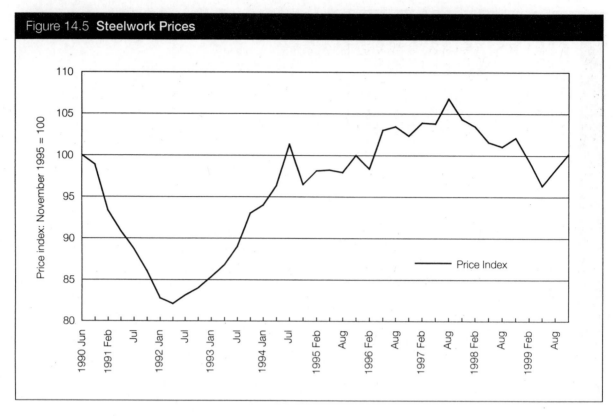

Source of Data: Table 14.7

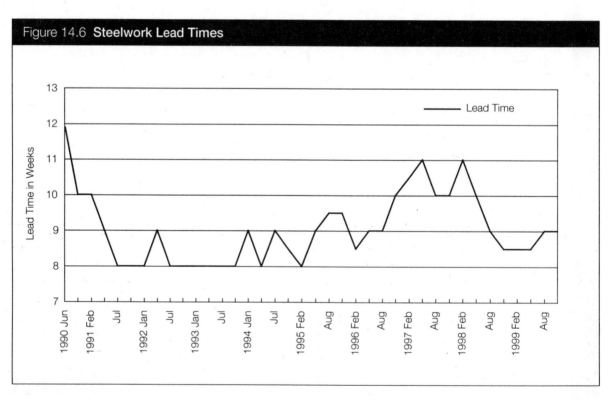

Source of Data: Table 14.7

Table 14.7 **Steelwork Prices and Lead Times**[1]

Regular shaped, braced 3/4 storey commercial structure of 150/200 tonnes

		Price per Tonne Index based on November 1995 = 100	Lead Time (Weeks)[2]
1990	June	100	12
	December	99	10
1991	February	93	10
	May	91	9
	July	89	8
	September	86	8
1992	January	83	8
	April	82	9
	July	83	8
	September	84	8
1993	January	85	8
	April	87	8
	July	89	8
	October	93	8
1994	January	94	9
	April	96	8
	July	101	9
	November	96	9
1995	February	98	8
	May	98	9
	August	98	10
	November	100	10
1996	February	98	9
	May	103	9
	August	103	9
	November	102	10
1997	February	104	11
	May	104	11
	August	107	10
	November	104	10
1998	February	103	11
	May	102	10
	August	101	9
	November	102	9
1999	February	99	9
	May	96	9
	August	98	9
	November	100	9

Notes

1. These figures are based on a representative sample of 30 BCSA member companies.
2. Time elapsed between placing an order and delivery starting on site.

Source of Data: British Constructional Steelwork Association
Contact: Gillian Mitchell MBE 020 7839 8566

CHAPTER 15

International Comparison

Table 15.1 Key Data

Country	Local Currency	Exchange Rate/£[1]	Pricing Key City Location	VAT General	Sales Tax (& the like) Building
EUROPE					
AUSTRIA	Schillings	21.84	Vienna	20.0%	20.0%
CYPRUS	Cyprus £	0.92	Nicosia	8.0%	8.0%
CZECH REPUBLIC	Koruna	57.20	Prague	22.0%	5.0%
DENMARK	Danish Krone	11.81	Copenhagen	25.0%	25.0%
FINLAND	Markka	9.44	Helsinki	22.0%	22.0%
FRANCE	French Franc	10.41	Paris	20.6%	20.6%
GERMANY	Deutsch-Mark	3.10	Berlin	16.0%	16.0%
HUNGARY	Forint	402.57	Budapest	25.0%	25.0%
IRELAND	Punt	1.25	Dublin	21.0%	12.5%
ITALY	Lira	3,072.88	Milan	20.0%	20.0%
NETHERLANDS	Guilder	3.50	Amsterdam	17.5%	17.5%
NORWAY	Nor. Krone	12.88	Oslo	23.0%	23.0%
POLAND	Zloty	6.83	Warsaw	22.0%	22.0%
ROMANIA	Leu	28,694.30	Bucharest	22.0%	22.0%
RUSSIA	Rouble	42,529.70	Moscow	20.0%	20.0%
SPAIN	Peseta	264.06	Barcelona	16.0%	16.0%
SWEDEN	Krona	13.62	Stockholm	25.0%	25.0%
SWITZERLAND	Swiss Franc	2.54	Zurich	7.5%	7.5%
UNITED KINGDOM	£ Sterling	1.00	London	17.5%	17.5%
INTERNATIONAL					
ARGENTINA	Peso	1.60	Buenos Aires	21.0%	21.0%
AUSTRALIA	Aus. $	2.53	Melbourne	–	–
CHINA	Yuan	13.29	Shanghai	17.0%	–
HONG KONG	HK $	12.48	Victoria	–	–
INDIA	Indian Rupee	69.67	New Delhi	0-20%	10-12%
INDONESIA	Rupiah	11,655.90	Jakarta	10.0%	10.0%
ISRAEL	Shekel	6.80	Tel Aviv	17.0%	17.0%
JAPAN	Yen	163.90	Tokyo	5.0%	5.0%
KENYA	Kenya Shilling	120.01	Nairobi	15.0%	15.0%
MALAYSIA	Ringgit	6.13	Kuala Lumpur	–	–
NEW ZEALAND	NZ $	3.14	Auckland	12.5%	12.5%
OMAN	Rial Omani	0.62	Muscat	–	–
PHILIPPINES	Peso	65.67	Manila	10.0%	10.0%
SAUDI ARABIA	Riyal	6.02	Riyadh	–	–
SINGAPORE	S $	2.68	Singapore	3.0%	3.0%
SOUTH AFRICA	Rand	9.85	Pretoria	14.0%	14.0%
THAILAND	Baht	62.35	Bangkok	10.0%	10.0%
UNITED ARAB EMIRATES	Dirham	5.90	Dubai	–	–
USA	US $	1.61	Various Cities	Varies	Varies
CANADA	priced US $	1.61			
CAYMAN ISLANDS	CI $	1.29	Grand Cayman	–	–

Notes

– = nil or less than half the final digit shown.

1. Exchange rates current at 26 November 1999 as published in the London Financial Times.

Source of Data: International Cost Survey, Gardiner & Theobald

Contact: David Hart 020 7209 3000

Table 15.2 Building Labour Rates[1] in 1999

Country	BASIC RATE[2]			ALL-IN RATE[3,4]		
	Unskilled £/hour	Semi-Skilled £/hour	Skilled £/hour	Unskilled £/hour	Semi-Skilled £/hour	Skilled £/hour
EUROPE						
AUSTRIA	4.70	5.14	6.07	9.60	10.33	11.76
CYPRUS	3.17	3.44	3.93	5.02	5.68	6.83
CZECH REPUBLIC	0.73	1.40	2.27	2.27	3.41	4.90
DENMARK	11.18	12.03	12.62	21.43	23.38	23.97
FINLAND	4.77	6.36	7.42	8.27	11.34	12.82
FRANCE	4.23	5.28	6.24	9.03	11.53	12.78
GERMANY – BERLIN	5.80	7.09	11.28	19.33	20.94	24.16
HUNGARY	0.87	1.12	1.61	2.48	3.23	4.60
IRELAND	5.13	5.83	6.40	9.19	10.58	11.50
ITALY	6.83	7.81	8.46	13.67	15.62	16.92
NETHERLANDS	5.72	6.58	7.15	12.87	14.01	16.01
NORWAY	5.44	0.00	9.94	16.31	0.00	21.74
POLAND	0.99	1.23	1.48	1.91	2.36	2.84
ROMANIA	0.81	1.56	1.96	1.06	2.09	2.65
RUSSIA	0.00	0.00	0.00	2.49	3.24	4.49
SPAIN	4.40	5.08	6.44	8.51	9.21	10.85
SWEDEN	0.00	5.87	8.08	0.00	12.48	16.89
SWITZERLAND	11.02	11.96	12.41	18.38	20.08	20.86
UNITED KINGDOM	4.55	5.20	6.05	6.01	6.83	7.91
INTERNATIONAL (EXCLUDING USA)						
ARGENTINA	1.33	1.56	1.81	2.87	3.36	3.99
AUSTRALIA	5.97	6.41	6.77	9.22	9.88	10.48
CHINA	0.19	0.27	0.44	0.22	0.31	0.51
HONG KONG	5.85	0.00	8.82	6.73	0.00	10.18
INDIA	0.11	0.13	0.19	0.14	0.17	0.23
INDONESIA	0.51	0.69	0.86	0.62	0.82	1.03
ISRAEL	3.53	4.41	5.89	4.12	5.59	7.65
JAPAN	11.98	15.49	16.78	17.97	23.23	25.17
KENYA	0.15	0.34	0.44	0.33	0.67	1.00
MALAYSIA	0.00	0.00	0.00	0.90	1.14	1.88
NEW ZEALAND	2.94	4.30	5.57	3.53	5.16	7.32
OMAN	1.02	1.25	1.36	1.46	1.78	1.94
PHILIPPINES	0.37	0.44	0.56	0.42	0.51	0.65
SAUDI ARABIA	0.66	0.83	1.66	1.33	1.66	2.49
SINGAPORE	1.34	1.79	2.24	2.20	2.27	2.42
SOUTH AFRICA	0.83	1.24	2.08	1.00	1.59	2.64
THAILAND	3.05	3.61	6.26	3.53	4.09	6.82
UNITED ARAB EMIRATES	0.71	0.75	0.78	1.02	1.07	1.12
USA						
BOSTON	9.34	13.08	16.19	16.19	20.24	24.60
HONOLULU	13.77	15.80	19.22	20.24	23.39	28.54
LAS VEGAS	10.90	11.52	16.19	16.51	18.69	21.80
LOS ANGELES	11.66	12.87	16.18	17.80	19.01	20.29
PHOENIX	9.03	11.83	14.33	16.19	19.31	22.73
PORTLAND	18.67	18.84	20.54	22.72	22.98	24.88
SAN FRANCISCO	13.70	16.19	18.69	26.78	28.65	38.62
SEATTLE	17.09	17.38	20.41	20.09	20.44	24.55
CARIBBEAN						
CAYMAN ISLANDS	7.32	10.28	13.39	8.88	12.30	19.00
BRITISH VIRGIN ISLANDS	3.74	4.98	7.47	4.98	6.54	9.97

Notes

1. The rates within these tables are representative of each country, and cannot be directly compared on a like-for-like basis.
2. The Basic Rate is the basic amount paid to a site operative, based upon figures promulgated by the national wage fixing body in the country. In many cases employers may be paying above or below the basic rate.
3. The All-In Rate is the gross hourly cost of employing the site operative, based upon the standard working week for the country including items such as insurances, statutory contributions and taxes.
4. All-In Rate for the Middle East includes for the cost of importing labour, food and accommodation.

Source of Data: International Cost Survey, Gardiner & Theobald
Contact: David Hart 020 7209 3000

Table 15.3 Building Material Supply Prices[1] in 1999

Country	Steel Rebar High Yield £/tonne	Structural Steel £/tonne	Ordinary Cement £/tonne	Concrete (20N/mm²) £/tonne	Aggregate All Grades £/tonne	Sand (Coarse) £/tonne	Plaster £/tonne	Carcassing Timber £/m³	Common Bricks £/1000	Concrete Blocks (100mm) £/m²	Glass (6mm) £/m²
EUROPE											
AUSTRIA	279	..	87	..	10	10	112	..	..	..	
CYPRUS	197	683	38	22	3	3	49	213	147	7	7
CZECH REPUBLIC	273	455	41	23	6	6	29	91	119	5	4
DENMARK	678	565	95	68	7	7	69	..	..	14	
FINLAND	307	646	65	34	6	6	..	127	..	6	21
FRANCE	576	576	67	61	9	9	81	144	173	4	14
GERMANY - BERLIN	242	515	64	42	11	8	119	177	209	14	23
HUNGARY	348	497	45	21	10	9	87	149	124	7	9
IRELAND	260	800	68	40	5	6	88	200	240	48	16
ITALY	179	586	50	42	7	11	75	146	49	4	18
NETHERLANDS	503	815	63	51	14	17	94	206	112	14	18
NORWAY	311	1,087	101	25	7	7	97	140	311	12	21
POLAND	199	224	31	26		2	51	143	98	6	3
ROMANIA	280	560	..	..	..	..	..	..	..	..	..
RUSSIA	198	494	46	43	7	4	..	56	..	20	28
SPAIN	360	379	40	37	8	9	..		68	4	12
SWEDEN	441	..	110	73	10	10	..	110	..	13	..
SWITZERLAND	595		106	68	13	20	138		218		
UNITED KINGDOM	359	554	76	57	11	11	127	202	173	7	21
INTERNATIONAL (EXCLUDING USA)											
ARGENTINA	293	498	74	56	4	6	78	50	75	6	19
AUSTRALIA	434	513	105	39	13	11	109	237	169	7	9
CHINA	169	211	26	26	4	4		98	21	2	3
HONG KONG	170	297	55	48	4	4	48	116	69	6	4
INDIA	244	309	37	34	11	3	..	..	20	4	11
INDONESIA	249	300	43	26	4	3	..	129	..	4	4
ISRAEL	..	..	..	..	..	..	..	..	..	..	..
JAPAN	177	427	53	73	26	28	..	323	..	9	12
KENYA	375	..	73	38	7	9	..	59	153	5	7
MALAYSIA	196	424	33	21	5	3	73	114	33	2	1
NEW ZEALAND	398	446	67	56	9	13	..	..	153	6	16
OMAN	324	..	47	21	3	5	36	170	..	5	..
PHILIPPINES	195	274	38	41	12	11	..	168	..	1	15
SAUDI ARABIA	241	498	43	33	4	2	58	125	..	3	20
SINGAPORE	186	261	30	17	4	4	..	268	50	3	9
SOUTH AFRICA	..	406	49	27	7	5	..	..	32	..	..
THAILAND	176	481	29	24	5	4	29	..	..	4	4
UNITED ARAB EMIRATES	178	305	41	30	6	11	170	271	..	5	11
USA											
BOSTON	436	872	125	40	11	11	218	184	187	12	36
HONOLULU	592	1,713	53	87	24	31	253	396	301	24	30
LAS VEGAS	498	965	100	50	5	5	92	125	436	6	26
LOS ANGELES	436	959	106	36	4	4	93	123	480	13	28
PHOENIX	440	1,134	130	26	12	9	214	174	175	12	33
PORTLAND	352	891	55	53	11	12	78	118	287	20	26
SAN FRANCISCO	343	810	55	73	37	11	193	146	218	26	40
SEATTLE	357	893	54	59	11	12	78	120	280	19	26
CARIBBEAN											
CAYMAN ISLANDS	523	2,647	106	128	24	24	..	315	..	8	24
BRITISH VIRGIN ISLANDS	473	1,869	75	90	34	34	..	279	..	8	49

Notes

.. = not available; no data have been provided by local offices

1. Unless stated, all rates include for delivery to site and local discounts, but exclude VAT and local taxes.

Source of Data: International Cost Survey, Gardiner & Theobald
Contact: David Hart 020 7209 3000

Table 15.4 Examples of Building Costs[1,2] in 1999

RESIDENTIAL, RETAIL & HOTELS

Country	HIGH RISE APARTMENTS £/m² Low	HIGH RISE APARTMENTS £/m² High	HIGH RISE APARTMENTS Typical Number of Floors	SHOPPING CENTRE £/m² Low	SHOPPING CENTRE £/m² High	SHOPPING CENTRE Typical Number of Floors	HIGH QUALITY CAPITAL CITY HOTEL £/m² Low	HIGH QUALITY CAPITAL CITY HOTEL £/m² High	HIGH QUALITY CAPITAL CITY HOTEL Typical Number of Floors	PROVINCIAL/ SUBURBAN HOTEL £/m² Low	PROVINCIAL/ SUBURBAN HOTEL £/m² High	PROVINCIAL/ SUBURBAN HOTEL Typical Number of Floors
EUROPE												
AUSTRIA	1,090	1,154	. .	. .	. .	. .	. .	. .	. .	. .	. .	. .
CYPRUS	360	513	4	306	546	2	546	874	5	350	492	3
CZECH REPUBLIC	306	434	1	367	490	1	577	829	8	395	504	3
DENMARK	905	1,006	5	1,062	1,154	1	1,339	1,431	5	973	1,052	3
FINLAND	763	1,113	4	530	646	2	1,049	1,166	5	816	933	3
FRANCE	432	720	6	384	624	2	817	1,153	6	432	576	2
GERMANY - BERLIN	644	838	5	805	1,128	3	1,289	1,450	10	805	1,128	5
HUNGARY	261	393	6	344	574	3	656	951	6	361	541	4
IRELAND	500	680	5	1,120	1,400	2	880	1,040	4	720	920	3
ITALY	488	716	4	814	976	1	651	1,139	4	521	911	2
NETHERLANDS	387	631	10	510	692	1	1,134	1,313	8	981	1,104	4
NORWAY	544	932	3	505	854	2	1,398	1,553	5	660	1,010	2
POLAND	334	585	. .	538	717	. .	684	854	. .	469	684	. .
ROMANIA	308	560	2	n/a	n/a	. .	616	934	5	. .	. .	. .
RUSSIA	n/a	n/a	. .	424	667	3	727	909	5	455	667	4
SPAIN	328	390	8	298	357	2	468	562	10	343	417	6
SWEDEN	808	918	3	698	808	3	734	881	4	698	771	3
SWITZERLAND	590	872	4	369	550		1,381	1,850	5	. .	. .	. .
UNITED KINGDOM	865	1,175	8	820	1,470	2	1,280	1,520	10	750	1,070	4
INTERNATIONAL (EXCLUDING USA)												
ARGENTINA	280	343	25	374	498	3	561	623	20	312	374	4
AUSTRALIA	502	673	15	409	553	2	900	1,204	15	717	1,011	2
CHINA	263	339	. .	564	677	. .	677	903	. .	527	752	. .
HONG KONG	850	1,162	45	1,002	1,242	4	994	1,459	25	n/a	n/a	
INDIA	86	115	14	100	129	2	86,119[3]	100,473[3]	. .	28,706[3]	43,060[3]	. .
INDONESIA	240	283	10	214	257		343	412	15	279	335	8
ISRAEL	368	405	16	294	405	2	441	552	12	324	441	3
JAPAN	1,666	1,849	20	671	1,666	5	2,221	3,325	20	1,574	2,032	10
KENYA	200	250	6	167	208	2	292	542	9	167	250	3
MALAYSIA	245	310	15	139	294	4	245	571	10	196	408	6
NEW ZEALAND	446	605	. .	271	446	. .	732	828	. .	573	669	. .
OMAN	417	712	. .	404	971	. .	445	801	. .	340	623	. .
PHILIPPINES	305	442	35	183	244	4	533	655	35	381	442	8
SAUDI ARABIA	581	797	7	498	830	2	913	1,246	7	747	1,079	4
SINGAPORE	447	596	25	410	484	2	783	894	10	596	745	7
SOUTH AFRICA	. .	. .	. .	183	264	1	304	365	7	264	325	5
THAILAND	265	321	20	200	233	4	401	481	20	192	289	3
UNITED ARAB EMIRATES	382	678	15	382	933	3	424	763	15	339	594	5

Table 15.4 Examples of Building Costs[1,2] in 1999 (continued)

RESIDENTIAL, RETAIL & HOTELS

Country	HIGH RISE APARTMENTS £/m² Low	High	Typical Number of Floors	SHOPPING CENTRE £/m² Low	High	Typical Number of Floors	HIGH QUALITY CAPITAL CITY HOTEL £/m² Low	High	Typical Number of Floors	PROVINCIAL/ SUBURBAN HOTEL £/m² Low	High	Typical Number of Floors
USA												
BOSTON	685	1,121	..	498	685	..	1,246	1,744	..	934	1,308	..
HONOLULU	498	903	..	554	841	..	1,308	1,619	..	810	1,277	..
LAS VEGAS	361	548	20	492	685	2	1,046	1,277	7	729	903	6
LOS ANGELES	386	685	25	623	747	2	1,090	1,308	7	685	872	6
PHOENIX	505	617	25	492	648	3	1,233	1,619	25	972	1,258	3
PORTLAND	536	1,071	15	424	1,071	2	903	1,370	10	505	841	5
SAN FRANCISCO	561	747	10	529	592	3	1,183	1,495	20	997	1,183	4
SEATTLE	573	1,071	..	436	1,071	..	941	1,339	..	505	841	..
CANADA												
CALGARY	529	747	..	374	561	..	903	1,277	..	467	754	..
CARIBBEAN												
CAYMAN ISLANDS	779	1,370	3	779	997	1	1,183	1,557	5	779	1,183	2
BRITISH VIRGIN ISLANDS	436	573	4	n/a	n/a	..	747	997	4	623	872	4

Notes

. . = not available; no data have been provided by local offices

n/a = not applicable

1. The above descriptions are indicative only, as construction specification and requirements will vary between countries. Therefore, building costs are for typical buildings in each country.

2. Typical number of floors have only been provided where identified by local offices.

3. These are costs per room

Source of Data: International Cost Survey, Gardiner & Theobald
Contact: David Hart 020 7209 3000

Table 15.5 Examples of Building Costs[1,2] in 1999

OFFICE & INDUSTRIAL

Country	HEATED OFFICES £/m² Low	High	Typical Number of Floors	AIR CONDITIONED OFFICES £/m² Low	High	Typical Number of Floors	FACTORIES, WAREHOUSES, INDUSTRIAL £/m² Low	High	Typical Number of Floors	HIGH TECHNOLOGY/ RESEARCH £/m² Low	High	Typical Number of Floors
EUROPE												
AUSTRIA	962	1,442	7	1,186	1,667	7	. .	. .	. .	. .	. .	. .
CYPRUS	300	579	5	339	644	6	197	415	1	437	874	1
CZECH REPUBLIC	358	455	6	455	664	6	192	378	1	350	498	1
DENMARK	983	1,083	2	1,242	1,371	2	877	969	1	1,246	1,357	2
FINLAND	699	880	5	763	933	5	435	530	1	646	763	2
FRANCE	576	817	4	672	961	4	192	240	1	480	576	2
GERMANY – BERLIN	773	1,128	5	966	1,450	5	322	548	2	902	1,450	4
HUNGARY	311	443	3	410	639	6	180	295	1	393	525	2
IRELAND	640	920	5	840	1,120	5	160	300	1	640	920	2
ITALY	521	651	8	651	976	8	260	521	1	651	1,302	2
NETHERLANDS	623	758	6	749	988	8	305	474	1	850	1,121	3
NORWAY	621	1,087	4	660	1,165	4	388	932	1	932	1,398	4
POLAND	373	606	. .	505	788	. .	233	385	. .	406	555	. .
ROMANIA	. .	. .	. .	560	747	6	311	436	2	n/a	n/a	. .
RUSSIA	424	727	5	515	909	6	424	667	2	n/a	n/a	. .
SPAIN	274	329	6	312	375	10	137	165	2	511	587	. .
SWEDEN	624	661	4	661	881	4	587	661	1	661	734	1
SWITZERLAND	. .	. .	. .	918	1,381	4	362	617	2	918	1,529	2
UNITED KINGDOM	795	1,080	4	1,195	1,690	8	235	380	1	700	1,030	2
INTERNATIONAL (EXCLUDING USA)												
ARGENTINA	n/a	n/a	. .	405	530	20	93	125	1			
AUSTRALIA	n/a	n/a	. .	590	731	17	192	247	1	592	790	2
CHINA	n/a	n/a	. .	489	640	. .	203	346	. .	579	715	. .
HONG KONG	n/a	n/a	. .	890	1,162	45	385	577	15	1,162	1,555	3
INDIA	n/a	n/a	. .	273	316	3	115	172	2	258	344	2
INDONESIA	n/a	n/a	. .	270	322	. .	195	236	. .	. .	. .	. .
ISRAEL	412	508	10	412	508	10	191	250	2	368	486	6
JAPAN	n/a	n/a	. .	1,617	2,636	15	488	1,147	1	1,147	1,611	3
KENYA	n/a	n/a	. .	292	375	12	167	250	1	375	458	2
MALAYSIA			. .	163	294	33	73	114	2	. .	. .	. .
NEW ZEALAND	509	573	20	573	669	25	96	143	. .	239	398	. .
OMAN	n/a	n/a	. .	356	623	. .	178	445	. .	566	971	. .
PHILIPPINES	n/a	n/a	. .	335	457	40	168	259	2	457	594	5
SAUDI ARABIA	n/a	n/a	. .	498	830	7	349	581	1	581	996	2
SINGAPORE	n/a	n/a	. .	559	708	25	287	373	2	317	410	2
SOUTH AFRICA	n/a	n/a	. .	203	284	2	162	223	1	244	304	1
THAILAND	n/a	n/a	. .	353	401	30	112	136	2	n/a	n/a	. .
UNITED ARAB EMIRATES	. . n/a	. . n/a	. . . .	. . 339	. . 594	. . 20	. . 170	. . 424	. . . .	. . 543	. . 933	. . 3

Table 15.5 Examples of Building Costs[1,2] in 1999 (continued)

OFFICE & INDUSTRIAL

Country	HEATED OFFICES			AIR CONDITIONED OFFICES			FACTORIES, WAREHOUSES, INDUSTRIAL			HIGH TECHNOLOGY/ RESEARCH		
	£/m² Low	£/m² High	Typical Number of Floors	£/m² Low	£/m² High	Typical Number of Floors	£/m² Low	£/m² High	Typical Number of Floors	£/m² Low	£/m² High	Typical Number of Floors
USA												
BOSTON	n/a	n/a	..	835	1,134	8	296	498	1	810	1,121	..
HONOLULU	n/a	n/a	..	716	1,121	..	234	436	..	747	1,121	..
LAS VEGAS	n/a	n/a	..	732	897	6	280	349	1	797	1,028	3
LOS ANGELES	n/a	n/a	..	747	997	6	311	424	1	841	1,121	4
PHOENIX	n/a	n/a	..	666	816	30	206	324	1	623	685	2
PORTLAND	n/a	n/a	..	523	1,227	..	280	511	..	704	934	..
SAN FRANCISCO	n/a	n/a	..	997	1,246	10	343	405	1	623	716	2
SEATTLE	n/a	n/a	..	536	1,239	..	305	505	..	704	934	..
CANADA												
CALGARY	n/a	n/a	..	529	903	..	386	498	..	760	934	..
CARIBBEAN												
CAYMAN ISLANDS	n/a	n/a	..	1,246	1,869	5	311	467	1	n/a	n/a	..
BRITISH VIRGIN ISLANDS	n/a	n/a	..	654	934	3	343	467	1	n/a	n/a	..

Notes

.. = not available; no data have been provided by local offices

n/a = not applicable

1. The above descriptions are indicative only, as construction specification and requirements will vary between countries. Therefore, building costs are for typical buildings in each country.

2. Typical number of floors have only been provided where identified by local offices.

Source of Data: International Cost Survey, Gardiner & Theobald

Contact: David Hart 020 7209 3000

Table 15.6 Inflation Rates and Forecasts[1,2,]

Country	RETAIL PRICE INFLATION								
	1993	1994	1995	1996	1997	1998	1999	2000	2001
EUROPE									
AUSTRIA	..	..	..	..	..	..	0.9%	1.2%	..
CYPRUS	..	..	..	..	..	..	..	..	..
CZECH REPUBLIC	22.2%	9.1%	9.0%	10.0%	9.2%	9.9%	2.1%	..	..
DENMARK	1.2%	2.0%	2.1%	2.1%	2.3%	3.4%	8.0%	8.0%	..
FINLAND	0.0%	5.0%	1.2%	-1.0%	2.4%	2.6%	2.6%	4.0%	..
FRANCE	2.1%	1.6%	2.1%	2.0%	1.1%	-1.0%	0.7%	1.0%	..
GERMANY – BERLIN	4.0%	5.0%	2.2%	1.5%	2.4%	1.0%	0.7%	0.6%	..
HUNGARY	15.0%	27.0%	29.8%	23.0%	18.0%	14.5%	10.5%	9.0%	..
IRELAND	2.9%	2.5%	1.6%	2.8%	1.7%	5.9%	..	..	..
ITALY	3.5%	5.9%	6.2%	3.9%	2.2%	3.5%	1.4%	2.0%	1.3%
NETHERLANDS	2.6%	2.6%	1.7%	2.5%	2.4%	1.8%	2.4%	2.4%	
NORWAY	1.9%	1.9%	2.1%	2.5%	1.6%	2.3%	3.0%	4.8%	4.7%
POLAND	35.4%	32.2%	28.0%	19.5%	15.3%	10.6%	8.5%	..	..
ROMANIA .	..	..	..	..	..	..	..	..	..
RUSSIA	..	..	..	..	..	52.0%	120.0%	..	..
SPAIN	.	..	..	..	..	..	2.7%	2.4%	.
SWEDEN	4.6%	2.2%	2.5%	0.5%	0.5%	2.9%	2.5%	..	..
SWITZERLAND	..	..	..	0.8%	0.8%	2.1%	1.6%	..	..
UNITED KINGDOM	1.6%	2.4%	3.3%	2.5%	3.1%	3.4%	1.3%	2.8%	3.0%

Country	BUILDING TENDER PRICE INFLATION								
	1993	1994	1995	1996	1997	1998	1999	2000	2001
EUROPE									
AUSTRIA	4.5%	3.5%	3.4%	1.5%	2.7%	2.2%	2.9%	..	..
CYPRUS	11.6%	6.7%	7.6%	6.5%	6.7%	8.0%	7.9%	10.3%	5.0%
CZECH REPUBLIC	17.9%	12.0%	9.3%	12.0%	10.7%	8.4%	2.0%	5.3%	..
DENMARK	2.9%	2.8%	5.4%	2.6%	4.2%	4.8%	9.5%	5.8%	..
FINLAND	-4.4%	15.0%	0.0%	-7.4%	9.0%	11.0%	12.5%	4.1%	..
FRANCE	3.6%	3.5%	1.9%	1.7%	1.1%	1.0%	1.7%	..	..
GERMANY – BERLIN	8.2%	5.6%	2.4%	2.1%	-2.5%	0.0%	..	..	..
HUNGARY	24.3%	30.2%	27.3%	19.0%	14.0%	19.0%	11.5%	10.0%	..
IRELAND	3.3%	8.0%	2.0%	2.6%	3.4%	3.2%	..	..	..
ITALY	-0.9%	1.8%	2.0%	2.2%	0.9%	3.3%	1.6%	2.4%	1.5%
NETHERLANDS	2.8%	0.9%	1.8%	1.8%	2.6%	1.7%	5.0%	3.2%	..
NORWAY	0.7%	4.1%	3.0%	1.7%	2.0%	4.1%	1.8%	2.3%	1.7%
POLAND	25.0%	19.6%	21.9%	21.1%	14.1%	7.0%	8.6%	..	..
ROMANIA	..	..	..	..	..	..	40.0%	7.1%	..
RUSSIA	..	..	..	..	..	..	..	..	..
SPAIN	2.8%	2.7%	3.5%	4.2%	3.2%	1.6%	5.4%	2.2%	..
SWEDEN	3.0%	1.6%	6.4%	1.2%	0.9%	2.5%	..	..	..
SWITZERLAND	1.0%	1.5%	..	..	..	..	..	..	..
UNITED KINGDOM	1.7%	6.0%	4.9%	3.7%	5.5%	4.6%	3.7%	3.7%	3.2%

Table 15.6 Inflation Rates and Forecasts[1,2] (continued)

Country	1993	1994	1995	1996	1997	1998	1999	2000	2001
				RETAIL PRICE INFLATION					
INTERNATIONAL (EXCLUDING USA)									
ARGENTINA	..	5.0%	4.0%	-3.0%	3.0%	2.0%	-3.8%	2.8%	..
AUSTRALIA	2.4%	2.1%	4.7%	1.2%	0.8%	0.8%	1.2%	3.0%	..
CHINA	..	..	..	..	1.8%	-1.5%	-1.3%	..	..
HONG KONG	10.8%	9.8%	8.2%	7.0%	5.3%	1.1%	2.8%	4.3%	..
INDIA	..	..	..	..	..	15.0%	3.1%	..	..
INDONESIA	4.9%	5.0%	3.9%	3.4%	6.9%	82.4%	..	..	..
ISRAEL	13.0%	8.0%	10.6%	6.7%	2.8%	4.7%	..	..	..
JAPAN	..	-1.1%	0.9%	-1.7%	-0.6%	-3.5%	0.3%	-0.2%	..
KENYA	..	..	..	..	..	..	..	..	..
MALAYSIA	..	..	..	..	..	..	..	..	..
NEW ZEALAND	2.2%	0.9%	2.9%	2.6%	0.0%	2.1%	2.4%	2.4%	..
OMAN	..	..	..	..	..	..	..	..	..
PHILIPPINES	..	..	..	..	5.3%	10.2%	5.7%	..	..
SAUDI ARABIA	0.8%	0.6%	5.0%	2.0%	2.0%	0.2%	1.9%	1.9%	2.3%
SINGAPORE	..	..	..	..	2.3%	-1.4%	0.9%	..	..
SOUTH AFRICA	..	..	..	6.0%	4.0%	4.1%	..	..	..
THAILAND	..	..	..	..	7.0%	5.8%	..	..	..
UNITED ARAB EMIRATES	-4.4%	0.0%	4.5%	0.0%	4.3%	8.3%	0.0%	0.0%	0.0%

Country	1993	1994	1995	1996	1997	1998	1999	2000	2001
				BUILDING TENDER PRICE INFLATION					
INTERNATIONAL (EXCLUDING USA)									
ARGENTINA	..	2.2%	0.8%	-5.0%	1.1%	0.0%	-0.8%	4.0%	..
AUSTRALIA	2.8%	1.8%	1.9%	2.8%	2.4%	3.2%	3.0%	3.0%	5.0%
CHINA	0.0%	-1.5%	-1.5%	-2.5%	-2.5%	-1.2%	-0.3%	-0.8%	..
HONG KONG	5.3%	13.0%	12.4%	10.2%	19.3%	-6.0%	-0.1%	..	..
INDIA	8.0%	7.7%	7.1%	6.6%	6.3%	5.9%	11.1%	10.0%	8.0%
INDONESIA	4.7%	1.4%	2.1%	2.3%	10.0%	100.0%	0.0%	..	..
ISRAEL	14.5%	11.9%	8.0%	8.0%	7.4%	2.1%	2.0%	4.6%	4.3%
JAPAN	-3.7%	-5.4%	-2.7%	-1.3%	-0.3%	-2.2%	-1.6%	..	..
KENYA	63.0%	15.5%	2.9%	8.1%	9.8%	8.2%	17.0%	17.0%	16.0%
MALAYSIA	3.4%	3.0%	2.0%	0.9%	3.2%	0.4%	4.0%	..	..
NEW ZEALAND	2.6%	2.1%	3.2%	0.4%	0.0%	0.1%	2.1%	2.0%	..
OMAN	..	..	..	..	..	..	..	..	..
PHILIPPINES	..	..	..	..	..	..	..	..	..
SAUDI ARABIA	..	..	..	..	..	..	..	..	..
SINGAPORE	-2.0%	0.0%	5.2%	8.9%	2.7%	-14.2%	-9.3%	..	..
SOUTH AFRICA	8.3%	8.4%	6.8%	5.7%	6.2%	7.2%	11.8%	6.5%	4.2%
THAILAND	5.4%	8.2%	5.7%	-0.9%	8.1%	-1.7%	-2.6%	4.3%	4.2%
UNITED ARAB EMIRATES	-3.5%	0.0%	4.5%	0.0%	4.3%	8.3%	0.1%	..	..

Table 15.6 Inflation Rates and Forecasts[1,2] (continued)

RETAIL PRICE INFLATION

Country	1993	1994	1995	1996	1997	1998	1999	2000	2001
USA	4.2%	2.7%	2.3%	2.1%	2.9%	4.5%	2.3%	2.6%	. .
BOSTON	. .	. .	. .	. .	. .	. .	. .	. .	. .
HONOLULU	. .	. .	. .	. .	. .	. .	. .	. .	. .
LAS VEGAS	. .	. .	. .	. .	. .	. .	. .	. .	. .
LOS ANGELES	. .	. .	. .	. .	. .	. .	. .	. .	. .
PHOENIX	. .	. .	. .	. .	. .	. .	. .	. .	. .
PORTLAND	. .	. .	. .	. .	. .	. .	. .	. .	. .
SAN FRANCISCO	. .	. .	. .	. .	. .	. .	. .	. .	. .
SEATTLE	. .	. .	. .	. .	. .	. .	. .	. .	. .

BUILDING TENDER PRICE INFLATION

Country	1993	1994	1995	1996	1997	1998	1999	2000	2001
USA									
BOSTON	1.9%	1.8%	4.5%	2.6%	5.0%	5.6%	8.3%	4.2%	. .
HONOLULU	1.3%	2.1%	3.2%	5.2%	3.4%	3.1%	2.9%	3.1%	. .
LAS VEGAS	3.3%	3.0%	1.9%	1.6%	2.4%	4.9%	2.8%	2.8%	. .
LOS ANGELES	1.6%	3.0%	1.6%	1.2%	2.2%	5.1%	4.3%	2.5%	. .
PHOENIX	2.9%	2.8%	2.7%	2.6%	2.6%	8.3%	3.1%	3.0%	. .
PORTLAND	2.8%	2.8%	6.2%	0.0%	5.9%	7.9%	1.4%	2.8%	. .
SAN FRANCISCO	2.9%	2.8%	2.7%	2.6%	2.6%	8.3%	4.7%	4.5%	. .
SEATTLE	2.8%	2.8%	3.6%	2.6%	5.9%	3.0%	3.7%	3.5%	. .

Notes

. . = not available; no data have been provided by local offices

1. Figures for 1999 to 2001 are forecasts.

2. Disclaimer: Forecasts are those of Gardiner & Theobald and in no way represent the views of DETR.

Source of Data: International Cost Survey, Gardiner & Theobald
Contact: David Hart 020 7209 3000

Table 15.7 Production of Constructional Steelwork (000 tonnes)

	Actual								Forecast	
	1991	1992	1993	1994	1995	1996	1997	1998	1999	2000
AUSTRIA	72	69	77	79	79	75	78	170	175	175
BELGIUM & LUXEMBOURG	385	300	304	308	310	285	305	. .	. .	. .
CZECH REPUBLIC	n/a	75	70	64	56	66	69	. .	. .	. .
DENMARK	97	92	78	88	95	99	110	77	71	71
FINLAND	125	95	90	95	100	106	116	85	91	91
FRANCE	697	641	550	548	605	636	630	668	695	695
GERMANY	1,621	1,706	1,637	1,698	1,945	1,704	1,815	1,832	1,877	1,877
WEST	1,349	1,395	1,270	1,250	. .	. .	. .	. .	. .	. .
EAST	272	311	367	448	. .	. .	. .	. .	. .	. .
ITALY	1,120	920	801	790	725	965	1,010	745	745	775
NETHERLANDS	495	512	449	479	536	575	560	n/a	n/a	n/a
NORWAY	40	39	40	44	48	47	75	64	57	56
SLOVENIA	n/a	17	16	15	14	11	22	27	31	35
SPAIN	960	900	720	750	760	749	883	. .	. .	. .
SWEDEN	78	72	67	61	86	107	132	97	108	110
SWITZERLAND	82	76	60	55	79	80	80	79	79	79
UNITED KINGDOM	903	833	858	937	989	1,004	1,099	1,074	1,057	1,011
TOTAL	6,675	6,347	5,817	6,011	6,427	6,509	6,984	4,918	4,986	4,975

Notes

. . = not available

Source of Data: The British Constructional Steelwork Association Limited
Contact: Gillian Mitchell MBE 020 7839 8566

Table 15.8: Outward Direct Investment in the Construction Industry

Direct investment of UK companies in overseas subsidiary and associate companies and earnings from UK direct investment overseas analysed by area and main country, 1995 to 1998

£ Million

Area and Main Country			Year	Net Investment[1,2]	Net Earnings[3]
EUROPE			1995	21	-20
			1996	-29	-60
			1997	262	-3
			1998	59	-2
	EU		1995	31	-22
			1996	-20	-51
			1997	292	-12
			1998	26	-4
		Austria	1995	–	–
			1996	–	–
			1997	–	–
			1998	–	–
		Belgium & Luxembourg (BLEU)	1995	1	0
			1996	0	0
			1997	2	..
			1998	..	..
		Denmark	1995	..	..
			1996	2	1
			1997	..	..
			1998	..	..
		Finland	1995	–	–
			1996	–	–
			1997	–	–
			1998	..	..
		France	1995	..	..
			1996	-31	-40
			1997	75	12
			1998	13	6
		Germany	1995	-10	-46
			1996	-35	-38
			1997	43	-35
			1998	5	-24
		Greece	1995	–	–
			1996	..	..
			1997	..	..
			1998	..	..
		Irish Republic	1995	..	0
			1996	3	-4
			1997	1	1
			1998	..	..
		Italy	1995	..	..
			1996	-8	-1
			1997	..	..
			1998	..	..
		Netherlands	1995	-1	3
			1996	39	24
			1997	-1	1
			1998	-6	4
		Portugal	1995	15	3
			1996	..	..
			1997	-2	1
			1998	0	0
		Spain	1995	..	..
			1996	13	9
			1997	..	3
			1998	4	1
		Sweden	1995	..	..
			1996	2	0
			1997	–	–
			1998	–	–

Table 15.8 Outward Direct Investment in the Construction Industry (continued)

Direct investment of UK companies in overseas subsidiary and associate companies and earnings from UK direct investment overseas analysed by area and main country, 1995 to 1998

£ Million

Area and Main Country	Year	Net Investment[1,2]	Net Earnings[3]
EFTA	1995	..	..
	1996	3	3
	1997	10	4
of which	1998	8	6
Norway	1995	-4	2
	1996	3	2
	1997	..	..
	1998	..	..
Switzerland	1995	..	..
	1996	..	..
	1997	..	..
	1998	..	..
OTHER	1995	..	..
EUROPEAN	1996	-11	-10
COUNTRIES	1997	-40	4
of which	1998	25	-3
Russia[4]	1995	..	..
	1996	-2	0
	1997	..	0
	1998	0	-1
UK Offshore	1995	–	–
Islands[5]	1996	–	–
	1997	..	..
	1998	..	..
AMERICA	1995	-32	..
	1996	291	43
	1997	-97	44
of which	1998	-115	91
Bermuda	1995	..	..
	1996	..	..
	1997	–	–
	1998	–	–
Brazil	1995	..	..
	1996	..	..
	1997	..	..
	1998	..	..
Canada	1995	1	0
	1996	32	20
	1997	66	26
	1998	..	13
Chile	1995	–	–
	1996	..	..
	1997	..	..
	1998	..	..
Colombia	1995	–	–
	1996	–	–
	1997	–	–
	1998	..	..
Mexico	1995	..	..
	1996	..	..
	1997	..	..
	1998	..	..
Panama	1995	–	–
	1996	–	–
	1997	–	–
	1998	–	–
USA	1995	-32	-13
	1996	258	21
	1997	-174	12
	1998	-102	71

Table 15.8 Outward Direct Investment in the Construction Industry (continued)

Direct investment of UK companies in overseas subsidiary and associate companies and earnings from UK direct investment overseas analysed by area and main country, 1995 to 1998

£ Million

Area and Main Country	Year	Net Investment[1,2]	Net Earnings[3]
ASIA	1995	23	..
	1996	..	20
	1997	-9	7
	1998	45	-3
NEAR & MIDDLE EAST COUNTRIES	1995	17	7
	1996	..	-4
	1997	7	-1
of which	1998	-10	1
Gulf	1995	16	8
Arabian	1996	-10	-5
Countries	1997	11	3
	1998	-10	5
OTHER ASIAN COUNTRIES	1995	6	..
	1996	16	26
	1997	-16	9
of which	1998	56	-4
Hong Kong	1995	14	4
	1996	18	26
	1997	-34	17
	1998	30	14
India	1995	..	..
	1996	2	1
	1997	..	..
	1998	..	..
Indonesia	1995	3	0
	1996	3	0
	1997	0	-1
	1998	-2	-2
Japan	1995	..	..
	1996	..	..
	1997	..	..
	1998	..	..
Malaysia	1995	1	7
	1996	-11	5
	1997	5	4
	1998	18	2
Singapore	1995	-10	-4
	1996	6	2
	1997	-1	1
	1998	0	0
South Korea	1995	–	–
	1996	..	..
	1997	..	..
	1998	–	–
Thailand	1995	0	0
	1996	0	0
	1997	0	..
	1998	1	..
AUSTRALASIA & OCEANIA	1995	5	7
	1996	..	20
	1997	23	15
of which	1998	27	8
Australia	1995	3	7
	1996	13	19
	1997	..	..
	1998	..	..
New Zealand	1995	..	..
	1996	..	..
	1997	..	..
	1998	..	..

Table 15.8 Outward Direct Investment in the Construction Industry (continued)

Direct investment of UK companies in overseas subsidiary and associate companies and earnings from UK direct investment overseas analysed by area and main country, 1995 to 1998

£ Million

Area and Main Country	Year	Net Investment[1,2]	Net Earnings[3]
AFRICA	1995	17	16
	1996	..	18
	1997	39	27
	1998	-15	16
of which			
Kenya	1995	..	..
	1996	–	–
	1997	–	–
	1998	–	–
Nigeria	1995	..	..
	1996	..	..
	1997	..	..
	1998	..	..
South Africa	1995	..	0
	1996	0	1
	1997	..	..
	1998	..	..
Zimbabwe	1995	-2	10
	1996	-6	2
	1997	0	..
	1998	0	0
WORLD TOTAL	1995	33	0
	1996	278	41
	1997	215	89
	1998	0	108
OECD	1995	-4	-26
	1996	282	10
	1997	204	45
	1998	-59	92
CENTRAL & EASTERN EUROPE	1995	-2	0
	1996	..	0
	1997	-1	6
	1998	..	-2

Notes

- = before nil or less than half the final digit shown

.. = not available

1. Net investment includes unremitted profits.
2. A minus sign indicates a net disinvestment (i.e. a decrease in the amount due to the UK).
3. A minus sign indicates net losses.
4. Prior to 1995 Russia covers other former USSR countries, the Baltic States and Albania.
5. The UK Offshore Islands consist of the Channel Islands and the Isle of Man, excluded from the definition of the economic territory of the UK from 1997.

Source of Data: Office for National Statistics
Contact: Simon Harrington 01633 813314

Table 15.9 Outward Direct Investment in the Construction Industry

Level of direct investment assets held overseas by the UK in subsidiary and associate companies at the end of 1998 analysed by area and main country

£ Million

Area and Main Country	Net Book Value[1,2]
EUROPE	688
EU	665
Austria	–
Belgium & Luxembourg	..
Denmark	..
Finland	..
France	39
Germany	140
Greece	..
Irish Republic	..
Italy	..
Netherlands	..
Portugal	29
Spain	..
Sweden	–
EFTA	18
Norway	..
Switzerland	..
OTHER EUROPEAN COUNTRIES	4
Russia[3]	6
UK Offshore Islands[4]	..
AMERICA	1,636
Bermuda	–
Brazil	..
Canada	93
Chile	..
Colombia	..
Mexico	..
Panama	–
USA	1,508
ASIA	360
NEAR & MIDDLE EAST COUNTRIES	38
GULF ARABIAN COUNTRIES	30
OTHER ASIAN COUNTRIES	323
Hong Kong	13
India	..
Indonesia	7
Japan	..
Malaysia	..
Singapore	..
South Korea	–
Thailand	..
AUSTRALASIA & OCEANIA	249
Australia	..
New Zealand	..
AFRICA	83
Kenya	–
Nigeria	..
South Africa	..
Zimbabwe	6

Table 15.9 Outward Direct Investment in the Construction Industry (continued)

Level of direct investment assets held overseas by the UK in subsidiary and associate companies at
the end of 1998 analysed by area and main country

£ Million

Area and Main Country	Net Book Value[1,2]
WORLD TOTAL	3,016
OECD	2,541
CENTRAL & EASTERN EUROPE	6

Notes

− = nil or less than half the final digit shown

.. = not available; data are disclosive

1. This is the overall level of outward direct investment at the end of 1998.

2. A negative figure indicates share capital adjusted by net inter-company loan balances.

3. Prior to 1995 Russia covers other former USSR countries, the Baltic States and Albania.

4. The UK Offshore Islands consist of the Channel Islands and the Isle of Man, excluded from the definition of the economic territory of the UK from 1997.

Source of Data: Office for National Statistics
Contact: Simon Harrington 01633 813314

Table 15.10 Inward Direct Investment in the Construction Industry

Direct investment by foreign companies in UK subsidiary and associate companies and earnings from overseas direct investment in the UK analysed by area and main country, 1995 to 1998

£ Million

Area and Main Country			Year	Net Investment[1,2]	Net Earnings[3]
EUROPE			1995	29	-15
			1996	977	-79
			1997	31	39
			1998	-29	51
	EU		1995	26	-18
			1996	10	-3
			1997	52	7
			1998	0	..
		Austria	1995	–	–
			1996	–	–
			1997	–	–
			1998	–	–
		Belgium & Luxembourg (BLEU)	1995	..	..
			1996	..	..
			1997	..	..
			1998	..	..
		Denmark	1995	..	..
			1996	..	..
			1997	..	..
			1998	..	..
		Finland	1995	–	–
			1996	–	–
			1997	–	–
			1998	–	–
		France	1995	..	..
			1996	..	..
			1997	..	..
			1998	..	..
		Germany	1995	..	..
			1996	..	..
			1997	..	..
			1998	..	..
		Greece	1995	–	–
			1996	–	–
			1997	–	–
			1998	–	–
		Irish Republic	1995	–	–
			1996	–	–
			1997	–	–
			1998	..	..
		Italy	1995	–	–
			1996	–	–
			1997	–	–
			1998	–	–
		Netherlands	1995	..	..
			1996	-1	5
			1997	..	..
			1998	25	17
		Portugal	1995	–	–
			1996	–	–
			1997	–	–
			1998	–	–
		Spain	1995	–	–
			1996	–	–
			1997	–	–
			1998	–	–
		Sweden	1995	..	..
			1996	2	0
			1997	0	..
			1998	0	..

Table 15.10 Inward Direct Investment in the Construction Industry (continued)

Direct investment by foreign companies in UK subsidiary and associate companies and earnings from overseas direct investment in the UK analysed by area and main country, 1995 to 1998

£ Million

Area and Main Country	Year	Net Investment[1,2]	Net Earnings[3]
EFTA	1995	..	..
	1996	..	..
	1997	..	..
of which	1998	..	..
Norway	1995	–	–
	1996	..	..
	1997	..	..
	1998	..	..
Switzerland	1995	..	..
	1996	..	..
	1997	..	..
	1998	..	..
OTHER EUROPEAN COUNTRIES	1995	..	..
	1996	..	..
	1997	..	..
of which	1998	..	–
Russia[4]	1995	–	–
	1996	–	–
	1997	–	–
	1998	–	–
UK Offshore Islands[5]	1995	–	–
	1996	–	–
	1997	..	..
	1998	–	–
AMERICA	1995	-271	-6
	1996	-6	..
	1997	86	95
of which	1998	-4	..
Canada	1995	..	..
	1996	..	..
	1997	..	..
	1998	..	..
USA	1995	11	..
	1996	80	..
	1997	57	..
	1998	..	..
ASIA	1995	-21	-11
	1996	25	-22
	1997	..	-5
	1998	9	-1
NEAR & MIDDLE EAST COUNTRIES	1995	–	–
	1996	–	–
	1997	..	..
	1998	..	..

Table 15.10 Inward Direct Investment in the Construction Industry (continued)

Direct investment by foreign companies in UK subsidiary and associate companies and earnings from overseas direct investment in the UK analysed by area and main country, 1995 to 1998

£ Million

Area and Main Country	Year	Net Investment[1,2]	Net Earnings[3]
OTHER ASIAN COUNTRIES	1995	-21	-11
	1996	25	-22
	1997	..	..
of which	1998	..	..
Hong Kong	1995	–	–
	1996	–	–
	1997	–	–
	1998	–	–
Japan	1995	-21	-11
	1996	..	..
	1997	..	:.
	1998	..	..
Singapore	1995	–	–
	1996	–	–
	1997	–	–
	1998	–	–
South Korea	1995	–	–
	1996	–	–
	1997	–	–
	1998	–	–
AUSTRALASIA & OCEANIA	1995	–	–
	1996	–	..
	1997	..	–
of which	1998	–	..
Australia	1995	–	–
	1996	–	–
	1997	–	–
	1998	–	–
New Zealand	1995	–	–
	1996	–	–
	1997	..	–
	1998	–	..
AFRICA	1995	–	–
	1996	–	–
	1997	–	–
of which	1998	–	–
South Africa	1995	–	–
	1996	–	–
	1997	–	–
	1998	–	–
WORLD TOTAL	1995	-264	-34
	1996	995	-35
	1997	159	128
	1998	-24	102
OECD	1995	31	2
	1996	1040	-53
	1997	109	96
	1998	65	86
CENTRAL & EASTERN EUROPE	1995	–	–
	1996	–	–
	1997	–	–
	1998	–	–

Notes

– = nil or less than half the final digit shown

.. = not available; data are disclosive

1. This is the overall level of outward direct investment at the end of 1998.

2. A negative figure indicates share capital adjusted by net inter-company loan balances.

3. Prior to 1995 Russia covers other former USSR countries, the Baltic States and Albania.

4. The UK Offshore Islands consist of the Channel Islands and the Isle of Man, excluded from the definition of the economic territory of the UK from 1997.

Source of Data: Office for National Statistics

Contact: Simon Harrington 01633 813314

Table 15.11 Inward Direct Investment in the Construction Industry

Level of direct investment assets held in the UK by foreign investors at the end of 1998 analysed by area and main country

£ Million

Area and Main Country	Net Book Value[1,2]
EUROPE	444
EU	162
Austria	–
Belgium & Luxembourg	..
Denmark	..
Finland	–
France	..
Germany	..
Greece	–
Irish Republic	..
Italy	–
Netherlands	80
Portugal	–
Spain	–
Sweden	11
EFTA	..
Norway	..
Switzerland	..
OTHER EUROPEAN COUNTRIES	..
Russia[3]	–
UK Offshore Islands[4]	–
AMERICA	..
Canada	..
USA	..
ASIA	26
NEAR & MIDDLE EAST COUNTRIES	..
OTHER ASIAN COUNTRIES	..
Hong Kong	–
Japan	..
Singapore	–
South Korea	–
AUSTRALASIA & OCEANIA	..
Australia	..
New Zealand	–
AFRICA	–
South Africa	–
WORLD TOTAL	1,018
OECD	880
CENTRAL & EASTERN EUROPE	–

Notes

– = nil or less than half the final digit shown

.. = not available; data are disclosive

1. This is the overall level of outward direct investment at the end of 1998.

2. A negative figure indicates share capital adjusted by net inter-company loan balances.

3. Prior to 1995 Russia covers other former USSR countries, the Baltic States and Albania.

4. The UK Offshore Islands consist of the Channel Islands and the Isle of Man, excluded from the definition of the economic territory of the UK from 1997.

Source of Data: Office for National Statistics
Contact: Simon Harrington 01633 813314

Table 15.12 Overseas construction activity by British companies[1]

£ Million Current Prices

	1989	1990	1991	1992	1993	1994	1995	1996	1997	1998 (r)	1999 (p)
(a) Value of New Contracts											
Europe	190	105	105	184	310	451	764	656	470	453	389
of which											
European Union	131	68	83	70	271	407	639	551	297	294	306
Middle East	141	200	238	235	403	607	406	371	464	446	524
Far East	104	353	568	897	750	975	1,607	983	931	902	915
of which											
Hong Kong	36	215	140	723	471	637	1,034	510	354	527	703
Africa	160	320	268	251	234	289	285	341	310	174	268
Americas	1667	1211	970	1271	1,669	1,384	2,244	2,026	1,617	2,017	1,225
of which											
North America	1620	1141	905	1158	1,540	1,275	2,135	1,978	1,538	1,904	1,150
Oceania	211	289	200	153	227	279	234	334	270	259	384
All Countries	**2,473**	**2,478**	**2,349**	**2,991**	**3,593**	**3,985**	**5,540**	**4,711**	**4,062**	**4,251**	**3,705**
(b) Value of Work Done											
Europe	77	140	179	181	253	373	419	602	538	496	434
of which											
European Union	40	59	120	137	178	275	365	500	443	388	325
Middle East	176	134	220	262	347	315	459	441	472	439	433
Far East	140	316	261	399	717	936	992	1,377	1,201	960	798
of which											
Hong Kong	57	214	141	190	409	564	587	890	694	549	416
Africa	162	219	180	215	183	233	265	363	151	202	200
Americas	813	955	1167	1180	1,218	1,296	1,567	1,626	1,660	1,891	1,276
of which											
North America	706	887	1109	1100	1,102	1,178	1,471	1,559	1,583	1,808	1,176
Oceania	159	260	310	222	227	227	268	286	278	270	283
All Countries	**1,527**	**2,024**	**2,317**	**2,459**	**2,945**	**3,380**	**3,970**	**4,695**	**4,300**	**4,258**	**3,424**
(c) Value of Work Outstanding											
Europe	264	258	170	151	212	301	657	790	687	631	537
of which											
European Union	194	188	137	47	144	275	567	698	517	413	345
Middle East	120	182	195	222	291	582	524	432	626	546	638
Far East	147	358	589	1117	1,208	1,235	1,830	1,443	1,124	1,021	1,115
of which											
Hong Kong	41	195	120	666	730	797	1,223	812	510	484	767
Africa	130	247	297	343	350	401	415	390	461	291	343
Americas	989	1211	1049	1270	1,589	1,772	2,452	2,791	2,656	2,773	1,837
of which											
North America	936	1160	993	1179	1,478	1,671	2,338	2,697	2,562	2,650	1,739
Oceania	131	246	147	87	106	160	125	172	161	153	251
All Countries	**1,781**	**2,502**	**2,447**	**3,190**	**3,756**	**4,451**	**6,003**	**6,018**	**5,715**	**5,415**	**4,721**

Notes
r = revised
p = provisional
1. Work of British construction companies and their overseas branches and subsidiaries.

Source of Data: Construction Market Intelligence, Department of the Environment, Transport and the Regions
Contact: David Williams 020 7944 5593

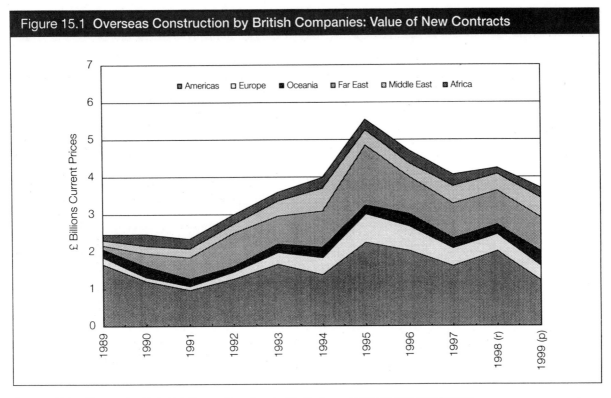

Figure 15.1 Overseas Construction by British Companies: Value of New Contracts

Source of Data: Construction Market Intelligence, Department of the Environment, Transport and the Regions
Contact: David Williams 020 7944 5593

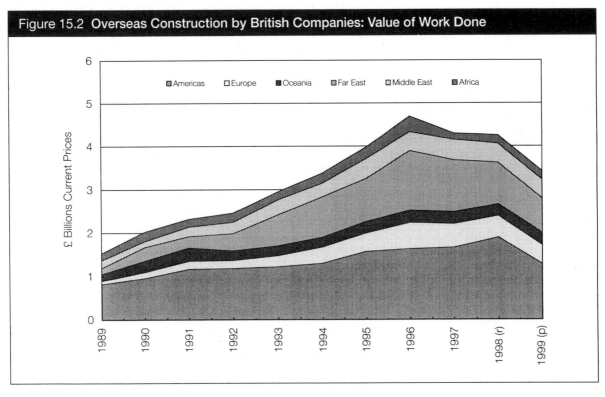

Figure 15.2 Overseas Construction by British Companies: Value of Work Done

Source of Data: Construction Market Intelligence, Department of the Environment, Transport and the Regions
Contact: David Williams 020 7944 5593

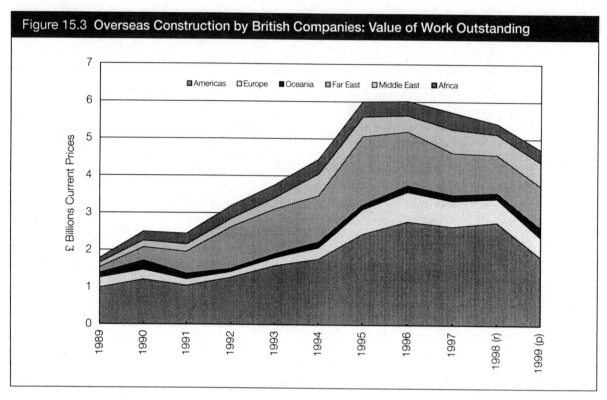

Figure 15.3 Overseas Construction by British Companies: Value of Work Outstanding

Source of Data: Construction Market Intelligence, Department of the Environment, Transport and the Regions
Contact: David Williams 020 7944 5593

EXPORT PROMOTION

The work of Export Promotion and Construction Materials Division (EPCM) supports one of the DETR's key objectives:

"To secure an efficient market in the construction industry, with successful UK firms that meet the needs of clients and society and are competitive at home and abroad."

To achieve this we:

- promote the construction industry's interests and capabilities in overseas markets;

- spread awareness of market opportunities; and

- provide appropriate, practical assistance.

As part of Trade Partners UK, we work closely with British Trade International, other Government departments and trade associations to ensure that our work is relevant and effective.

We arrange Ministerial led trade missions that take companies to markets we believe hold significant opportunities. A Ministerial presence opens doors and enables companies to meet key decision-makers. So far in 2000 we have taken successful trade missions to South Africa, Lebanon and Kazakhstan and are taking trade missions to Chile, Brazil, Indonesia and Romania later on in the year. We also arrange inward missions and seminars/conferences.

We have commissioned, from British Embassies, a number of targeted market information reports on the construction industry. These construction reports contain information on possible opportunities, threats and useful contacts both in the UK and overseas.

We are keen to get across what the UK construction industry can offer other markets and for the second year running we have organised an Industry Briefing Course for commercial officers from British Embassies around the world. The aim of the course was to give the commercial officers a better understanding of the construction industry so that they can promote it more effectively in their home markets.

If you are interested in finding out more about how we can help you please contact us at: EPCM, DETR, Zone 3/J3, Eland House, Bressenden Place, London, SW1E 5DU; Tel: 020 7944 5682; Fax: 020 7944 5669; E-mail: mona_shah@detr.gsi.gov.uk. Alternatively access our webpage at www.construction.detr.gov.uk/epcm/conaims.htm

CHAPTER 16

DETR Construction Innovation and Research Business Plans

The Government's aim for construction is to secure an efficient market in the industry, with innovative and successful UK firms that meet the needs of clients and society and are competitive at home and abroad.

In support of these aims the Department of the Environment, Transport and the Regions (DETR) invests upwards of £22m each year in a Programme of construction-related innovation and research. The purpose of the Programme is to develop and make available information and knowledge which will assist the UK construction industry improve it's performance, profitability and competitiveness, provide enhanced value for it's clients and contribute more widely to sustainable development. The Programme also aims to develop the innovation and research necessary to underpin changes in the Building Regulations, which promote healthier, safer and more sustainable buildings and easier use of buildings by disabled people.

FOCUSING ON RESEARCH PRIORITIES

The approach adopted in pursuing these aims has been to identify and focus the Programme on a number of areas of strategic importance, rather than spreading resources thinly across a wider research canvas. The research priorities are reviewed each year to ensure that the Programme continues to respond to industry requirements, to address the issues of most relevance in meeting its strategic aims and to ensure a sound scientific base for regulatory reviews.

Table 16.1 The Plan	
Overall Aim: **To help secure an efficient market in the construction industry, with innovative and successful UK firms that meet the needs of clients and society and are competitive at home and abroad.**	
Business Plan:	**Themes:**
Sustainable Construction *To help the industry meet the obligations of the sustainable construction strategy*	• Supporting clients • Measuring performance • Improving design • Improving the construction process • More sustainable occupation and use • Communication
Safety and Health in Buildings *To enable policy with respect to safety and health in buildings to be based on a firm scientific footing*	• Materials and technology • Human factors • Development • Regulation • Communication
Technology and Performance *To improve the technological performance of UK construction*	• Improving buildability • Investigating whole life costs • Developing new technologies and techniques • Better codes/standards; alternatives and associated guidance • Communication
Construction Process *To develop new and improved approaches to construction processes*	• Integrating design and construction • Improving supply chain management • Improving productivity and performance • Techniques for improving value • Communication
Best Practice *To promote beneficial changes in management practice and business process in the construction industry*	• Understand the benefits • Raise awareness of the benefits • Provide advice and guidance

Source of Data: CIRM, Department of the Environment, Transport and the Regions
Contact: John Troughton 020 7944 5689

CHANGING EMPHASES

The Plan is an evolving one and there are new emphases which reflect the changing business and social environment, the 'Rethinking Construction' agenda flowing from Sir John Egan's review of the construction industry, developing technologies and the response to earlier calls from the research community.

The time horizons for implementation of research strategies can be lengthy. It often takes many years between the commissioning of research and a resulting impact on procedures within the industry.

The process of achieving successful and widespread application of new ideas and new ways of working can be accelerated by effective dissemination of new knowledge. The Business Plan places increased emphasis on knowledge dissemination. It aims to do this through the development of strategies and mechanisms for the publication and communication of innovation and research. These are set out in the sections below and in the specific communications strategies contained within the five individual Plans.

The **Best Practice Business Plan** will concentrate on preparing material about management best practice for dissemination by the Construction Best Practice Programme (CBPP). The CBPP aims to raise awareness of current and emerging best practice, help construction organisations implement best practice and foster a climate of innovation in construction. Both the CBPP and other mechanisms will be used to increase awareness and impact of the rest of the Programme. The CBPP is currently funded from resources available within the Construction Innovation and Research Programme.

One important business priority has been to improve construction processes; this is reflected in the **Construction Process Business Plan**. Many process themes, upon which research portfolios are now well established within the Programme, were highlighted as key areas for industry change in 'Rethinking Construction'. Following the publication of 'Rethinking Construction' the newly established 'Movement for Innovation' (M4I) identified a series of demonstration projects to illustrate to the industry how innovations can work in practice. Research work under the Process, and other Plans, will inform and complement the M4I agenda. As new knowledge gaps are identified, priorities in the Process Business Plan may be adjusted.

The Government is working with the construction industry to develop a strategy for achieving more sustainable construction – this will be published later this year. Many of the changes in business process, materials and technology, best practice and the Regulatory environment – which the Business Plans seek to inform – will assist directly in achieving the aims of the strategy for sustainable construction. To further underpin the strategy, the **Sustainable Construction Business Plan** places increased emphasis on (i) the production of practical tools and guidance for implementing environmental best practice, and (ii) techniques for quantifying the resulting business and social benefits.

The need to ensure that the Building Regulations are in tune with modern technology, and the accumulation of data on the performance of buildings, has resulted in the identification of a number of new areas for research under the **Safety and Health in Buildings Business Plan**.

As in previous years, the largest element of the Plan relates to **Technology and Performance**. In directing resources to this area, DETR is particularly concerned to focus on projects which give most scope for improving the performance and competitiveness of the industry at home and abroad.

PUBLICITY AND COMMUNICATIONS

Greater emphasis is being placed on achieving wider communication and take-up of R & I findings. The overarching communications strategy, which has been put in place over the last year, will continue to be developed and refined. The CBPP is the prime dissemination mechanism for ensuring that the industry is aware of, and is able to apply management best practice. To ensure focused communication for all outputs of the Business Plans, the strategy will:

- ensure that communication of R & I findings is integral to each Business Plan, so that every research project has robust dissemination activity built in from the outset;

- at the proposal assessment stage, give increased weight to projects which have set out clear communication and implementation strategies, including clear routes to target audiences;

- use the CBPP fully to promote improvements in management and business process;

- encourage proposals within Business Plans for more effective dissemination of R&I findings: for example bringing several projects together under a wider theme or for a particular audience; repackaging research for use by a particular audience; or helping to facilitate networks for practitioners to take forward research findings;

- expand the range of information on the Construction section of the DETR website and ensure that information on all completed and ongoing DETR funded projects is included.

The Department is determined to ensure that the research and innovation it funds is exploited properly, that follow up work is properly identified and that target audiences for outputs are properly engaged. The strategy outlined above should allow for this. The Department expects to draw on expertise from consultants who help manage the Business Plans, as well as encouraging proposals from others with workable ideas for enhancing the dissemination of particular areas of the Plan.

RESOURCES

Ministers have agreed that the resources within the Construction R & I Programme should be managed to obtain a gradual increase in investment in the Construction Process and Best Practice areas, whilst reducing funding on issues covered by Technology & Performance (which still remains the major area of funding). Safety & Health and Sustainable Construction remain areas of high Ministerial priority. A small amount of funding is available for Cross Plan activities.

The chart below shows the Baseline Trend Scenario of how funding has evolved, and is planned, from 1996/97 over a five year period.

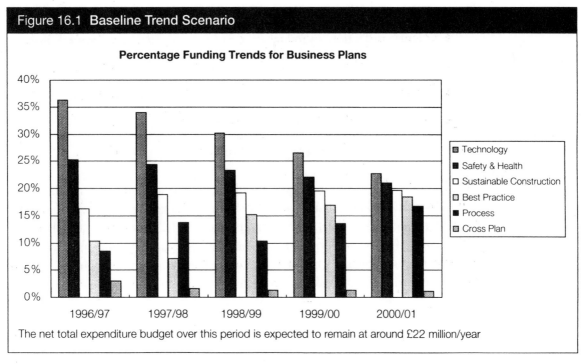

Figure 16.1 Baseline Trend Scenario

Percentage Funding Trends for Business Plans

The net total expenditure budget over this period is expected to remain at around £22 million/year

Source of Data: CIRM, Department of Environment, Transport and the Regions
Contact: John Troughton 020 7944 5689

CHAPTER 17

Construction Industry Key Performance Indicators and Benchmarking

KEY PERFORMANCE INDICATORS (KPIS)

The Construction Industry Key Performance Indicators are produced by a partnership of the Department of the Environment, Transport and the Regions (DETR) the Construction Industry Board and the Construction Best Practice Programme using data from the DETR, Building Cost Information Service, Construction Clients Forum, Health and Safety Executive, and Dun & Bradstreet and other third-party financial analysts. Further information is available from the website address http://www.cbpp.org.uk.

The second annual Construction Industry KPIs, relating to performance in 1999, were published in April 2000. These adopt the format of an overall industry wallchart and a KPI pack containing a handbook and a number of sector specific wallcharts, a book of additional performance indicators and an industry progress report. The KPIs are intended for use by individual firms wishing to measure and compare their performance.

The ten KPIs are as follows:

- Client Satisfaction – Product
- Client Satisfaction – Service
- Defects
- Cost Predictability (Design, Construction)
- Time Predictability (Design, Construction)

- Profitability
- Productivity
- Safety
- Cost
- Time

The KPI wallcharts are arranged as a 'family tree'. At the highest (or headline) level, the 'All Construction wallchart includes data from all the major construction industry sectors excluding material suppliers. More detail is given in the sector-specific wallcharts for new-build housing (public and private), new-build non-housing (public and private), infrastructure, and repair & maintenance and refurbishment (see Figure 17.1). The public category includes central and local government, health, schools and defence.

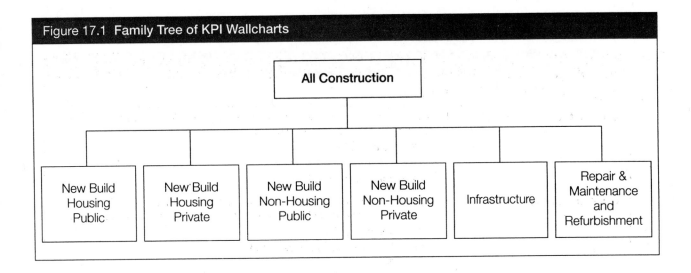

Figure 17.1 Family Tree of KPI Wallcharts

In order to define the Project Performance KPIs, three key project stages have been identified:

A. Commit to Invest — the point at which the client decides in principle to invest in a project, sets out the requirements in business terms and authorises the project team to proceed with the conceptual plan.

B. Commit to Construct — the point at which the client authorises the project team to start the construction project.

C. Available for Use — the point at which the project is available for substantial occupancy or use. This may be in advance of the completion of the project.

These stages are shown diagramatically in Figure 17.2.

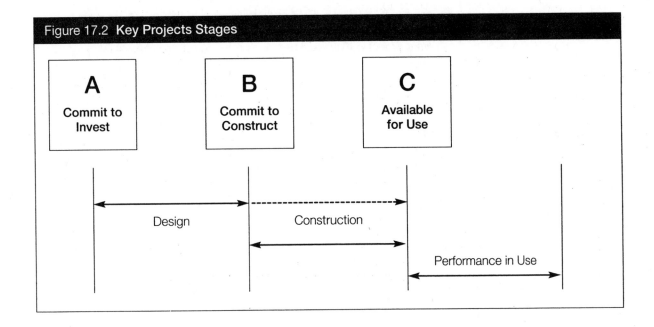

Figure 17.2 Key Projects Stages

The KPI Definitions are shown in Table 17.1

Table 17.1 Definitions used for the Project and Company Performance KPIs	
Project KPIs	**Definition**
Client Satisfaction – Product	How satisfied the client was with the finished product/facility, for projects completed in 1999, using a 1 to 10 scale, where: **10** = Totally satisfied **5** = Neither satisfied or dissatisfied **1** = Totally dissatisfied
Client Satisfaction – Service	How satisfied the client was with the service of the consultants and main contractor, for projects completed in 1999, using a 1 to 10 scale, where: **10** = Totally satisfied **5** = Neither satisfied or dissatisfied **1** = Totally dissatisfied
Defects	Condition of the facility with respect to defects at the time of handover for projects completed in 1999, using a 1 to 10 scale, where: **10** = Defect Free **8** = Some defects and no significant impact on client **5** = Some defects and some impact on client **3** = Major defects with major impact on client **1** = Totally defective
Predictability – Cost	There are two indicators for projects completed in 1999 compared with 1998 – one for design cost and one for construction cost. **1** Design Cost – actual cost at Available for Use less the estimated cost at Commit to Invest, expressed as a percentage of the estimated cost at Commit to Invest. **2** Construction Cost – actual cost at Available for Use less the estimated cost at Commit to Construct, expressed as a percentage of the estimated cost at Commit to Construct.
Predictability – Time	There are two indicators for projects completed in 1999 compared with 1998 – one for design phase and one for construction phase. **1** Design Time – actual duration at Commit to Construct less the estimated duration at Commit to Invest, expressed as a percentage of the estimated duration at Commit to Invest. **2** Construction Time – actual duration at Available for Use less the estimated duration at Commit to Construct, expressed as a percentage of the estimated duration at Commit to Construct.
Construction Cost	The normalised[2] construction cost of a project in 1999 less the normalised cost of a similar project in 1998, expressed as a percentage of the normalised project cost in 1998.
Construction Time	The normalised[2] time to construct a project in 1999 less the normalised time to construct a similar project in 1998, expressed as a percentage of the normalised time to construct a project in 1998.
Company KPIS	**Definition**
Profitability	Company profit before tax and interest as a percentage of sales, reported in 1999.
Productivity	Company value added[3] per employee (£), reported in 1999.
Safety	Reportable accidents[4] per 100,000 employed per year in 1998/99.

Notes
1. 'Normalisation' is a statistical method for removing the effects of location, function, size and inflation.
2. Value added is turnover less all costs subcontracted to, or supplied by, other parties.
3. Reportable accidents are defined in Health and Safety Statistics 1998/99 published by the Health and Safety Commission as fatalities, major injuries, and over 3 day injuries to employees, self employed people and members of the public.

The Industry Average Performance for 'All Construction' and for each construction sector is shown in Table 17.2. The value given is the median of the sample used except for 'Safety' where the arithmetic mean is used.

Table 17.2 Industry Average Performance

KPIs	All Construction	New Build Housing Public	New Build Housing Private	New Build Non Housing Public	New Build Non Housing Private	Infra-structure	Repair & Maintenance and Refurbishment
Client Satisfaction – Product	8	8◇	8*	8◇	8◇	8*	8
Client Satisfaction Service	8	8◇	8*	8◇	8◇	8*	8
Defects	8	8◇	8*	8◇	8◇	8*	8
Predictability – Cost							
Design	0%	0%◇	+3%◇	0%	0%	0%*	0%
Construction	+1%	0%◇	+2%◇	+1%	+1%	+1%*	+2%
Predictability – Time							
Design	+20%	+58%	+11%◇	+19%	+10%	+20%*	+21%
Construction	0%	-10%	+2%◇	0%	0%	0%*	-1%
Profitability	+4.7%	+4.0%♦	+4.0%♦	+3.4%♦	+3.4%♦	+3.4%	+4.7%
Productivity (£k)	27	26♦	26♦	24♦	24♦	29	27*
Safety (No.)	1037†	1037†	1037†	1037†	1037†	1037†	1037†
Construction Cost	-2%	-2%	+2%	+2%	0%	-15%	-3%
Construction Time	+3%	+3%	+8%	+2%	-2%	+5%	0%

Notes

.   Insufficient data available in this sector, therefore All Construction average used.

◇   Limited data – use with caution.

♦   The data in this sector is not split between public and private.

†   No sector breakdown is available.

A summary of the performance of the industry in 1998 and 1999 is shown in Table 17.3. Figure 17.3 below then converts these performance figures to an index, with the 1998 performance for each KPI normalised to a base of 100.

Table 17.3 Summary of Industry Performances 1998 & 1999

KPI	Measure	1998	1999
Client Satisfaction – Product	% scoring 8/10 or better	72%	73%
Client Satisfaction – Service	% scoring 8/10 or better	58%	63%
Defects*	% scoring 8/10 or better*	70%	65%
Cost Pedictability – Design	% on target or better	65%	64%
Cost Predictability – Construction	% on target or better	37%	45%
Time Predictability – Design	% on target or better	27%	37%
Time Predictability – Construction	% on target or better	34%	62%
Profitability*	median profit before interest & tax*	3.2	4.7
Productivity*	median turnover/employed*	60	59
Safety	mean accident incident rate	997	1037
Cost	change compared with one year ago	–3%	–2%
Time*	change compared with one year ago*	–1%	+3%

* The measure for this KPI was changed this year, comparison with the previous year is not strictly valid

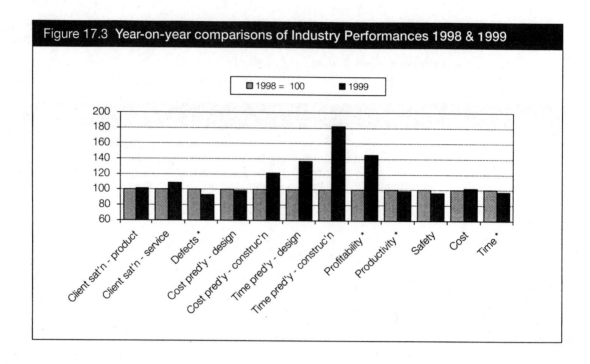

Figure 17.3 Year-on-year comparisons of Industry Performances 1998 & 1999

BENCHMARKING

Sir John Egan's report, *Rethinking Construction* concerned itself directly with improving both the effectiveness and efficiency of the UK Construction Industry. It challenged the industry to meet some ambitious improvement targets and to measure its performance over a range of its activities.

The creation of **Key Performance Indicators** and the release of the first KPI Pack in early 1999, were the first steps in the process of answering those challenges. The Construction Industry KPIs present the construction industry's range of performance using ten headline measures, but do not cover the more diagnostic elements of performance measurement. The KPI Report for the Minister for Construction produced by the KPI Working Group addresses the latter issue by presenting organisations with a framework with which to benchmark and analyse their activities.

Identified within the report are both operational and diagnostic indicators to promote a better understanding of the headline KPI results. The former relates to the specific activities of the firm, while the latter provide information on why changes may have occurred in the headline or operational indicators. Both allow more thorough analysis of how and where improvements can be made.

The indicators are intended to be applied at the project or company level; however, it is also possible for them to be applied at other stages of the construction supply chain so as to capture the network of clients and suppliers. Figure 17.4 shows the inter-relationships between the KPIs and the parts of the supply chain.

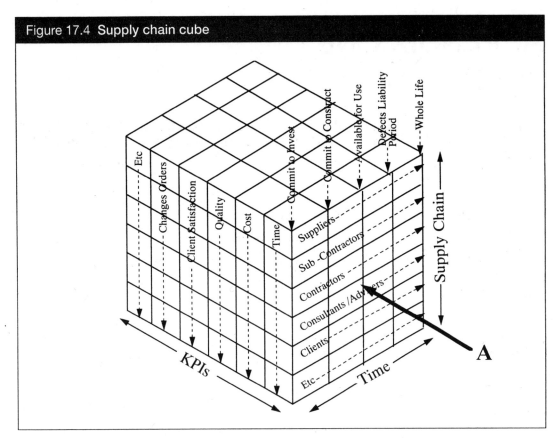

Figure 17.4 **Supply chain cube**

Each individual building block represents a different component of the KPIs. The cube can be "sliced" to allow analysis from any of the three dimensions pictured: supply chain component, time or by KPI. For example, in the diagram, the first face of the cube if analysed in the direction of arrow A would be time. Here the materials supplier is above the contractors, so the latter is the client. The relevant indicators to consider are specific to the particular supplier-client relationship under examination. By slicing the cube along a different dimension, one can analyse a separate part of the supply chain or a different point in the project cycle.

The adaptability of the framework allows companies to take individual KPIs and adapt them to their specific needs and therefore, prompts the industry to integrate it throughout the whole supply chain.

A full copy of the report is contained
in the DETR website at www.construction.detr.gov.uk/cmi/kpi.htm

Source: Construction Market Intelligence, Department of the Environment, Transport and the Regions
Contact: Bob Packham 020 7944 5764 or e-mail robert_packham@detr.gsi.gov.uk

APPENDIX ONE

List of DETR, DOE and Standard Regions

Standard Regions	DETR Regions (after 8/2/96)	DOE Regions (before 8/2/96)
North Cleveland Cumbria Durham Northumberland Tyne & Wear	**North East** Cleveland Durham Northumberland Tyne & Wear	**Northern** Cleveland Durham Northumberland Tyne & Wear
Yorkshire & Humberside Humberside North Yorkshire South Yorkshire West Yorkshire	**Yorkshire & The Humber** Humberside North Yorkshire South Yorkshire West Yorkshire	**Yorkshire & Humberside** Humberside North Yorkshire South Yorkshire West Yorkshire
North West Cheshire Lancashire Greater Manchester Merseyside	**North West** Cheshire Cumbria Lancashire Greater Manchester **Merseyside** Merseyside	**North West** Cheshire Cumbria Lancashire Greater Manchester Merseyside
East Midlands Derbyshire Leicestershire Lincolnshire Northamptonshire Nottinghamshire	**East Midlands** Derbyshire Leicestershire Lincolnshire Northamptonshire Nottinghamshire	**East Midlands** Derbyshire Leicestershire Lincolnshire Northamptonshire Nottinghamshire
West Midlands Hereford & Worcester Shropshire Staffordshire Warwickshire West Midlands	**West Midlands** Hereford & Worcester Shropshire Staffordshire Warwickshire West Midlands	**West Midlands** Hereford & Worcester Shropshire Staffordshire Warwickshire West Midlands
East Anglia Cambridgeshire Norfolk Suffolk	**Eastern** Bedfordshire Cambridgeshire Essex Hertfordshire Norfolk Suffolk	**Eastern** Bedfordshire Buckinghamshire Cambridgeshire Essex Hertfordshire Norfolk Suffolk
South West Avon Cornwall Devon Dorset Gloucestershire Somerset Wiltshire	**South West** Avon Cornwall Devon Dorset Gloucestershire Somerset Wiltshire	**South West** Avon Cornwall Devon Dorset Gloucestershire Somerset Wiltshire
South East Bedfordshire Berkshire Buckinghamshire East Sussex Essex Greater London Hampshire Hertfordshire Isle of Wight Kent Oxfordshire Surrey West Sussex	**South East** Berkshire Buckinghamshire East Sussex Hampshire Isle of Wight Kent Oxfordshire Surrey West Sussex **Greater London** London	**South East** Berkshire East Sussex Hampshire Isle of Wight Kent Oxfordshire Surrey West Sussex **London** All London Boroughs

APPENDIX TWO

List of SSI Inspection Regions

SSI Inspection Regions

North West
Bolton
Bury
Cheshire
Knowsley
Lancashire
Liverpool
Manchester
Oldham
Rochdale
Salford
Sefton
St. Helens
Stockport
Tameside
Trafford
Wigan
Wirral

North East
Barnsley
Bradford
Calderdale
Cleveland
Cumbria
Doncaster
Durham
Gateshead
Humberside
Kirklees
Leeds
Newcastle Upon Tyne
North Tyneside
North Yorkshire
Northumberside
Rotherham
Sheffield
South Tyneside
Sunderland
Wakefield

Central
Birmingham
Cambridgeshire
Coventry
Derbyshire
Dudley
Hereford & Worcester
Leicestershire
Lincolnshire
Norfolk
Nottinghamshire
Sandwell
Shropshire
Solihull
Staffordshire
Suffolk
Walsall
Warwickshire
Wolverhampton

Southern & South West
Avon
Berkshire
Buckinghamshire
Cornwall
Devon
Dorset
Gloucestershire
Hampshire
Isle of Wight
Isle of Scilly
Northamptonshire
Oxfordshire
Somerset
Wiltshire

London East
Barking
Bexley
Bromley
Camden
City of London
East Sussex
Enfield
Essex
Greenwich
Hackney
Haringey
Havering
Islington
Kent
Lambeth
Lewisham
Newham
Redbridge
Southwark
Tower Hamlets
Waltham Forest

London West
Barnet
Bedfordshire
Brent
Croydon
Ealing
Hammersmith
Harrow
Hertfordshire
Hillingdon
Hounslow
Kensington
Kingston Upon Thames
Merton
Richmond Upon Thames
Surrey
Sutton
Wandsworth
West Sussex
Westminster

APPENDIX THREE

Notes and Definitions

DEFINITION OF THE CONSTRUCTION INDUSTRY

The industry is defined in accordance with Division 42 of the Revised 1992 Standard Industrial Classification (SIC), i.e.

General Construction and Demolition Work

Establishments engaged in building and civil engineering work, not sufficiently specialised to be classified elsewhere in Division 42 and demolition work. Direct labour establishments of local authorities and Government Departments are included.

Construction and Repair of Buildings

Establishments engaged in the construction, improvement and repair of both residential and non-residential buildings, including specialists engaged in sections of construction and repair work such as bricklaying, building maintenance and restoration, carpentry, roofing, scaffolding and the erection of steel and concrete structures for buildings.

Civil Engineering

Construction of roads, car parks, railways, airport runways, bridges and tunnels. Hydraulic engineering e.g. dams, reservoirs, harbours, rivers and canals. Irrigation and land drainage systems. Laying of pipelines, sewers, gas and water mains and electricity cables. Construction of overhead lines, line supports and aerial towers. Construction of fixed concrete oil production platforms. Construction work at oil refineries, steelworks, electricity and gas installations and other large sites. Shaft drilling and mine sinking. Laying out of parks and sports grounds. Contractors responsible for the design, construction and commissioning of complete plants and manufacturers of constructional steelwork are not classified to construction.

Installation of Fixtures and Fittings

Establishments engaged in the installation of fixtures and fittings, including gas fittings, plumbing, heating and ventilation plant, sound and heat insulation, electrical fixtures and fittings.

Building Completion Work

Establishments specialising in building completion work such as painting and decorating, glazing, plastering, tiling, on-site joinery and carpentry, flooring (including parquet floor laying), installation of fire-places etc. Builders' joinery and carpentry manufacturers and shop and office fittings manufacturers are not classified to construction.

The main differences between the 1980 and 1968 versions of the SIC are that establishments engaged in open-cast coal-mining and plant hire without operatives are not classified to construction in the new classification.

Construction work carried out by direct employees of Government departments, Local authorities, new towns and nationalised industries in the transport sector, where information is available, is included in the published construction output and employment statistics. Output by direct employees of utilities is excluded from construction. Such work is classified to the same SIC heading as the establishment, which they serve.

New construction work includes extensions, major alterations (i.e. improvements), site preparation and demolition, except for housing where work done on improvements, extensions and alterations and house/flat conversions is included under repair and maintenance. New construction work includes houses converted to other uses.

Public and Private Sector

Public work is for any public authority such as government departments, public utilities, nationalised industries, universities, the Post Office, new town corporations, housing associations, etc.

Private work is for a private owner or organisation or for a private developer, and includes work carried out by firms on their own initiative. It includes work where the private sector carries the majority of the risk/gain, i.e. in principle, all PFI contracts are private.

Contractors

DETR maintains a statistical register of private contracting 'firms', from which samples are drawn for regular enquiries into orders, output and employment. These 'firms' are strictly 'reporting units' – some large firms prefer to report the operations of parts of their companies separately, e.g. regional divisions. Other companies may make single returns covering associated companies each of which are legally separate firms. In 1974 there were about 93,000 reporting units on the register. By late 1977 the number on the register had declined to about 78,000, due in part to a fall in construction activity, but also due to the difficulty experienced in picking up new entrants to the industry. Arrangements were made to obtain names and addresses of firms newly registered for VAT purposes as construction firms: with this additional source of information, the statistical register expanded by 1981 to about 115,000. In 1982, the VAT based register and the Department's register were 'matched' and 45,000 more construction firms were identified and added to the Department's listing of construction companies. Most of the new additions were very small firms. In 1992, the VAT based registers maintained by the Department of the Environment and by the Office for National Statistics (ONS) were 'matched' so that the overall coverage improved. There were 195,000 firms in the register in mid-1993. Following further acceptance of information from the ONS register, the firms fell to 163,000 in 1996. Estimation procedures have been used to avoid any discontinuities in the Department's published series. The register does not include the majority of self-employed workers.

There is now a regular exchange of information between DETR and the Interdepartmental Business Register held by ONS. All firms are asked each year for information required to bring the statistical register up to date. Most firms with manpower of less than 8, are sent a 'Small Firms Return' covering only the register information. The remainder receive an 'annual return' which includes in respect of third quarter output and October employment, the more detailed questions which do not appear for the other three quarters of the year; the usual quarterly enquiries are sent to samples of about 12,000 firms, including all those with 60 or more employees. To reduce the burden on firms, the information required in these 'firm-based' quarterly enquiries was reduced in 1980, and a new 'project-based' enquiry covering a sample of 12,000 new construction projects was initiated. In this enquiry each project is surveyed, to completion, and new projects identified in the New Orders enquiries are added each quarter.

New Orders questionnaires are sent to a sample of 5,500 main contractors firms every month.

CONSTRUCTION NEW ORDERS AND OUTPUT (CHAPTERS 1 & 2)

Value of New Orders (Tables 1.1-1.7)

This information relates to contracts for new construction work awarded to main contractors by clients in both the public and private sectors, including extensions to existing contracts and construction work in 'package deals'. Also included is speculative work, undertaken on the initiative of the firm, where no contract or order is awarded; the value of this work is recorded in the period when foundation works are started, e.g. on houses or offices for eventual sale or lease. The site value and architects' or consultants' fees are excluded from the value of the order.

The figures are collected in current price terms. These are revalued at constant (1995) prices, then seasonally adjusted and converted into index form based on 1995=100.

Regional classification of orders is based on the location of the site work, not the location of the firm, by using regional boundaries.

Contracts are allocated to types of work according to the type of construction involved, not the type of client. (Before the second quarter of 1976 the client was taken into account e.g. some offices for industrial groups were classified as 'industrial' instead of 'offices', and roads at airfields were allocated to 'airfields' not, as now, to 'roads'. This change in approach and certain revisions of the descriptions have led to minor discontinuities in the published series.)

Value of Output (Tables 2.1-2.5, 2.8-2.9)

Contractor's output is defined as the amount chargeable to customers for building and civil engineering work done in the relevant period. Contractors are asked to include the value of work done on their own initiative on buildings such as dwellings or offices for eventual sale or lease, and of work done by their own operatives on the construction and maintenance of their own premises. The value of goods made by the contractors themselves and used in the work is also included. In the 'firm-based' returns, work done for the firm by sub-contractors is excluded to avoid double counting, since sub-contractors are also sampled. The 'project-based' enquiry includes only main contractors who are asked to report all work on the site, including sub-contracted work. The results of the 'project-based' enquiry are used to distribute the overall estimate of contractors' output on new work (based on the firm-based returns and estimates of unrecorded output) between the different types of new work carried out. The regional classification of new work is therefore based on the location of site work, while for repair and maintenance it is based on the location of the firm.

The figures collected are at current values. These are revalued at constant prices and then seasonally adjusted and converted into index form, based on 1995=100.

The value of output results in both current and constant prices include estimates of 'unrecorded output' of firms and individuals not on the statistical register (mainly self-employed workers). These estimates are based on the difference between the number of employees and working proprietors reported by firms on the Department's register and the estimates of total employment (employees and self-employed) produced by the Office for National Statistics.

When VAT was introduced in April 1973, contractors were asked to include this, where appropriate, in the value of work done. From the third quarter of 1977, they were asked to exclude VAT and this change has led to minor discontinuities in the current price series for repair and maintenance.

Direct Labour Output of Public Sector (Tables 2.6-2.7)

Government Departments, local authorities, new towns and national industries in the transport sector (where information is available) are surveyed quarterly, in respect of work done by employees engaged on building and civil engineering work. Output by direct employees of other public bodies e.g. NCB, UKAEA is excluded from construction. Such work is classified to the same heading as the establishment, which they serve.

Type of New Work: Detailed Descriptions

Orders and output have been classified in accordance with revised descriptions given below from 1st quarter 1980. Prior to 1st quarter 1980 there were differences in definition; see *Housing and Construction Statistics: Notes and Definitions Supplement, 1991.*

Prior to 1st quarter 1985, telephone exchanges and cabling work for British Telecom were classified as communications work for the public sector. From 1st quarter 1985 this work has been classified to the private sector. From 1st quarter 1987 construction work for British Gas has been classified to the private sector. From 1st quarter 1990, construction work for water companies in England and Wales has been classified to the private sector. From 1st quarter 1991, construction work for electricity companies in England and Wales has been classified to the private sector. From 2nd quarter 1996 construction work for rail companies has been classified to the private sector.

Type of Work	Examples of Kind of Work Covered[1]
(a) Public Sector Housing	Local authority housing schemes, hostels (except youth hostels), married quarters for the services and police; old peoples' homes; orphanages and childrens' remand homes; and the provision within housing sites of roads and services for gases, water, electricity, sewage and drainage.
(b) Private Sector Housing	All privately owned buildings for residential use, such as houses, flats and maisonettes, bungalows, cottages, vicarages, and provision of services to new developments.
(c) Infrastructure	
Water	Reservoirs, purification plants, dams (except for hydro-electric schemes), aqueducts, wells, conduits, water works, pumping stations, water mains, hydraulic works, etc
Sewerage	Sewerage disposal works, laying of sewers and surface drains.
Electricity	All buildings and civil engineering work for electrical undertakings such as power stations, dams and other works on hydro-electric schemes, sub-stations, laying of cables and the erection of overhead lines.
Gas	Gas works, gas mains and gas storage.

---

[1] Mixed development schemes are included in the category, which describes the major part of the scheme.

Communications	Post offices, sorting offices, telephone exchanges, switching centres, cables etc.
Air Transport	Air terminals, runways, hangars, reception halls, radar installations, perimeter fencing, etc, which are for use in connection with airfields.
Railways	Permanent way, tunnels, bridges, cuttings, stations, engine sheds, etc, and electrification of both surface and underground railways.
Harbours (Including Waterways)	All works and buildings directly connected with harbours, wharves, docks, piers, jetties (including oil jetties), canals and waterways, dredging, sea walls, embankments, and water defences.
Roads	Roads, pavements, bridges, footpaths, lighting, tunnels, flyovers, fencing, etc.

(d) Non-Housing Excluding Infrastructure[2]

Factories	Factories, shipyards, breweries, chemical works, coke ovens and furnaces (other than at steelworks), skill centres, laundries, refineries (other than oil), workshops, Royal Mint (in public sector).
Warehouses	Warehouses, wholesale depots.
Oil	Oil installations including refineries, distribution pipelines and terminals, production platforms (but not modules or rigs).
Steel	Furnaces, coke ovens and other buildings directly concerned with the production of steel (excludes offices and constructional steelwork).
Coal	All new coal mine construction such as sinking shafts, tunnelling, etc, works and buildings at the pithead which are for use in connection with the pit. Open cast coal extraction is excluded.
Schools and Colleges	Schools or colleges (including technical colleges and institutes of agriculture) except medical schools and junior special schools which are classified under 'Health'. Schools and colleges in the private sector are considered to be those financed wholly from private funds e.g. some religious colleges including their halls of residence.
Universities	Universities including halls of residence, research establishments, etc.

2 Private work is classified between industrial and commercial as follows:

Industrial – Factories, Warehouses, Oil, Steel, Coal.

Commercial – Schools and Colleges, Universities, Health, Offices, Entertainment, Garages, Shops, Agriculture, Miscellaneous.

Health	Hospitals including medical schools; clinics; surgeries (unless part of a house); medical research stations (except when part of a factory, school or university); welfare centres; centres for the handicapped and for rehabilitation; adult training centres and junior special schools.
Offices	Office buildings, banks, embassies. Police HQ's, local and central government offices (including town halls) are classified to the public sector.
Entertainment	Theatres, concert halls, cinemas, film studios, bowling alleys, clubs, hotels, public houses, restaurants, cafes, holiday camps, yacht marinas, dance halls, swimming pools, works and buildings at sports grounds, stadiums and other places of sport or recreation and for commercial television, betting shops, youth hostels and centres; service areas on motorways are also classified in this category as the garage is usually only a small part of the complex which includes cafes, restaurants, etc.
Garages	Buildings for storage, repair and maintenance of road vehicles; transport workshops, bus depots, road goods transport depots and car parks.
Shops	All buildings for retail distribution such as shops, department stores, retail markets, showrooms, etc.
Agriculture	All buildings and work on farms, market gardens and horticultural establishments such as barns, animals' houses, fencing, stores, greenhouses, boiler houses, agricultural and fen drainage, veterinary clinics, etc, but not houses (see category (c)), or buildings solely or mainly for retail sales which are included under 'shops'.
Miscellaneous	All work not clearly covered by any other heading, such as: fire stations; barracks for the forces (except married quarters, classified under 'Housing'); naval dockyards; RAF airfields; police stations; prisons; reformatories, remand homes; borstals; civil defence work; UK Atomic Energy Authority work; council depots; public conveniences; museums, conference centres; crematoria; libraries; caravan sites, except those at holiday resorts, exhibitions; wholesale markets; Royal Ordnance factories.

STRUCTURE OF THE CONSTRUCTION INDUSTRY (CHAPTER 3)

Private Contractors Construction Work (Tables 3.1, 3.3-3.15)

The information from the 'Annual returns', sent to about 30,000 firms, and the 'Small firms returns' sent to all others on the statistical register, forms the basis of several analyses of the structure of the private construction industry by size of firm, by trade of firm and by region of registration. The results are adjusted to allow for non-response but, unlike the quarterly output and employment series, no adjustment is made to allow for firms not on the statistical register. Because of this, changes between the years may simply reflect changes in the register not structural changes in the industry. For example the increase in the number of firms in 1982 and 1983 is due mainly to the addition of firms identified in the industry following a 'matching' of the VAT based register and the Department's construction register. There is also a discontinuity between 1991 and 1992 as a result of improvements to the coverage of the register resulting from the matching exercise carried out with the Inter-Departmental Business Register (IDBR) maintained by the Office for National Statistics. Likewise the reduction between 1995 and 1996 was due to further acceptance of information from the IDBR.

Although the term 'firms' is used throughout, strictly these are 'reporting units' since some large firms, instead of reporting as single units, prefer to report the operations of parts of their companies (e.g. regional divisions) separately. Other companies may make single returns covering associated companies each of which are legally separate firms.

The size of a firm is determined by total direct employment – including working proprietors as well as employees. Each firm self-classifies trade on the basis of the activity which forms the most significant part of its turnover. The regional classification is by reference to the address of the reporting office

Insolvencies (Table 3.2)

Figures for bankruptcies include receiving orders, administration orders and deeds of arrangement made during the year, excluding any rescinded before the end of the year. The debtor may be an individual or a partnership: if orders are made against each member of a partnership, these are consolidated into a single order before being administered. The analysis of occupations relates only to the self-employed, not to employees or company directors who account for 20-25 per cent bankruptcies and who are included in the total of all orders. The minimum debt to support a creditor's petition for bankruptcy was raised from £50 to £200 in 1977 and from £200 to £750 on 1 October 1984.

Liquidations of companies include both those, which are compulsory following court orders and creditors' voluntary liquidations when debtor companies wind-up, their affairs after agreeing terms with their creditors.

Analyses by industry are not available for sequestrations in Scotland.

CONSTRUCTION PRICE INDICES (CHAPTER 4)

Public Sector Building (Non-Housebuilding) Tender Price Index (Tables 4.1, 4.3-4.4)

This index measures the movement of prices in competitive tenders for building contracts in the public sector in Great Britain. It does not include contracts for housing work, work of a mainly civil engineering nature, mechanical and electrical work nor repair and maintenance work. Contribution to the data from which this index is calculated is by the submission of priced bills of quantities. The principal contributors are the Department of Education and Employment, the Department of Health, the Department of Social Security, the Ministry of Defence, the Home Office, the Lord Chancellors Department, the Ministry of Agriculture Fisheries and Foods and the Local Authorities of England, Scotland and Wales.

Each project index is calculated from the price levels established by comparing the price of items to a minimum value of 25 per cent for each trade or section of the bills of quantities with standard base prices. The resultant factors are combined by applying weights representing the total value of each trade or section. Preliminaries and other general charges are spread proportionally over each item of the bill of quantities.

For each quarter, adjustment factors for location and building function are calculated using the project indices within that quarter and the past eleven quarters.

The quarterly index is calculated by taking the median value of the project indices within that quarter. A three-quarter moving average is used to smooth the published indices.

The published index is an all-in index and is published along with the adjustment factors for location and building function which should be applied before the index is used.

Public Sector Housebuilding Tender Price Index (Tables 4.1-4.2)

This index measures the movement of prices in competitive tenders for social housebuilding contracts in the England and Wales. It includes new build contracts for single dwellings and those built in blocks, of up to four storeys high. Contribution to the data from which this index is calculated is by way of the completion of survey forms. The contributors are the Local Authorities and Housing Associations of England and Wales

The survey form requires the specification and unit prices, taken from the bills of quantities, of 18 well-defined items chosen to represent the price movement of the significant trades or sections of the bills of quantities. The table below shows the 18 trades or sections and their weightings.

For each quarter, adjustment factors for location are calculated using the project prices for that quarter and the past eleven quarters. The quarterly index is calculated as a laspeyres price index in which the weighted arithmetic average of the adjusted (by the above factor) prices for the 18 items, within that quarter, is taken in relation to the base year. Occasionally it may happen that a particular item or items is (or are) not represented in any scheme for a particular quarter. When this occurs it is assumed that the price of such item(s) changes in the same manner as the weighted arithmetic average for the remaining items. A three-quarter moving average is used to smooth the published indices. The published index is an all-in index and is published along with the adjustment factors for location, which should be applied before the index is used.

Trades or sections and their weightings			
		Percentage Weights	
		New Build	**Rehab**
1	Excavation and Earthworks	6.2	2.1
2	Insitu Concrete	4.1	1.1
3	Brickwork	12.7	9.1
4	Blockwork	7.5	5.7
5	Structural Timber	9.9	4.3
6	Roof Coverings	4.6	6.5
7	Dry Lining and Boarding	2.2	4.8
8	Windows	4.7	5.9
9	Doors and Internal Joinery	6.7	11.5
10	Wet Plaster Finishes	5.5	13.2
11	Decorations	3.9	5.2
12	Sanitary Fittings	4.3	5.3
13	Thermal Insulation	1.1	2.2
14	Hard Pavings	5.4	1.8
15	Soft Landscaping/Furniture	3.2	1.1
16	Below Ground Services	6.9	2.3
17	Mechanical Installation	5.9	10.5
18	Electrical Installation	5.4	7.6

Road Construction Tender Price Index (Tables 4.1, 4.5-4.7)

This index measures the movement of prices in competitive tenders for road construction contracts in Great Britain. It includes new road construction and major maintenance works of a value exceeding £250,000. Contribution to the data from which this index is calculated is by way of the completion of survey forms and submission of priced bills of quantities. The contributors are the Highways Agency and the Highways Departments of the Local Authorities of England, Scotland and Wales.

Each project index is calculated from the price levels established by comparing the price of items to a minimum value of 25 per cent for each trade or section of the bills of quantities with standard base prices. The resultant factors are combined by applying weights representing the total value of each trade or section. Preliminaries and other general charges are spread proportionally over each item of the Bills of Quantities.

For each quarter, adjustment factors for road type, location and contract size are calculated using the project indices for that quarter and the past eleven quarters.

The quarterly medians of the adjusted (by the above factors) project indices divided through by the public sector building tender price index for that quarter, are smoothed by a kalman filter after transformation to normality. The quarterly index is obtained by reversing the transformation on the result of the smoothing and then multiplying by the public sector building tender price index.

The published index is an all-in index and is published along with the adjustment factors for road type, location and contract size, which should be applied before the index is used.

Construction Output Price Indices (Tables 4.8-4.11)

These indices measure the movement of prices of construction work being carried out, they are derived from tender price indices and are used as 'deflators' to convert contractors' output of new construction work from current prices to constant prices. Repair and maintenance output and all direct labour outputs are deflated using indices based on costs of material and labour.

Output of new construction work in a quarter is made up of work done on contracts let during or before that quarter: the deflator can be constructed from the value and volume of orders placed in previous quarters once adjustments have been made to tender prices for changes in materials and labour costs for which reimbursement is allowed under 'variation of price' clauses.

The output price indices are calculated as follows:

Let the quarter for which the output is considered be T, and let t be any calendar quarter before that quarter. Let the value at current prices of orders placed in quarter t be X_t. Let the proportion of orders containing 'variation of price' clauses be P_t and, of these contracts, let the proportions adjustable for changes in costs be M_t for materials and L_t for labour. The proportion of fixed price orders is then $(1-P_t)$. The value of output at current prices in quarter T, Y_T is made up of F_T from fixed price contracts and V_T from contracts adjustable for cost changes.

Then:

$$F_T = \sum_{t=-\infty}^{T} \alpha_{tT} \cdot X_t \cdot (1-P_t)$$

where α_{tT} is the proportion of work done on contracts placed in quarter t being performed in quarter T.

If the indices of material costs and labour costs are $l_t{}^M$ and $l_t{}^L$ respectively, then

$$V_t = \sum_{t=-\infty}^{T} \alpha_{tT} \cdot X_t \cdot P_t \cdot \left[1 + M_t \cdot \frac{\left(l_T{}^M - l_t{}^M\right)}{l_t{}^M} + L_t \cdot \frac{\left(l_T{}^L - l_t{}^L\right)}{l_t{}^L} \right]$$

Thus the total current price output $Y_T = F_T + V_T$

$$= \sum_{t=-\infty}^{T} \alpha_{tT} \cdot X_t \cdot \left[1 + P_t \cdot \left[M_t \cdot \frac{\left(l_T{}^M - l_t{}^M\right)}{l_t{}^M} + L_t \cdot \frac{\left(l_T{}^L - l_t{}^L\right)}{l_t{}^L} \right] \right]$$

If the appropriate tender price index is A_t the volume Z_t of orders in

quarter t is $\dfrac{X_t}{A_t}$

Thus the volume of output in quarter T, H_T is given by

$$H_T = \sum_{t=-\infty}^{T} \alpha_{tT} \cdot Z_t = \sum_{t=-\infty}^{T} \alpha_{tT} \cdot \frac{X_t}{A_t}$$

Hence the appropriate deflator for output in a quarter T is given by

$$D_t = \frac{Y_T}{H_T}$$

Separate output price indices are calculated for each of the five new work sectors, public housing, private housing, public non-housing, private industrial and private commercial. Within each sector contracts are further sub-divided into short, medium and long contracts, depending upon the expected duration, and the values of α_{tT} and P_t are estimated at this lower level.

A fuller description of the output price indices was published in *Economic Trends No 297 (July 1978)*. These indices superseded the Cost of New Construction (CNC) index described in *Notes and Definitions Supplement 1977*.

CONSTRUCTION INVESTMENT (CHAPTER 6)

These tables give expenditure on fixed assets and cover replacement of, or addition to, existing fixed assets. For more details see *National Income and Expenditure* (HMSO).

LOCAL AUTHORITY EXPENDITURE (CHAPTER 9)

Local Authorities and New Towns (Table 9.3)

Analyses of the local authorities' direct labour construction output and employment are published, by size of authority, by type of authority and by region. Classification of size is determined by the number of operatives only, not including APTC staff as is the case for private firms.

Administrative, professional, technical and clerical staff statistics relate to employees engaged on the design, management, control, etc of building and civil engineering work irrespective of whether that work is to be carried out by the authorities' own direct labour or by contractors.

TRENDS IN EMPLOYMENT AND THE PROFESSIONS (CHAPTER 12)

Employment by Contractors (Tables 12.1-12.2)

Employment figures collected in the enquiries are the numbers on firms' payrolls in a specified week, and cover all employees, differentiating between operatives and administrative, professional, technical and clerical (APTC) staff, and also any working proprietors. (See below.)

Administrative, professional, technical and clerical staff are defined as those employees, including directors who receive a definite wage, salary or commission, who do not do manual work.

Operatives are all employees who do manual work, including for example, transport drivers, stores staff, canteen workers, workers making goods for sale.

Direct Labour Employed in the Public Sector

Employment statistics are collected quarterly from local authorities, new towns, transport authorities and government departments. From these returns estimates are made of the total direct labour force employed on construction work in the public sector. The returns cover all employees engaged mainly on construction work, whether or not they are contained within a separate building department of the authority concerned. Such employees are quite distinct from the operatives of private contractors who may be working for the authority under contract.

Employees' Earnings and Hours (Tables 12.3-12.5)

Information on earnings and hours is obtained from the New Earnings Survey conducted by the Office for National Statistics.

The New Earnings Survey collects information from the employers of a sample of 1 per cent of employees in Great Britain. Since 1975, selection has been made from Pay-as-you-earn records held by Inland Revenue tax offices, covering employees paying National Insurance contributions through PAYE schemes. Prior to 1975 selection was made when National Insurance contribution cards were exchanged each year at Department of Health and Social Security offices.

The results of this survey relate to a pay-period in April of each year and are not necessarily representative of pay and hours over a longer period. If earnings for the pay-period are increased retrospectively after the survey returns have been completed, because of delayed pay settlement, these increases will not be reflected in the results. Figures included in tables 12.4-12.5 are for full-time employees: i.e. employees generally expected to work for more than 30 hours in a normal week (excluding main meal-breaks and all overtime), or, if normal hours are not specified because of the nature of the job, employees regarded as full-time by their employers. Full results of the April surveys appear in the publication *New Earnings Survey*. Key results of this survey are published in the *Labour Market Trends*.

CONSTRUCTION BUILDING MATERIALS (CHAPTER 14)

The following notes show that in a number of cases information is collected more frequently than quarterly. This additional information is included in another of the Department's publications, '*Monthly Statistics of Building Materials and Components*'.

(Tables 14.1-14.5)

Bricks

Information on the production, deliveries and stocks of bricks is collected monthly from each brickworks in Great Britain. Bricks are classified as (i) facing, where appearance is important, (ii) engineering, where strength is important or (iii) common, where neither appearance nor strength is a requirement.

Regional figures for brick deliveries refer to deliveries from brickworks in the region to destinations both within and outside the United Kingdom.

Concrete Building Blocks

Information is collected monthly from the major producers and quarterly from all producers. The inquiry is voluntary for the first two months of each quarter whereas the quarterly inquiry is covered by the Statistics of Trade Act. The results of the quarterly inquiry are used to adjust those for the first two months of each quarter.

Sand and Gravel

Information is collected quarterly under the Statistics of Trade Act from a sample of producers and the results are grossed using the latest available Annual Mineral Raised Inquiry (AMRI) which covers all producers of sand and gravel. The annual figures given in the quarterly volumes do not always agree with those given in the annual volume since the former is the total of the quarterly grossed figures whereas the latter comes from AMRI. The figures for 1998 in this volume come from the quarterly inquiry.

Crushed Rock

Information on crushed rock is collected annually in AMRI and is available about ten months after the end of the latest year being collected. Detailed geographical analysis of the production of both crushed rock and sand and gravel are published each year in *Business Monitor PA 1007* (HMSO).

Cement

From the beginning of 1996, all the UK manufacturers are covered, with cement production being extended to included exports, and cement deliveries being extended to include all imports rather than only manufacturers' imports.

Fibre Cement Products

Information on fibre cement products is collected quarterly.

Concrete Roofing Tiles

The figures relate to roof area covered and are derived from the quarterly returns of individual producers.

Ready Mixed Concrete

Figures are of production in the United Kingdom and are based on a quarterly summary of production provided by the Quarry Products Association.

Slate

Information on the production, deliveries and stocks of slate is collected quarterly from individual producers. The figures exclude slate waste used for fill.

Imports and Exports of Building Materials and Components (Table 14.6)

The figures are derived from the total import and export values for the commodities, which are published in *Overseas Trade Statistics (TSO)*. As many of the commodities are not used solely in construction, estimates have been made of the imports and exports for constructional use.

INTERNATIONAL COMPARISON (CHAPTER 15)

British Construction Work Overseas (Table 15.12)

The Department collects annual information about building and civil engineering work undertaken overseas by British contractors and their subsidiaries. This covers the amount of new contracts won and the work done in the year, and of work outstanding at the end of the year, all at current prices, for each country.

INDEX

A

Aggregates
 exports 14.6
 imports 14.6
 price index 14.1
 region 14.4
 sales 14.4-14.5
APTCs
 construction industry 12.2
 occupation 3.12
Architects' Workload
 new commissions 12.9
 production drawings 12.10

B

Bankruptcies 3.2
Benchmarking 17
Brick Deliveries 14.2
 exports 14.6
 imports 14.6
 price index 14.1
 production 14.2
 region 14.2
 stocks 14.2
Building Costs 15.4-15.5
Building Labour Rates 15.2
Building Materials 14.1-14.7, 15.3
 aggregates 14.1, 14.4-14.6
 bricks 14.2
 cement 14.5
 clinker 14.5
 concrete blocks 14.3
 concrete roofing tiles 14.5
 crushed rock 14.4-14.5
 exports 14.6
 fibre cement products 14.5
 imports 14.6
 price index 14.1
 ready mixed concrete 14.5
 roofing tiles 14.5
 sand and gravel 14.1-14.5
 slate 14.5
 supply prices 15.3

C

Cement
 deliveries 14.5
 exports 14.6
 imports 14.6
 price index 14.1
 production 14.5
 stocks 14.5

Clinker
 production 14.5
 stocks 14.5
Concrete
 blocks 14.3
 exports 14.6
 imports 14.6
 price index 14.1
 ready mixed 14.5
 roofing tiles 14.5
Construction New Orders
 See New Orders
Construction Output
 See Output
Consulting Engineers
 fees for work 12.11
 new commissions 12.12
Contractors
 See New Orders
 See Output
 See Private Contractors
Cost Indices
 Building Non-Housing 5.1
 House Building 5.2
 Infrastructure 5.4
 Maintenance for Building Housing 5.6
 Maintenance for Building Non-Housing 5.5
 Road Construction 5.3

D

Deflators
 contractors' output 4.11
 direct labour output 4.11
Direct Labour 2.6-2.7, 4.10

E

Earnings
 all industries and services 12.5
 construction craft 12.4
 construction industry 12.3-12.5
 overtime pay 12.3
 payments by results 12.3
 premium payments 12.3
Employees 12.1-12.2, 12.6
Employment 3.4-3.5, 3.11-3.15, 12.1-12.2
 APTCs 3.12, 12.2
 ethnic origin 12.7
 gender 12.6
 operatives 3.5, 3.11, 12.2
 working proprietors 3.13

F

Fees for Work
consulting engineers 12.11
Floorspace
availability 7.1
take-up 7.1
new office designs 7.2

G

Government Expenditure
local authorities 9.1-9.3
MoD non-PPI spend 8.13
non-PFI health building projects 8.4
non-PFI road contracts 8.7
See Private Finance Initiative
Wales 13.10
Gross Domestic Fixed Capital Formation
construction industry 6.1-6.3

H

Health & Safety 12.22-12.25
Hours
all industries and services 12.5
construction industry 12.3-12.5
overtime 12.3
Housing Starts
Northern Ireland 13.2-13.4

I

Inflation Rates
building tender price 15.6
retail 15.6
Innovation 16
Insolvencies 3.2
Investment 6.1-6.3, 15.8-15.11
assets held in UK by foreign investors 15.11
assets held overseas by UK companies 15.9
by foreign companies in UK 15.10
earnings from overseas investment in UK 15.10
earnings from UK investment overseas 15.8
of UK companies overseas 15.8

K

Key Performance Indicators 17

L

Labour
building rates 15.2
disputes 12.21
See Employment
See Unemployment

Lead Times 14.7
Local Authority Expenditure
construction, conversion and renovation 9.1
housing, repair & maintenance revenue 9.2
PFI projects 8.9
work done 9.3

M

Materials 14.1-14.7
See Building Materials

N

National Lottery Funded Projects
fund application 10.1
funding received 10.2
New Commissions
architects 12.9
consulting engineers 12.12
New Orders
contractors 1.1-1.3
duration 1.7
region 1.5
type of work 1.4
value range 1.6
Northern Ireland
housing association starts 13.4
housing executive starts 13.2
index of construction 13.1
private sector housing starts 13.3

O

Operatives
construction industry 12.2
occupation 3.5, 3.11
Output
all agencies 2.1-2.3
contractors 2.4-2.5
direct labour 2.6-2.7
region 2.9
type of work 2.8
Output Price Indices
all industries 4.8
new construction 4.8, 4.9
public works 4.10
Overseas
building costs 15.4-15.5
building labour rates 15.2
building material supply prices 15.3
construction activity 15.12
constructional steelwork 15.7
inflation rates 15.6
investment 15.8-15.11

key data 15.1
service trade 12.18-12.19

P
Planning Applications and Decisions 11.1-11.5, 13.8-13.9
Price Indices
 See Output Price Indices
 See Tender Price Indices
Private Contractors
 APTCs 3.12
 employment 3.4-3.5, 3.11-3.15
 number of firms 3.1, 3.6-3.7
 operatives 3.5, 3.11
 region of registration 3.1, 3.3-3.5, 3.7, 3.15
 size of firm 3.1, 3.3-3.8, 3.10-3.13
 trade of firm 3.1, 3.3-3.6, 3.9-3.12, 3.14
 work done 3.3, 3.8-3.10
Private Finance Initiative (PFI)
 DfEE 8.1
 DETR 8.2
 Department of Health 8.3
 Highways Agency 8.6
 Home Office 8.8
 Local Authorities 8.9
 Inland Revenue 8.10
 Lord Chancellor's Department 8.11
 Ministry of Defence 8.12
 Scottish Executive 8.14
 National Assembly for Wales 8.15
 Northern Ireland 8.16
Professional Services
 number of businesses 12.20
 overseas service trade 12.18-12.19
 turnover 12.20
 See Architects
 See Consulting Engineers

R
Research Business Plans 16
Resource Cost Indices
 See Cost Indices
RICS
 Workload 12.8

S
Scotland
 new orders 13.12
 output 13.13
Self-Employed 12.1-12.2, 12.6
Size of Firm
 See Private Contractors

Steelwork
 prices 14.7
 production 15.7
Stoppages 12.21
Supply Chain Cube 17.2

T
Tender Price Indices
 Social Housing 4.1-4.2
 Public Sector Building (Non-Housing) 4.1, 4.3-4.4
 Road Construction 4.1, 4.5-4.7
Trade of Firm
 See Private Contractors
Training 12.13-12.14

U
Unemployment
 skilled trades 12.15
 carpenters and joiners 12.17

V
Vacancies in Skilled Trades 12.16

W
Wales
 capital account expenditure 13.10
 household projections 13.11
 index of construction 13.7
 mineral planning applications 13.9
 new orders 13.5
 output 13.6
 planning applications 13.8
Working Proprietors
 construction industry 3.13
Workload
 Architects 12.9-12.10
 RICS 12.8